Handbook of ELISPOT

METHODS IN MOLECULAR BIOLOGY™

John M. Walker, SERIES EDITOR

METHODS IN MOLECULAR BIOLOGY™

Handbook of ELISPOT

Methods and Protocols

Edited by

Alexander E. Kalyuzhny

R&D Systems Inc., Minneapolis, MN

HUMANA PRESS ✳ TOTOWA, NEW JERSEY

© 2005 Humana Press Inc.
999 Riverview Drive, Suite 208
Totowa, New Jersey 07512

www.humanapress.com

All papers, comments, opinions, conclusions, or recommendations are those of the author(s), and do not necessarily reflect the views of the publisher.

This publication is printed on acid-free paper. ∞
ANSI Z39.48-1984 (American Standards Institute)

Permanence of Paper for Printed Library Materials.

Production Editor: Nicole E. Furia
Cover design by Patricia F. Cleary
Cover Illustration: (Foreground) Figure 2 from Chapter 2, "Chemistry and Biology of the ELISPOT Assay" by Alexander E. Kalyuzhny; (Background) Figure 5 from Chapter 20, "A Gel-Based Dual Antibody Capture and Detection Method for Assaying of Extracellular Cytokine Secretion: *EliCell*,"by Lisa A. Spencer, Rossana C. N. Melo, Sandra A. C. Perez, and Peter F. Weller.

For additional copies, pricing for bulk purchases, and/or information about other Humana titles, contact Humana at the above address or at any of the following numbers: Tel.: 973-256-1699; Fax: 973-256-8341; E-mail: humana@humanapr.com; or visit our Website: www.humanapress.com

Printed in the United States of America. 10 9 8 7 6 5 4 3 2 1

eISBN 1-59259-903-6

ISSN 1064-3745

Library of Congress Cataloging in Publication Data

Handbook of ELISPOT methods and protocols / edited by Alexander E.Kalyuzhny.
     p. ; cm. -- (Methods in molecular biology ; 302)
  Includes bibliographical references and index.
  ISBN 1-58829-469-2 (alk. paper)
  1. Enzyme-linked immunosorbent assay.
  [DNLM: 1. Enzyme-Linked Immunosorbent Assay--methods. 2.
Cytokines--immunology. 3. Interferon Type II--immunology. 4.
T-Lymphocytes--immunology. QW 525.5.E6 H236 2005] I. Kalyuzhny, Alexander
E. II. Series: Methods in molecular biology (Clifton, N.J.) ; v. 302.
  QP519.9.E48H36 2005
  616.07'9--dc22
                                    2004016928

Preface

Enzyme-linked immunospot (ELISPOT) assay is a molecular tool for accurate quantification of cytokine-secreting immune system cells that is widely used in many fields of biomedical research including vaccine development, transplantation studies, and HIV, cancer, and allergy research. Although ELISPOT has been known to researchers for more than two decades, it still remains a state-of-the-art technique requiring solid knowledge and skills to run the assay. Being a combination of both bioassay and immunoassay techniques, ELISPOT consists of a set of multiple sequential procedures, each playing an important role in the outcome of the assay. It is well known that setting up an ELISPOT assay that produces recognizable and quantifiable spots may be very difficult not only for beginners, but also for experienced researchers. The *Handbook of ELISPOT: Methods and Protocols* is the first book dedicated entirely to the ELISPOT technique and is written for researchers who wish to learn about this assay and excel in performing it.

Handbook of ELISPOT: Methods and Protocols provides a comprehensive collection of ELISPOT protocols covering topics from vaccine development, tuberculosis research on animal models (mice, rat, and monkey), and for human studies. The book begins with a chapter on the history of ELISPOT technique written by one of the inventors of the ELISPOT assay (Chapter 1) and is followed by chapters on chemical and biological aspects of ELISPOT assays (Chapter 2), use of membrane-backed plates (Chapter 3), standardization and validation procedures (Chapter 4), removal of cells from ELISPOT plates (Chapter 5), cell separation techniques (Chapter 6), and quantification of ELISPOT data (Chapters 7 and 8). Chapters 9–12 cover the application of ELISPOT assays on animal models including rhesus macaque (Chapter 9), feline (Chapter 10), and mouse (Chapters 11 and 12) animal models. Application of the ELISPOT assay to human cells is covered in Chapters 13–16, which focus on using this assay to study measles (Chapter 13), multiple sclerosis (Chapter 14), monitoring immune responses (Chapter 15), and studying autoimmune sensorineural hearing loss (Chapter 16). Chapters 17–20 describe modifications of the ELISPOT assay, including development of multicytokine detection systems (Chapters 17–19) and combination of ELISPOT assay with immunocytochemistry (Chapter 20).

This book will serve both as a convenient reference manual for beginners and as a troubleshooting guide for experienced ELISPOT users. Methods and protocols are written by the leading researchers in their fields and presented in such a way that they can be easily adapted and modified for different research projects. The *Handbook of ELISPOT: Methods and Protocols* contains detailed technical reviews on many different aspects of ELISPOT assays with emphasis on their merits and shortcomings.

I would like to thank all of the authors who dedicated a significant amount of their time to prepare high quality manuscripts. Their efforts will contribute to understanding the principles of ELISPOT assays and allow the use of this technique in many diverse fields of biomedical science. I thank Sarah Palzer for her editorial assistance, and I am particularly thankful to Dr. Monica Tsang for her advice and support with editing this book. I also wish to thank R&D Systems Inc. for supporting ELISPOT projects and inspiring my editorial work.

Alexander E. Kalyuzhny

Contents

vii

Contributors

TANYA BAILEY • *University of Minnesota Medical School, Minneapolis, MN*

GEOFFREY A. COLE • *Point Therapeutics Inc., Boston, MA*

JOSEPHINE H. COX • *The US Military HIV Research Program, Henry M. Jackson Foundation, Rockville, MD*

GREGG A. DEAN • *Department of Molecular Biomedical Sciences, College of Veterinary Medicine, North Carolina State University, Raleigh, NC*

ALLEN C. EAVES • *StemCell Technologies Inc.; Director, Terry Fox Laboratory; and Professor of Medicine, University of British Columbia, Vancouver, British Columbia, Canada*

GUIDO FERRARI • *Department of Surgery, Duke University, Durham, NC*

JOANNE FLYNN • *Department of Molecular Genetics and Biochemistry University of Pittsburgh, Pittsburgh, PA*

WOLF H. FRIDMAN • *Unité d'Immunologie Biologique, Hopital Européen Georges Pompidou, Université René Descartes, Paris, France*

AGNES GAZAGNE • *Unité d'Immunologie Biologique, Hopital Européen Georges Pompidou, Université René Descartes, Paris, France*

ANGELA GRANT • *R&D Systems Inc., Minneapolis, MN*

CHRIS HARTNETT • *R&D Systems Inc., Minneapolis, MN*

SYLVIA JANETZKI • *ZellNet Consulting Inc., Fort Lee, NJ*

ALEXANDER E. KALYUZHNY • *R&D Systems Inc., Minneapolis, MN*

ALORA S. LAVOY • *Department of Molecular Biomedical Sciences, College of Veterinary Medicine, North Carolina State University, Raleigh, NC*

VANJA LAZAREVIC • *Department of Molecular Genetics and Biochemistry University of Pittsburgh, Pittsburgh, PA*

PAUL VIKTOR LEHMANN • *Cellular Technology Ltd., Cleveland, OH*

WOLF MALKUSCH • *Carl Zeiss Vision Zeppelinstr., Hallbergmoos, Germany*

MARK MATIJEVIC • *MGI Pharma Biologics, Lexington, MA*

ROSSANA C. N. MELO • *Department of Biology, Federal University of Juiz de Fora, Juiz de Fora, Brazil and Department of Medicine, Harvard Thorndike Laboratories, Charles A. Dana Research Institute, Beth Israel Deaconess Medical Center, Harvard Medical School, Boston, MA*

IOANA R. MOLDOVAN • *Department of Neurosciences, Lerner Research Institute, The Cleveland Clinic Foundation, Cleveland, OH*

BHARTI NEHETE • *Department of Immunology, University of Texas MD Anderson Cancer Center, Houston, TX*

PRAMOD N. NEHETE • *Department of Immunology, University of Texas MD Anderson Cancer Center, Houston, TX*

MIKIO NISHIDA • *Division of Health Care Pharmacy, Faculty of Pharmacy, Meijo University, Tempaku-ku, Nagoya, Aichi, Japan*

SUSHILA K. NORDONE • *Department of Molecular Biomedical Sciences, College of Veterinary Medicine, North Carolina State University, Raleigh, NC*

NEAL ODEN • *The EMMES Corporation, Rockeville, MD*

YOSHIHIRO OKAMOTO • *Division of Health Care Pharmacy, Faculty of Pharmacy, Meijo University, Tempaku-ku, Nagoya, Aichi, Japan*

INNA G. OVSYANNIKOVA • *Mayo Vaccine Research Group, Mayo Clinic and Foundation, Rochester, MN*

SARAH PALZER • *R&D Systems Inc., Minneapolis, MN*

SANTOSH PAWAR • *Department of Molecular Genetics and Biochemistry University of Pittsburgh, Pittsburgh, PA*

CLARA M. PELFREY • *Department of Neurosciences, Cleveland Clinic Foundation Lerner Research Institute and Institute of Pathology, Case Western Reserve University, Cleveland, OH*

SANDRA A. C. PEREZ • *Department of Medicine, Harvard Thorndike Laboratories, Charles A. Dana Research Institute, Beth Israel Deaconess Medical Center, Harvard Medical School, Boston, MA*

CARRIE E. PETERS • *StemCell Technologies Inc., Vancouver, British Columbia, Canada*

GREGORY A. POLAND • *Mayo Vaccine Research Group, and Program in Translational Immunovirology and Biodefense, Mayo Clinic and Foundation, Rochester, MN*

JENNA E. RYAN • *Mayo Vaccine Research Group, Mayo Clinic and Foundation, Rochester, MN*

K. JAGANNADHA SASTRY • *Department of Immunology, University of Texas MD Anderson Cancer Center, Houston, TX*

JONATHON D. SEDGWICK • *Stress and Immune Responses, Lilly Research Laboratories, Lilly Corporate Center, Indianapolis, IN*

C. ARTURO SOLARES • *Department of Immunology, Lerner Research Institute, and the Head and Neck Institute, The Cleveland Clinic Foundation, Cleveland, OH*

LISA A. SPENCER • *Department of Medicine, Harvard Thorndike Laboratories, Charles A. Dana Research Institute, Beth Israel Deaconess Medical Center, Harvard Medical School, Boston, MA*

ROSEMARY STEVENS • *Department of Molecular Biomedical Sciences, College of Veterinary Medicine, North Carolina State University, Raleigh, NC*

ERIC TARTOUR • *Unité d'Immunologie Biologique, Hopital Européen Georges Pompidou, Université René Descartes, Paris, France*

MONICA TSANG • *R&D Systems Inc., Minneapolis, MN*
VINCENT K. TUOHY • *Department of Immunology, Lerner Research Institute, The Cleveland Clinic Foundation, Cleveland, OH*
ROBERT G. URBAN • *MGI Pharma Biologics, Lexington, MA*
BENOIT VINGERT • *Unité d'Immunologie Biologique, Hopital Européen Georges Pompidou, Université René Descartes, Paris, France*
ALAN J. WEISS • *Life Sciences Division, Millipore Corporation, Danvers, MA*
PETER F. WELLER • *Department of Medicine, Harvard Thorndike Laboratories, Charles A. Dana Research Institute, Beth Israel Deaconess Medical Center, Harvard Medical School, Boston, MA*
STEVEN M. WOODSIDE • *StemCell Technologies Inc., Vancouver, British Columbia, Canada*

Color Plates

Color Plates 1-11 appear as an insert following p. 50.

Color Plate 1:
Chapter 2, Figure 3. Two-cytokine ELISPOT assay (custom-made kit; R&D Systems). IL-2 release from human peripheral blood lymphocytes is detected using Alkaline phosphatase-BCIP/NBT reagents, whereas IFN-γ is detected using HRP-AEC detection system. *See* discussion on p. 19.

Color Plate 2:
Chapter 2, Figure 7. High degree of programmed cell death (apoptosis) in cells plated into the ELISPOT plate may result in lack of spots. Cells attached to membranes (green fluorescence) were labeled immunocytochemically for an apoptosis marker active Caspase-3 using R&D Systems anticaspase-3 antibodies (red color). Note the high number of apoptotic cells. *See* discussion on pp. 25–26.

Color Plate 3:
Chapter 8, Figure 10. ELISPOT specimen with similar number of spots per well. Some wells are labeled with a red marker (left), others with a blue marker (middle), and some with both markers resulting in violet spots (right). Result of counts per well with teaching on each color and cross check: Setting 1: color 88–97 (red), setting 2: color 57–62 (blue), setting 3: color 82–95 (violet). *See* discussion on p. 148.

Color Plate 4:
Chapter 8, Figure 11. ELISPOT specimen labeled with fluorescence markers for IFN-γ (fluorescein isothiocyanate, green spots) and IL-5 (rhodamine, red spots). Spots from cells expressing both cytokines appear yellow. Count results: red spots, 229; green spots, 291; yellow spots, 64. *See* discussion on p. 149.

Color Plate 5:
Chapter 19, Figure 1. Dual-color immunoenzymatic ELISPOT for the detection of IFN-γ and IL-4 producing cells. Left: B.EBV cells and a TH2 T-cell clone were mixed and activated with PMA–ionomycin. IFN-γ and IL-4 derived from secreting cells were detected using substrates specific for horseradish peroxidase (3-amino-9-ethylcarbazole, $C_{14}H_{14}N_2$) or alkaline phosphatase (5-bromo-4-chloro-3-indolylphosphate/Nitroblue tetrazolium chloride), respectively. Enzymes were linked to detection antibodies for IFN-γ and IL-4. Red spots corresponded to IFN-γ secreting cells, whereas blue spots belong to IL-4 producing cells. Right: Greater enlargement of a quadrant from the left. The arrows showed the difficulties in the interpretation of mixed color spots. *See* discussion on p. 290.

Color Plate 6:
Chapter 19, Figure 2. Dual-color fluorospot for the detection of IL-2 and/or IFN-γ-producing cells. Peripheral blood mononuclear cells were stimulated with PMA and ionomycin in PVDF plates. IFN-γ- and/or IL-2-producing cells were characterized by a dual-color fluorospot assay. Green spots corresponded to IFN-γ secreting cells, whereas red spots belong to IL-2-producing cells. Yellow spots corresponded to cells coexpressing IFN-γand IL-2. No spots were observed when non-stimulating cells were used for the dualcolor fluorospot. See discussion on pp. 292–293.

Color Plate 7:
Chapter 20, Figure 3. Positive staining results using EliCell system. Detection of cytokine release fromeosinophils stimulated with physiologic (**A**) or nonphysiologic (**B**) stimuli. In (**B**), eosinophils were stimulated with 0.5 µ*M* A23187. Panel 2 illustrates simultaneous detection of 2 cytokines labeled with Alexa 488 or Alexa 546. Digital pictures were taken using 100X magnification objective. *See* discussion on pp. 306–307,309.

Color Plate 8:
Chapter 20, Figure 4. Staining artifacts using EliCell system. Phase-contrast and fluorescence microscopy of identical fields of eosinophils incubated in EliCell preparations. Damaged (**A**, **B**, and **C**) or permeabilized (**D**) cells show nonspecific staining. In **D**, the image was overlaid. Digital pictures were taken using 100X magnification objective. *See* discussion on pp. 307–309.

Color Plate 9:
Chapter 20, Figure 5. Viability of cells after EliCell assay. EliCell preparation of eosinophils stained with acridine orange/ethidium bromide mixture after fixation. Most cells show green fluorescent nucleus indicative of cell viability. Digital pictures were taken using 100X magnification objective. *See* discussion on p. 309.

Color Plate 10:
Chapter 20, Figure 6. Morphology of eosinophils during EliCell assay. Light micrographs of eosinophils observed in the EliCell system before (**A**) and after stimulation with eotaxin (**B–D**). Morphological changes characterized by cell elongation are clearly seen in stimulated cells. Cells were stained with Hema 3 (**A–C**) or fast green/hematoxylin (**D**). n, nucleus. Digital pictures were taken using 100X magnification objective. *See* discussion on p. 310.

Color Plate 11:
Chapter 20, Figure 7. Intracellular lipid body staining of eosinophils using EliCell system. Phasecontrast and fluorescence microscopy of identical field of chemokine-stimulated eosinophil in an EliCell preparation. Cytoplasmic lipid bodies are indicated (arrows). Cells were stained with BODIPY. Digital pictures were taken using 100X magnification. *See* discussion on p. 312.

I

ELISPOT ASSAY SETUP AND DATA ANALYSIS

1

ELISPOT Assay

A Personal Retrospective

Jonathon D. Sedgwick

"And yet the true creator is Necessity, who is the Mother of our Invention."

—Plato, *The Republic*

Summary

In 1983, papers describing the enzyme-linked immunospot (ELISPOT) technique were published by two groups, the first description from a team in Perth, Western Australia, and the second, soon thereafter, from a group in Gothenburg, Sweden. Described here is my recollection of the background and circumstances that lead to the assay's development within the Perth group. Included are the early studies in 1981 through early 1982 that were conducted to bring the assay to fruition—both setbacks and solutions, and finally some generally unknown but amusing insights into the naming of the ELISPOT assay by the Gothenburg group.

Key Words: Historical; retrospective; discovery; invention; inhalation tolerance; hemagglutination assay; hemolytic plaque assay; ELISA; ELISA-plaque; Spot-ELISA; ELISPOT; antibody-secreting cell; IgE-secreting cell; cytokine-secreting cell.

1. Introduction

I had spent 1980 as a Bachelor of Science Honors student in the laboratories of the then Clinical Immunology Research Unit of the Princess Margaret (Hospital) Children's Medical Research Foundation in my hometown of Perth, Western Australia, pursuing a very challenging project on a new histamine-releasing activity *(1)*. While enthusiastically supervised by Kevin Turner, the Director of the Unit, and Patrick Holt, a Senior Research Fellow, it became clear as the work progressed that there was little core expertise in the Unit to push this area further, to remain competitive, and provide the breadth to support me through a PhD program. Fortunately, other work ongoing in both Kevin's and Patrick's laboratories was much more appropriate and interesting.

From: *Methods in Molecular Biology, vol. 302: Handbook of ELISPOT: Methods and Protocols*
Edited by: A. E. Kalyuzhny © Humana Press Inc., Totowa, NJ

Their studies had shown that mice exposed weekly to aerosolized antigen during a 6- to 7-week time frame exhibited only a transient and self-limiting serum immunoglobin E (IgE) response after 3–4 weeks of exposure but a sustained and increasing serum IgG response. However, after completion of the respiratory antigen exposure regimen and parenteral antigen/adjuvant challenge, mice showed antigen-specific resistance to the induction of an IgE response. This was an unexpected outcome, the aerosol model having been established in an attempt to sensitize the animals and create a model of atopy or asthma. In any event, the concept of "inhalation tolerance" had been born. These findings were eventually published in March 1981 *(2)*. Another group working in the United States also had made this observation and published their findings in June 1981 *(3)*.

Based on my now-proven ability to work 24/7 and survive through an impossible project, Patrick asked me to join his group to pursue my PhD, beginning in early 1981, the project being to understand the cellular mechanisms of this new type of immunological tolerance. This line of research was successful and productive *(4,* reviewed in **ref. 5**) and contributed to the formulation of Patrick's now widely -accepted hypotheses concerning the role of neonatal antigen exposure in modulating development of atopy and asthma in children *(6,7)*. The enzyme-linked immunospot (ELISPOT) assay was developed essentially as a side-line project to the core inhalation program. As is always the case with invention, the ELISPOT assay was conceived out of need.

2. A Need Arises—in the Right Place, at the Right Time, and With the Right People

One of the problems we needed to resolve early in the program was the lack of high-throughput, sensitive, and specific assays for antigen-specific as well as total immunoglobulins, particularly for the IgE isotype. We now take the availability of such assays for granted, but 20 years ago, monoclonal antibodies (mAbs) were rare, the enzyme-linked immunosorbent assay (ELISA) was still a very new technique, and the specificity of most polyclonal antibodies was highly questionable. This was a major problem for very low concentration isotypes such as IgE and IgA, where even minor crossreactivity to IgG or shared light-chain reactivity invalidated the assays. Based on my notebook from this era, I started working on the inhalation program on February 19, 1981, and did almost nothing else for the next 5 months then try to develop specific ELISAs for IgE and IgG and compare these to the standard assays of the time—the I^{125}-based radioabsorbent test and in vivo passive cutaneous anaphylaxis for IgE, and the sheep erythrocyte-based hemagglutination (HA) assay for IgG and IgM. Where I was physically located to do this work was, as it turned out, fortuitous for these studies as well as developing the ELISPOT assay.

For my Honors year I was placed in one of the larger laboratories that was run by Kevin Turner and Geoff Stewart, another Research Fellow in the Unit and, because Patrick's laboratory was small and crowded, I did not move to his area for at least one year after beginning my PhD, rather remaining in Kevin's and Geoff's laboratory. Sitting opposite me was Patricia Price, a PhD student with Kevin Turner who taught me much about HA assays. She and I, with total disregard for our colleagues, polluted the environment with β-mercaptoethanol, used to differentiate between IgG and IgM in the HA assay. Geoff and a PhD student, Andrew McWilliams, in this laboratory were skilled protein chemists and familiar with the techniques for antibody characterization and purification. Of particular note was their expertise with immunodiffusion in-gel assays that I used to characterize antibody specificities. The classic "Ouchterlony" double immunodiffusion test was the technique most widely used. Dr. Ouchterlony also played a part in the ELISPOT story—more of that later.

Reasonable progress was made in the development of assays for the detection of immunoglobulins secreted into serum and other fluids through a combination of many techniques, but especially the ELISA, which I had by this time explored carefully and modified to my needs. The holy grail, however, was to define where antibody was being made during respiratory exposure, especially the transient IgE response. I recall Patrick and I discussing this issue at length and the difficulties surrounding the detection of very small numbers of IgE antibody-secreting cells (ASCs). The Jerne hemolytic plaque assay *(8)* was the most widely used assay at that time for the detection of ASC and especially the Cunningham modification *(9)*, where two glass slides are joined with double-sided tape to create a thin chamber into which the mixture of ASCs and erythrocytes are placed. This assay was suitable for the IgG isotype, for which the number of ASCs was high and thus the development of "direct" (or IgM-secreting cell plaques) could be allowed for. We also were aware that Rector and colleagues *(10)* had used this approach for the detection of IgE-secreting cells in a mouse system. Nevertheless, my records show that I did not make any attempts to develop hemolytic plaque assays until after the ELISPOT was conceived.

With my by now substantial familiarity with the standard ELISA and background discussions regarding our need to detect ASCs, it was perhaps a small intellectual leap to pose the question whether it would be possible to use an ELISA approach to detect antibody secreted from cells directly rather than for the measurement of antibody in solution. I recall having the idea and immediately initiating discussions with Patrick, Geoff, Patricia, Kevin, and probably many others. That the assay ever came to fruition is entirely the result of the input from these colleagues and especially Patrick. First, Patrick immediately advised me that the concept of solid-phase "capture" of secreted antibody was not new and pointed me to published studies to help me develop the concept. The most relevant was from Don Mason (who I later joined for my postdoctoral Fellowship in 1985) in the then Medical Research Council Cellular Immunology Unit in Oxford. Don had coated one glass slide with antigen, placed another on top, maintaining separation of the two slides using thin coverslips to create the cell incubation chamber, introduced ASCs into the narrow chamber and then, after incubation, the slides were separated, iodinated secondary antibody added, and spots of secreted antigen-specific antibody detected using autoradiography *(11)*. Distinct to this approach, the ELISA-based technique would be dependent upon localization of a colored substrate, so the likely need to develop spots within a gel matrix to "trap and localize" the color was discussed as was the use of low-melting temperature gels given that the assay would be performed in plastic wells. My only experience with ELISA was with soluble substrates, especially the yellow alkaline phosphatase (AP) substrate, *p*-nitrophenylphosphate, and I recall discussing with both Patrick and Geoff whether the gel would hold this in place. No one knew, but we decided to try it anyway.

The breakthrough in terms of substrate to detect spots of antibody came from parallel discussions with Geoff Stewart. I recall this very clearly. Geoff had been experimenting with separation of some house-dust mite allergens using in-gel techniques and needed a method to detect the allergens with AP-labeled antibodies. He had obtained a Sigma Aldrich kit that was designed to separate alkaline phosphatase isoforms in-gel using electrophoresis followed by addition of a direct AP substrate that turned blue upon AP enzymatic cleavage. The substrate used in the kit was 5-bromo-4-chloro-3-indolylphosphate (BCIP). The beauty of BCIP was its intense color, insolubility, and stability. Unlike many substrates, especially horseradish peroxidase substrates, it was not light sensitive and did not fade.

The first test of the concept was performed on August 31, 1981 **(Fig. 1)**. It was not done with cells, but rather by adding small (1 μL) volumes of rat IgG to plastic wells, followed by incubation to adsorb the antibody to the plastic to create a "spot" of antibody. This was followed by an anti-rat IgG, a second AP-conjugated detecting antibody, and then the addition of BCIP or *p*-nitrophenylphosphate in agarose. The concept worked in this first trial. I noted the following (*see* **Fig. 1,** at the bottom).

"Results: It worked: Yellow dots spread rapidly, however, so this is not very suitable. The BCIP substrate, however, gave clear, blue dots & did not spread."

During the next 3 weeks I used this same artificial spot system to optimize the system, especially the concentration of agarose to use, BCIP substrate buffers, best plastics to use for coating, and concentrations of coating antigen. We discovered early on that the ELISPOT required a much higher coating antigen concentration than standard ELISA, probably because of a need to have high localized density of substrate to observe individual spots by eye. It was not

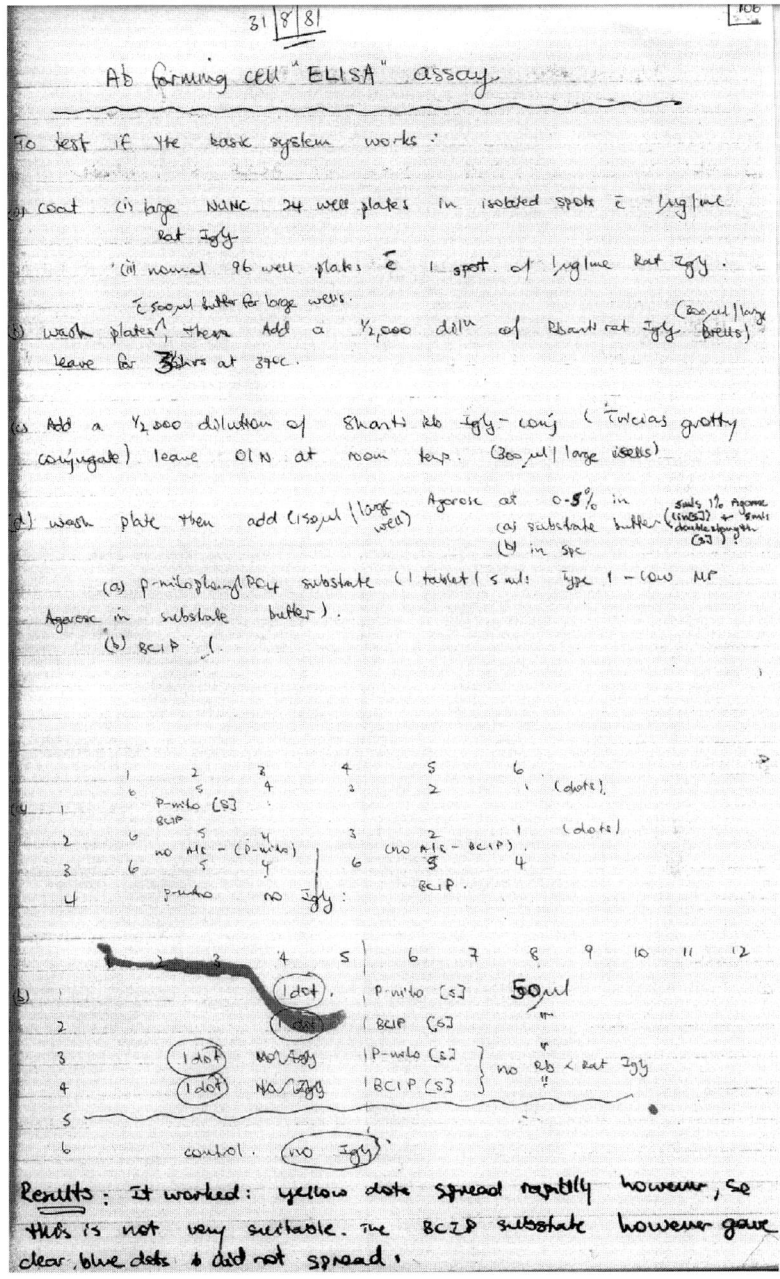

Fig. 1. Page from notebook dated August 31, 1981, describing the first test to assess the feasibility of detecting small spots of immunoglobulin in an ELISA format. It was described as an 'Ab-forming cell "ELISA" assay.' The BCIP substrate was also tested for the first time here.

until September 18, 1981, that I tried actual cells in the assay (splenocytes from ovalbumin-immunized mice), and antigen-specific IgG secreting cells were detected in this first study. The quality, however, was poor, and a problem arose that sidetracked us for at least three months.

I was using a range of plastic dishes for the assay, all of them round wells, including small bacterial agar dishes and 24-well tissue-culture dishes. In every case, cells redistributed to the edge of the wells during incubation, probably because of natural swirling actions as the media warmed during incubation. For reasons that are not entirely obvious to me now, we never tried smaller wells (96-well format) in any systematic way. We now know that the problem is less in smaller wells, especially if volumes are kept low (<50 µL). The problem was such that I discarded the use of plastic wells entirely and turned to a combination of the "Cunningham" chamber technique, coupled with the "Don Mason" variation using coated slides but using plastic slides rather than glass that would better adsorb coated antigen. After cell incubation in the chamber, the slides were broken apart and spots developed by addition of antibodies and substrates as in the ELISA. This worked quite well, including for the detection of IgE-secreting cells, but was very tedious. In parallel, to prove that the ELISPOT was quantitative, it was now necessary to compare it directly to the gold standard of the time, namely the hemolytic plaque assay. This technique too did not come easily to me, especially in terms of the modifications needed to enable the detection of protein antigen-specific ASCs. The techniques for coating of erythrocytes with antigen such as ovalbumin were not consistent from one study to the next and, more often than not, unsuccessful even for IgG and IgM-ASC. I never achieved satisfactory detection of IgE-secreting cells using the hemolytic plaque assay but rather compared the IgE ELISPOT to a method known as heterologous adaptive cutaneous anaphylaxis, in which cells containing the IgE-ASC are injected id into rat skin, the cells secrete IgE locally, and this is detected by the measurement of local mast cell degranulation as in the passive cutaneous anaphylaxis. We quickly determined that the ELISPOT was quantitative and almost always much more sensitive than existing in vitro or in vivo techniques. Even if only rare ASCs were present in a cell suspension, the ELISPOT would detect them.

The next breakthrough came early in 1982 through discussions with Patrick about the continuing problem of movement of cells to the periphery of round wells and the tedious nature of the slide-chamber approach. Patrick suggested (I don't know why exactly but it was very insightful) to try square wells instead of round ones. Some old 25-well virology repli dishes from Dyos Corp., UK were located, and I tried the assay in those. It worked immediately. Cells were distributed evenly across the wells, and the sensitivity was excellent. Presumably, the square well configuration discouraged the swirling effects that occurred in round wells. Results were gather rapidly after that, and a paper describing the assay was submitted to the *Journal of Immunological Methods* in the first week of April 1982. At the time, we did not have a catchy name for the technique and so it was referred to only as a "solid-phase immunoenzymatic technique." the cartoon from this manuscript outlining the concept (*see* **Fig. 2**) is still relevant today, the only notable difference from most diagrams one sees of this technique today is that it shows square wells in step 6!

The manuscript, although received mid-April, was not reviewed in a timely way. I recall Patrick contacting the Editors repeatedly but with little success. Finally, we were advised in August 1982 that the paper was accepted—without change!? It was published 10 months after submission, in February 1983 *(12)*. We followed this up rapidly with a Brief Report in the *Journal of Experimental Medicine (13)* using the ELISPOT assay to detect, for the first time, the site of production of IgE-secreting cells in a primary immune response, insights that would have been difficult to derive using previously available technologies. In that study we also used an anti-IgE antibody as coating protein rather than antigen to enable capture of all IgE and thus, detection of total IgE-secreting cells. This "reverse" technology enabled, in principle, the detection of secreted products of virtually any type so long as an antibody could be generated against that product, and this ideas was discussed for detection of T-cell products *(14)*, although we never developed that concept further ourselves. This "reverse" approach has in fact become the most common mode of use for the ELISPOT assay, especially for the detection of cytokine-secreting cells.

3. Concurrent Developments in Sweden

That new ideas and technologies are often developed simultaneously by a number of laboratories unaware that there are similar activities ongoing elsewhere is, perhaps surprisingly, not unusual (cf., inhalation tolerance, **refs.** *2* and *3*) and probably reflects a global convergence of ideas and development of enabling technologies. The development of the ELISPOT is a classic case. In December 1983 Cecil Czerkinsky and colleagues based in Gothenburg, Sweden, described in the *Journal of Immunological Methods*, an identical technique to the one we described. "ELISPOT" was used in the title *(15)*, but this is not the first use of this name. Having now become very firm friends and coauthored a number of reviews on the ELISPOT technique, Cecil and I have discussed the paths that lead to the development of this assay in two laboratories on opposite sides of the world almost simultaneously. It transpired that Cecil's ELISPOT paper for the detection of ASC *(15)* was actually preceded by a paper published in June of 1983 in the *Journal of Clinical Microbiology* in which he described an "enzyme-linked immunospot" assay for the detection of bacteria secreting enterotoxin. The term ELISPOT is first used in this publication *(16)*.

1. Antigen conjugated to plate (ꙮ)

   Wash

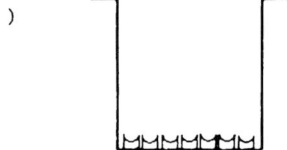

2. Add cell suspension; incubate
   1 hr. Secreted antibody binds
   to antigen (◗).

   Wash

3. Add Rb-anti-Rat Ig;
   incubate 2 hrs. (⋏)

   Wash

4. Add Sh-anti-Rb IgG-Enzyme
   conjugate; incubate overnight (⋏)

   Wash

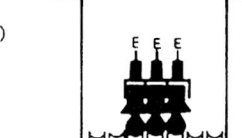

5. Add soluble substrate in
   Agarose (☆); incubate to
   allow formation of insoluble
   product (★).

6. Invert chamber over light
   source and enumerate macroscopic
   blue spots.

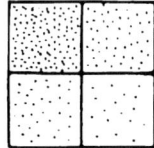

He had been working on the concept directed to bacterial secreted products for many months prior. The application of this technology to the detection of ASCs was an obvious next step. Cecil and his colleagues have been a major force in the development of novel ELISPOT technologies, especially in their early descriptions of the use of ELISPOT for the detection and enumeration of cytokine-secreting lymphocytes *(17)*.

4. The Legacy

Now, just more than 20 years after the first description of the technique, we see here a book devoted completely to ELISPOT. It is very gratifying for me to see how the technique has matured and developed in directions that were unimaginable at the time Patrick and I described it in the early 1980s. Since our first description, key advances have included: the first use by Moller and Borrebaeck of nonplastic (originally nitrocellulose)-based wells that enabled increased sensitivity and without the requirement for in-gel substrates *(18)*, improvements in substrates and use of multicolored detection systems *(19,20)*, and application of computer-assisted spot-counting technologies *(21)* and the commercialization of these technologies by numerous instrumentation companies *(22)*. Probably most fulfilling is the fact that the ELISPOT assay is now the method of choice, including in an increasing number of clinical immunology, clinical virology, and clinical pathology laboratories, for the monitoring of cytotoxic responses (through measurement of interferon-γ-secreting s) in cancer immunotherapy *(23)* and viral-vaccine trials, including in most HIV vaccine trials *(24)*.

5. "What's in a Name"?

As noted above, the Perth team did not coin the term ELISPOT. In January 1984 Patrick, Kevin, other colleagues, and I published a study *(25)* in which the term "ELISA-plaque" was first used to describe the technique, based on the fact that this was *ELISA*-based and was a technique to replace the hemolytic *plaque* assay for the detection of ASCs. Almost no one outside the Perth group used this name, but that team and I (after leaving for Oxford in 1985) continued to call the assay ELISA-plaque until the early 1990s. Through this period, the assay was called by a range of names by different investigators, including spot-ELISA, ELISA-spot, ELISA immunospot and, of course, ELISPOT. Ultimately,

---

Fig. 2. Figure 1 from the original publication. Note in part 6 the depiction of square wells that were used in the early ELISPOT studies. Reproduced from **ref.** *12* with permission from J. D. Sedgwick, P. G. Holt, and Elsevier. A solid-phase immunoenzymatic technique for the enumeration of specific antibody-secreting cells.

ELISPOT has been chosen by the scientific community and so it should—it is by far the best and most descriptive.

I will close with an amusing story that relates to the development of the name ELISPOT. In around 1986 I was invited by Cecil Czerkinsky to visit Gothenburg and present a seminar. On touring Cecil's laboratory, I was introduced to an older man who turned out to be the famous Dr. O. Ouchterlony responsible for the development of the double immunodiffusion assay that bares his name *(26,27)*. Dr. Ouchterlony did not know I had anything to do with the development of the ELISPOT assay. Rather, he had heard that I was an Australian and wanted to relate to me his experiences as an expert witness for the defense in the trial of Lindy Chamberlain accused of killing her infant daughter, Azaria, in a camp-ground in central Australia in the vicinity of Ayers Rock, or Uluru. Chamberlain had maintained her innocence, claiming that her daughter was taken by a dingo, a wild Australian dog. The story and the eventual trial was sensationalized, was covered internationally, and eventually was made into a film (*A Cry in the Dark* starring Meryl Streep and Sam Neill.) In any event, a key piece of data used by Chamberlain's accusers was that they claimed identification of fetal blood in the Chamberlain's car, based on the use of antibodies supposedly specific for fetal hemoglobin and tested using the Ouchterlony technique. The quality control was poor, the antibodies were non-specific, the original gels had not been saved, and Dr. Ouchterlony testified that the data was uninterpretable. Chamberlain was convicted nevertheless and jailed for life with hard labor (although she was released and exonerated some years later). Dr. Ouchterlony presented me with a signed copy of a manuscript he had written and published in which he detailed all the issues surrounding the trial and especially those pertaining to the evidence of fetal hemoglobin *(28)*. He was about to leave when Cecil told him about my role in the development of the ELISPOT assay. Dr. Ouchterlony was a co-author on Cecil's ELISPOT paper and turned out to be a close friend and mentor to Cecil. Dr. Ouchterlony turned to me and said something to the effect that he had advised Cecil that almost more important than the value of the technique was to ensure that it had a good name!

Acknowledgments

Thanks to Patrick Holt for his support and friendship during the past 20-plus years and for his review of this manuscript. The assistance of Deborah Jones (Eli Lilly and Company) with manuscript preparation is greatly appreciated.

References

1. Sedgwick, J. D., Holt, P .G., and Turner, K. J. (1981) Production of a histamine-releasing lymphokine by antigen-stimulated human peripheral T cells. *Clin. Exp. Immunol.* **45**, 409–418.

2. Holt, P. G., Batty J. E., and Turner K. J. (1981) Inhibition of specific IgE responses in mice by pre-exposure to inhaled antigen. *Immunology* **42,** 409–417.
3. Fox, P. C., and Siraganian R. P. (1981) IgE antibody suppression following aerosol exposure to antigens. *Immunology* **43,** 227–234.
4. Sedgwick, J. D., and Holt, P. G. (1983) Induction of IgE-isotype specific tolerance by passive antigenic stimulation of the respiratory mucosa. *Immunology* **50,** 625–630.
5. Holt P. G., and Sedgwick, J. D. (1987) Suppression of IgE responses following inhalation of antigen: a natural homeostatic mechanism which limits sensitization to aero allergens. *Immunol. Today* **8,** 14–15.
6. Holt, P. G., and Macaubas, C. (1997) Development of long-term tolerance versus sensitization to environmental allergens during the perinatal period. *Curr. Opin. Immunol.* **9,** 782–787.
7. Holt, P. G., Macaubas, C., Stumbles, P. A., and Sly, P. D. (1999) The role of allergy in the development of asthma. *Nature.* **402** (Suppl.), B12–B17.
8. Jerne, N. K., and Nordin, A. A. (1963) Plaque formation in agar by single antibody-producing cells. *Science*, **140**, 405.
9. Cunningham, A. J., and Szenberg, A. (1968) Further improvements in the plaque technique for detecting single antibody-forming cells. *Immunology*, **14**, 599–600.
10. Rector, E. S., Lang, G. M., Carter, B. G., Kelly, K. A., Bundesen, P. G., Bottcher, I., et al. (1980) The enumeration of mouse IgE secreting cells using plaque-forming cell assays. *Eur. J. Immunol.* **10**, 944–949.
11. Mason, D. W. (1976) An improved autoradiographic technique for the detection of antibody-forming cells. *J. Immunol. Methods.* **10**, 301–306.
12. Sedgwick, J. D., and Holt, P. G. (1983) A solid-phase immunoenzymatic technique for the enumeration of specific antibody secreting cells. *J. Immunol. Methods* **57**, 301–309.
13. Sedgwick, J. D., and Holt, P. G. (1983) Kinetics and distribution of antigen-specific IgE-secreting cells during the primary antibody response in the rat. *J. Exp. Med.* **157**, 2178–2183.
14. Sedgwick, J. D., and Holt, P. G. (1986) The ELISA-plaque assay for the detection and enumeration of antibody secreting cells: an overview. *J. Immunol. Methods*, **87**, 37–44.
15. Czerkinsky, C. C., Nilsson, L. A., Nygren, H., Ouchterlony, O. and Tarkowski, A. (1983) A solid-phase enzyme-linked immunospot (ELISPOT) assay for enumeration of specific antibody-secreting cells. *J. Immunol. Methods.* **65**, 109–121.
16. Czerkinsky, C. C., and Svennerholm, A. M. (1983) Ganglioside GM1 enzyme-linked immunospot assay for simple identification of heat-labile enterotoxin-producing Escherichia coli. *J. Clin. Microbiol.* **17**, 965–969.
17. Czerkinsky, C., Andersson, G., Ferrua, B., Nordström, I., Quiding, M., Eriksson, K., et al. (1991) Detection of human cytokine-secreting cells in distinct anatomical compartments. *Immunol. Revs.* *119*, 5–22.
18. Moller, S. A. and Borrebaeck, C. A. (1985) A filter immuno-plaque assay for the detection of antibody-secreting cells in vitro. *J. Immunol. Methods* **79**, 195–204.

19. Franci, C., Ingles, J., Castro, R., and Vidal, J (1986) Further studies on the ELISA-spot technique. Its application to particulate antigens and a potential improvement in sensitivity. *J. Immunol. Methods* **88**, 225–232.
20. Czerkinsky, C., and Sedgwick, J. (1993) Enzyme-linked immunospot (ELISPOT) assays for detection of specific antibody-secreting cells, in *Methods of Immunological Analysis, Volume 3, Cells and Tissues* (Masseyeff, R. F., Albert, W. H., and Staines, N. A. eds.), VCH Verlagsgesellschaft mbH, Weinheim Germany, pp 504–540.
21. Cui, Y., and Chang, L. J. (1997) Computer-assisted, quantitative cytokine enzyme-linked immunospot analysis of human immune effector cell function. *Biotechniques*, **22**, 1146–1149.
22. Karulin, A. Y., Hesse, M. D., Tary-Lehmann, M., and Lehmann, P. V. (2000) Single cytokine-producing CD4 memory cells predominate in type 1 and type 2 immunity. *J. Immunol.* **164**, 1862–1872.
23. Romero, P., Cerottini, J-C., and Waanders, G. A. (1998) Novel methods to monitor antigen-specific cytotoxic T-cell responses in cancer immunotherapy. *Mol. Med. Today.* **4**, 305–312.
24. Mwau, M., McMichael, A. J., and Hanke, T. (2002) Design and validation of an enzyme-linked immunospot assay for use in clinical trials of candidate HIV vaccines. *AIDS Res. Hum. Retroviruses.* **18**, 611–618.
25. Holt, P. G., Cameron, K. J., Stewart, G. A., Sedgwick, J. D., and Turner, K. J. (1984) Enumeration of human immunoglobulin secreting cells by the ELISA-plaque method: IgE and IgG isotypes. *Clin. Immunol. Immunopathol.* **30**, 159–164.
26. Ouchterlony, O. (1949) Antigen-antibody reactions in gels. *Acta Pathol. Microbiol. Scand.* **26**, 507–517.
27. Ouchterlony, O. (1958) Diffusion-in-gel methods for immunological analysis. *Prog. Allergy.* **5**, 1–78.
28. Ouchterlony, O. (1987) Carl Prausnitz memorial lecture. Immunoprecipitation in court—the Chamberlain case. *Int. Arch. Allergy Appl. Immunol.* **82**, 233–237.

2

Chemistry and Biology of the ELISPOT Assay

Alexander E. Kalyuzhny

Summary

Enzyme-linked immunospot, or ELISPOT, assay allows the detection of low frequencies of cells secreting various molecules. ELISPOT can be used in many areas of research and, because of its high sensitivity, has the potential to become a valuable diagnostic tool. Based on the same "sandwich" immunochemical principles as enzyme-linked immunosorbent assay, ELISPOT is easy to perform and quantify the results. At the same time ELISPOT remains a state-of-the-art technique that requires accuracy, thorough selection of antibodies and detection reagents, and an understanding of the principles of data analysis. This review covers various technical aspects of the ELISPOT assay, including immunochemical principles of the assay, selection of reagents and plates, and troubleshooting recommendations.

Key Words: ELISPOT; detection antibodies; capture antibodies; spot-forming cells; quantification of spots; spot artifacts.

1. Historic Overview

In 1983, Sedgewick and Holt *(1)* published a paper in the *Journal of Immunological Methods* describing a novel technique for the enumeration of antibody-secreting cells. The new technique was built on the same solid-phase immunoenzymatic principles as the enzyme-linked immunosorbent assay (ELISA): antigen was immobilized to a solid support (plastic dish) to bind antibodies released by cultured splenocytes. Later, in 1983, another article describing a similar antibody detection technique was published in the same journal by Czerkinsky and colleagues *(2)*, who coined the name for this assay "enzyme-linked immunospot," or ELISPOT. Later, the original ELISPOT technique was modified in that the solid phase was coated with antibodies (rather than the antigen) to capture antigens (for example, cytokines) released by cultured cells *(3)*. As modified, reversed ELISPOT has become very popular and appears to be used more frequently than its predecessor. Some researchers call it "reversed

From: *Methods in Molecular Biology, vol. 302: Handbook of ELISPOT: Methods and Protocols*
Edited by: A. E. Kalyuzhny © Humana Press Inc., Totowa, NJ

ELISPOT," whereas most truncated this name to just "ELISPOT," In this chapter, I will cover various technical aspects of the reversed ELISPOT assay and, like most researchers, also will call it simply ELISPOT.

2. Fields of Application of ELISPOT Assay

As it has been reported by Tanguay and Killion, ELISPOT appears to be 200 times more sensitive than ELISA in detecting secreted cytokines *(4)*. These authors have shown that it was below delectability level of ELISA to detect cytokines released by less than 10^4 cells, whereas as many as 10–100 cells per well was sufficient for the detection of cytokine-releasing cells. Such a high sensitivity makes ELISPOT a technique of choice for the detection of spontaneous and antigen-induced secretion of cytokines (e.g., interferon [IFN]-γ, tumor necrosis factor [TNF]-α, interleukin [IL]-2, IL-4) from peripheral blood lymphocytes *(5,6)*. ELSIPOT is widely used for vaccine development *(7–9)*, AIDS research *(10,11)*, cancer research (for review. *see* **ref**. *12*), infectious diseases monitoring *(13)*, autoimmune disease studies *(14)*, and allergy and transplantation research *(15,16)*.

3. Immunochemical Principles of ELISPOT Assay

Even though ELISPOT uses the same immunochemical "sandwich" principles as ELISA (**Fig. 1**) there are two main differences between these two assays. First, ELISA measures the real concentration of the cytokine *(17)* and thus answers the question "how much is secreted?", whereas ELISPOT enumerates secreting cells answering the question "what is the frequency of secreting cells?" *(1,2)*. Therefore, one assay should be used not "instead of," but rather "in addition to" the other. Second, ELISA is an immunoassay designed to analyze mostly cell-free media *(17)*, whereas ELISPOT is a combination of both immunoassay and bioassay because live cells are cultured directly in ELISPOT plates. It appears that the quality of spots depends on both immunoassay and bioassay components (*see* examples in troubleshooting in **Subheading 7.1.**).

4. Nuts and Bolts of ELISPOT Assay

The performance of ELISPOT assay depends on the quality of four major components: (1) antibodies (both capture and detection), (2) enzyme conjugates, (3) chromogenic substrates, and (4) membrane-backed plates. Because the secretion activity of cells in ELSIPOT is determined by the number of spots on the bottom of the plate *(1,2)*, it appears that all four components should be optimized to facilitate the formation of detectable spots. Spots should have strong staining intensity (high signal-to-noise ratio) and have well-defined edges. It also is desirable that spots have a small diameter to

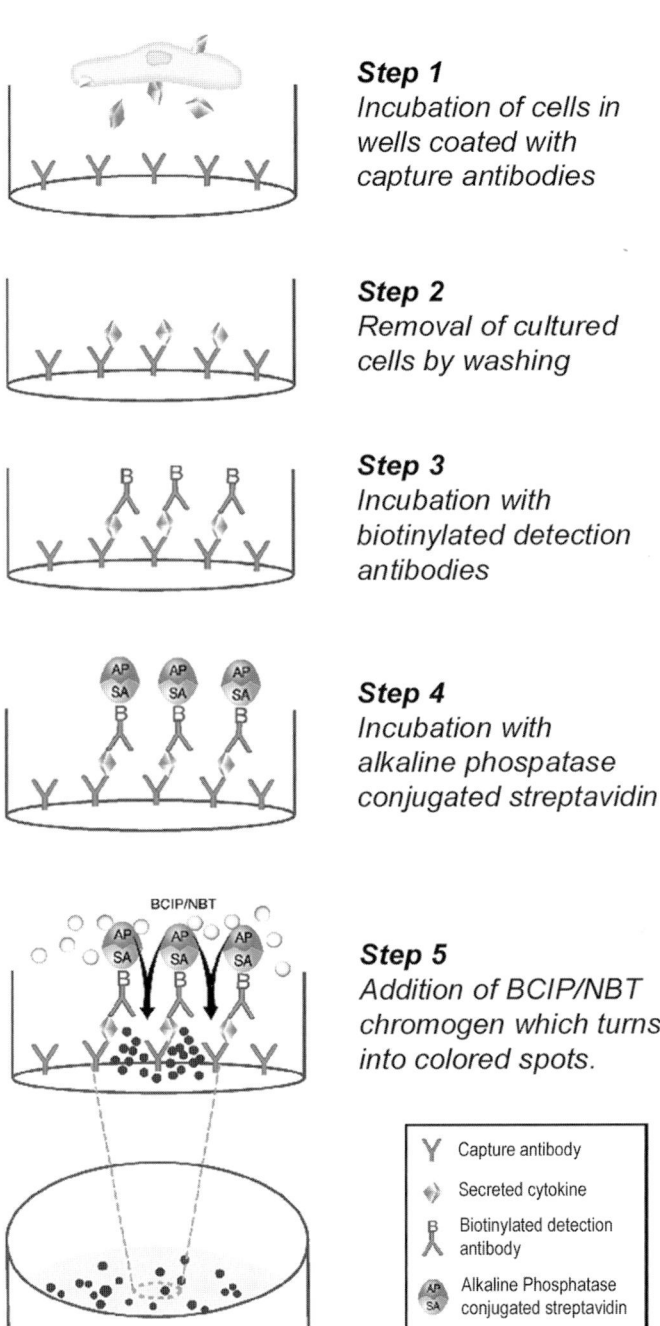

Step 1
Incubation of cells in wells coated with capture antibodies

Step 2
Removal of cultured cells by washing

Step 3
Incubation with biotinylated detection antibodies

Step 4
Incubation with alkaline phospatase conjugated streptavidin

BCIP/NBT

Step 5
Addition of BCIP/NBT chromogen which turns into colored spots.

Y Capture antibody

◆ Secreted cytokine

B Biotinylated detection antibody

AP SA Alkaline Phosphatase conjugated streptavidin

○ BCIP/NBT substrate

● Color precipitate (forms spots)

Fig. 1. Typical ELISPOT assay procedure.

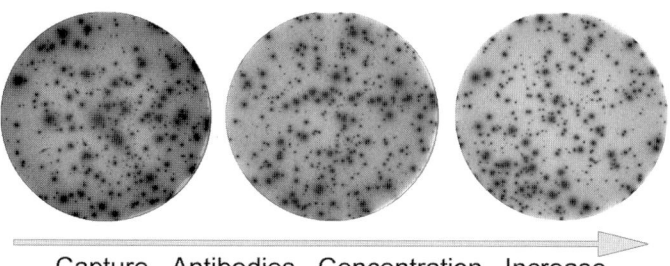

Capture Antibodies Concentration Increase

Fig. 2. Effect of capture antibodies' concentration on the size of spots and background staining (human IL-8 ELISPOT kit; R&D Systems).

avoid their merging with each other: a few merged spots may be erroneously counted as a single spot.

4.1. Antibodies

Both monoclonal and polyclonal antibodies can be used in ELISPOT assays for either antigen capture or antigen detection. ELISPOT can use capture and detection antibodies that were raised either against the entire antigen molecule (e.g., antirecombinant protein antibodies) or against a portion of the antigen (e.g., antipeptide antibodies). The critical factor in choosing capture and detection antibodies is their ability to recognize nonoverlapping epitopes of the target antigen *(17)*. For these reasons it is not recommended to use the same monoclonal antibody for both capture and detection in the same ELISPOT assay. Suitability of antibodies for such applications as immunohistochemistry and western blotting and even ELISA does not necessarily guarantee that these antibodies will also work in ELISPOT (A. Kalyuzhny, personal observations). The only reliable method to identify the best capture and detection antibody combinations is to test antibodies directly in an ELISPOT assay. The concentration of capture antibodies has to be optimized to obtain intensely stained spots with well-defined edges: **Fig. 2** illustrates the effect of coating antibody concentration on the size of spots, intensity of their staining, and the background. Detection antibodies used in ELISPOT need to be conjugated to biotin to make possible their reaction with streptavidin-enzyme conjugates *(18)*. The reason detection antibodies need to be biotinylated is to avoid crossreactivity: if both capture and detection antibodies are raised in the same species (e.g., mouse), antibodies (e.g., anti-mouse) conjugated to enzyme will bind to both capture and detection antibodies rather binding to detection antibodies only. Alternatively, detection system may use detection antibodies directly conjugated to enzyme (so-called direct conjugate). Unfortunately the sensitivity of

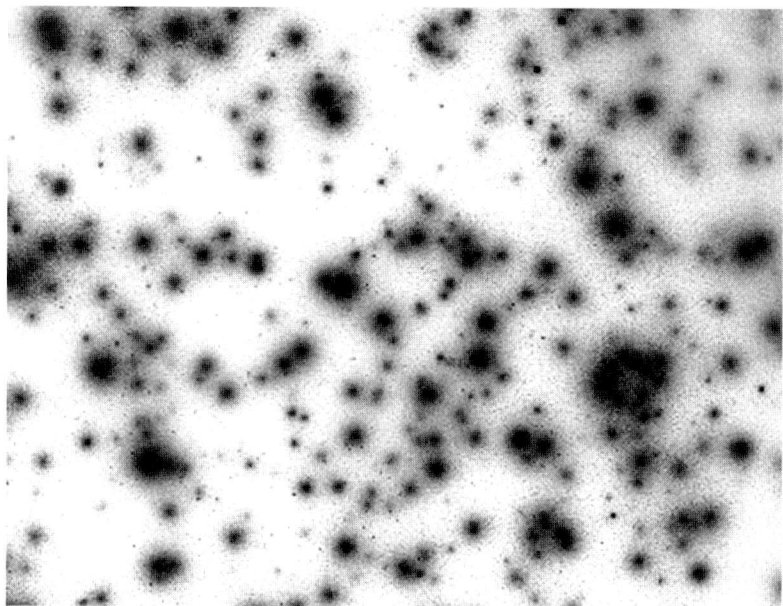

Fig. 3. Two-cytokine ELISPOT assay (custom-made kit; R&D Systems). IL-2 release from human peripheral blood lymphocytes is detected using Alkaline phosphatase-BCIP/NBT reagents, whereas IFN-γ is detected using HRP-AEC detection system. *See* **Color Plate 1** following page 50.

ELISPOT assays that use direct conjugates may be lower than that of avidin-biotin ones.

4.2. Enzyme Conjugates

Horseradish peroxidase (HRP) or alkaline phosphatase (AP) can be used as streptavidin conjugates *(18)*. HRP (optimum pH 7.6) in the presence of hydrogen peroxide (H_2O_2) catalyzes the oxidation of substrates, which change color with the loss of electrons. The advantage of using HRP is its high turnover rate (spots develop faster), whereas the drawback is increased background. Unlike HRP, AP (optimum pH 9.0–9.6) has a linear reaction rate (spots develop slower), allowing for longer incubations with chromogenic substrates *(18)* without a risk of developing background staining. Longer incubation may be performed if it is necessary to increase the sensitivity of AP-based assay. By combining HRP and AP, it is possible to develop an ELISPOT assay for simultaneous detection of two different cell-secreted molecules (**Fig. 3**; **refs.** *19* and *20*). The major drawback of mulianalyte systems is the loss of sensitivity for each of the antigens. I have observed that a number of spots formed by IL-2 and IFN-γ secreted form peripheral blood mononuclear cells in the plate coated with anti-IL-2 and anti-IFN-γ antibodies was noticeably lower in comparison with corresponding single-cytokine assays (A. Kalyuzhny, personal observation). I have found that the drop in sensitivity becomes even more profound if ELISPOT plate is coated with more than two capture antibodies (*see* Chapter 18). The mechanism underlying this phenomenon is not known, and additional research is needed to find the ways of building high-sensitivity multianalyte ELISPOT assays.

4.3. Enzyme Substrates

Regardless of which enzyme conjugate is used, their corresponding substrates should produce intense and stable colors. HRP substrate such as AEC (3-amino-9-ethylcarbazole, $C_{14}H_{14}N_2$) forms intense red color spots *(18)*. However, AEC is unstable *(18)*, and spots will bleach in a short period of time. This, in turn, will result in irrecoverable loss of primary data. Another HRP substrate, DAB (3,3′-diaminobenzidine, $C_{12}H_{14}N_4$), produces brown color spots that are less intense than their AEC counterparts *(18)* and, although stable, DAB is poisonous and potentially carcinogenic. One of the most frequently used substrates for AP is a mixture of BCIP (5-bromo-4-chloro-3-indolylphosphate *p*-toluidine salt) and NBT (Nitroblue tetrazolium chloride) which forms intense black–blue spots *(18)*. Because of the high stability of BCIP/NBT, spots do not fade, and ELISPOT plates can be re-analyzed after being stored for several years.

4.4. Assay-Developing Procedures

The secretion capacity of cells may be tested in two ways: (1) cells are cultured in a designated plate and then transferred into ELISPOT plates *(21–23)*, or (2) cells are stimulated and cultured directly in ELISPOT plates *(24)*. Depending on the research project, cells may be stimulated one way or the other, but it should be kept in mind that cells cultured and stimulated outside ELISPOT plate need to be transferred into a fresh culture medium before being plated into an ELISPOT plate to avoid background staining.

4.5. Membrane-Backed Microplates

ELISPOT assays can be performed using either 96-well clear plastic plates *(4,25)* or plates backed with membranes such as polyvinylidene diflouride *(26,27)* and nitrocellulose *(25,28)*. Unlike lateral flow and flow-through assays, membranes in ELISPOT assay are used for other reasons: they support the growth of cells and have a much higher retaining capacity for capture antibodies (because higher surface area) than conventional plastic plates. In ELISPOT assay the flow of reagents through or across the membrane is not required, but rather a diffusion of cell-secreted molecules towards capture antibodies immobilized on the membrane. Membrane plates are manufactured by different vendors, including the Millipore Corporation, Pall Corporation, and Whatman. All vendors manufacture comparable plates, but it appears that Millipore plates are more popular for ELISPOT assay. This may be attributed to the fact that membranes with spots can be easily removed from Millipore plates for compact filing and protection purposes (*see* membrane removal systems in **Subheading 6.**).

4.6. Types of ELISPOT Assays

There are two major commercial formats of ELISPOT assay: (1) fully developed and optimized ready-to-use kits (RTU) and (2) so-called do-it-yourself (DIY) kits which, include reagents and uncoated 96-well plates to develop an assay. RTU kits may include precoated 96-well plates and all necessary reagents to run the assay. DIY kits need to be optimized by the researcher, which can be a very laborious procedure. RTU kits are more expensive than DIY ones, but RTU kits are the best choice for large-scale clinical trial experiments requiring convenience and a high degree of accuracy (*29*). R&D Systems, Inc. was the first company to design and introduce completely optimized RTU ELISPOT kits, which include dry precoated membrane microplates, wash buffers, detection antibodies, and AP-BCIP/NBT detection reagents.

5. ELISPOT Data Analysis

In ELISA assays, the concentration of the molecules in the sample is determined by measuring the optical density of the color substrate solution filling the wells (*17*), whereas in ELISPOT, cell-secretion capacity is measured by counting colored spots on the bottom of the well (*1,2*). The term "spot-forming cells," or SFC, is used as a quantitative measure of the cell secretion activity in ELISPOT assay (*30,31*).

5.1. Quantification of Spots

After finishing the assay, spots can be counted either manually or by using computer-aided image analysis (*32*). Manual counting is performed under the stereomicroscope using, for example, a hand-held tally counter. Manual counting is very tedious and time-consuming but appears to be of higher sensitivity, allowing investigator to identify faint spots of smaller sizes and decide whether spot is "real" or an artifact. Computer-aided quantification can be performed using either inexpensive semiautomated (MVS Pacific; www.mvspacific.com) or more expensive but fully automated systems offered by such vendors as Zeiss (www.zeiss.com; *see* Chapter 8 in this book), Cellular Technology (www.immunospot.com; *see* Chapter 7 in this book), and AutoImmun Diagnostika (www.elispot.com)

Regardless of the system used, a 96-well ELISPOT plate is mounted onto microscopy stage and moved in front of the lens to capture images of individual wells. When using semiautomated systems, the operator moves the ELISPOT plate either by hand or by using a joystick connected to the stage, whereas on fully automated readers the plate is moved automatically according to the image collection sequence set by the operator. The main component of each ELISPOT reader is its software allowing capture of images and quantification of spots: the better the design of the spot-recognition and image-processing algorithm, the higher the value of the software. Fully automated systems are faster and more expensive but not necessarily more accurate than semiautomated ones. It appears that customers are paying more for convenience of automation rather than for higher accuracy of quantification.

5.1.1. Manual Quantification of Spots

The typical set-up for manual spot counting would be a 4X objective lens with 10X eyepieces. The main concern with manual counting is the human error and subjective bias, for example, very small spots may go unnoticed, whereas two close spots may be counted as a single spot. Interestingly, that even with these limitations, the human visual system has higher resolution than the existing computerized analyzers.

5.1.2. Computer-Aided Quantification of Spots

Computer-aided quantification of spots is thought to be more reliable than the manual counting *(32)*. Computerized systems use a charged-coupled device (CCD) camera to visualize and capture digital images of each well. Many image-processing algorithms are designed to detect and count spots in each captured image automatically. Unfortunately, the finite pixel size of the CCD camera poses serious limitations on both resolution and detectability for both smaller spots and for clusters of closely-spaced spots of all sizes. These limitations are usually summarized by quoting a Nyquist-limited resolution of two-pixels' width *(33)*. Pixel size is an important consideration in computerized ELISPOT readers. For accurate quantification pixels need to be at least half the size of the smallest spots.

5.1.3. Types of ELISPOT Readers

Current systems can be further grouped as "macro-imagers" or "image-tiling systems."

5.1.3.1. MACRO-IMAGERS

These readers capture an image from an entire well. The camera is moved from well to well, either manually or automatically. The limitation of the

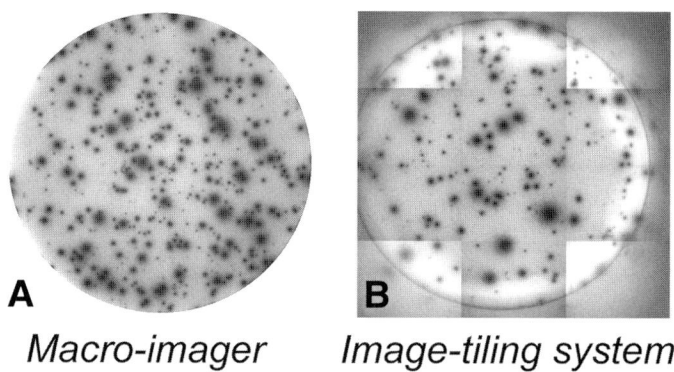

A **Macro-imager**　　**B** **Image-tiling system**

Fig. 4. Typical single well ELISPOT images. (**A**) image captured with macro-imager (ImageHub; MVS Pacific: www.mvspacific.com). (**B**) image captured using the image-tiling system (KS ELISPOT reader; Carl Zeiss).

macro approach is the number of pixels in the CCD's focal plane array. A typical 1000×1000 pixel CCD imager would have 6-µm pixels when focused on a 6-mm diameter membrane on the bottom of the well. As a result, small spots may go undetected, whereas clusters of spots will be counted as a single spot. **Figure 4A** shows the typical image of an ELISPOT well captured using a macro-imager.

5.1.3.2. IMAGE-TILING SYSTEMS

Higher resolution can be achieved by using higher magnification and capturing multiple image "tiles," each from a small portion of a single membrane. These individual image tiles are then "stitched" or "seamed" together into a larger image that can be analyzed (for example, U.S. Patent 4,760,385 discloses the principles of image tiling). Image-tiling systems are more expensive than macro-imagers because they require a fully automated microscope that moves the 96-well plate while a computer and a video-formatted CCD camera automatically coordinate the capture of many individual images or tiles. **Figure 4B** shows the typical image collected by such a tiling system, the KS Elispot reader (Zeiss). Tiling systems are not only expensive, but they become prohibitively slow at higher resolutions. An additional drawback of tiling systems is that they often sacrifice image quality at tile boundaries where the combination of imperfect tile alignment and optical distortion may result in image artifacts (refer to **Fig. 4B**).

6. Archiving of Primary ELISPOT Data

After finishing the experiment, stained 96-well plates become primary experimental data, and it may be required to store them in a safe place. Unfortunately

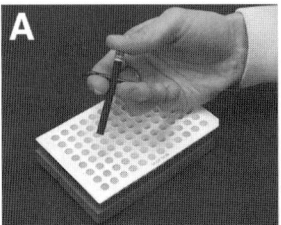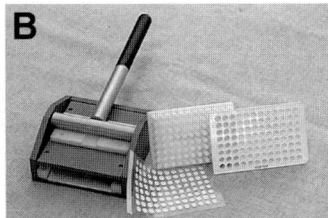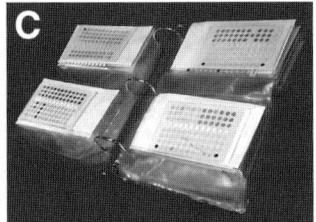

Fig. 5. Archiving stained membranes from ELISPOT plates. (**A**) single-membrane removal tool (Zellnet; www.zellnet.com). (**B**) membrane-removal device for simultaneous removal of all 96 membranes from the plate and their transfer onto adhesive film (MVS Pacific; www.mvspacific.com). (**C**) removed membranes can be stored in a regular photo album and reanalyzed when needed.

96-well plates are bulky, and their storage requires a lot of space, especially during large-scale clinical trials. To solve this problem, membranes with spots can be punched out of the plates, laminated, barcoded, and stored in a regular photo-album. **Figure 5** depicts two types of membrane removal systems: single-well puncher made by Zellnet (**Fig. 5A**; www.zellnet.com) and 96-well membrane removal device (**Fig. 5B**) designed by MVS Pacific (US Patent 6,631,649; www.mvspacific.com). The latter device allows for simultaneous removal and transferring of all 96 membranes from the plate onto adhesive film in less than a minute. Adhesive film with attached membranes can be laminated to protect membranes with spots from damage during their handling. Removed and laminated membranes can be stored in a regular photo album as shown on **Fig. 5C**. If needed, removed membranes can be reanalyzed using ELISPOT readers.

7. Troubleshooting ELISPOT Assays

7.1. Staining

The quality of staining has a strongest impact on the accuracy of the quantification of spots in an ELISPOT assay. There are two major staining problems that require troubleshooting: background staining and staining of spots. Background in an ELISPOT assay is defined as a staining that covers either a part of or the entire membrane. Backgrounds may be further categorized as either specific or nonspecific. Specific background is formed as the result of specific binding of cell-secreted molecules by capture antibodies: molecules that are released from the cell dissociate from capture antibodies surrounding the releasing cell, diffuse, and bind to capture antibodies in the cell-free zone. A specific background may occur, for example, if an ELISPOT plate with cells is disturbed during the incubation. Once an ELISPOT plate is placed into the incubator, it should not be touched or moved for the entire incubation period.

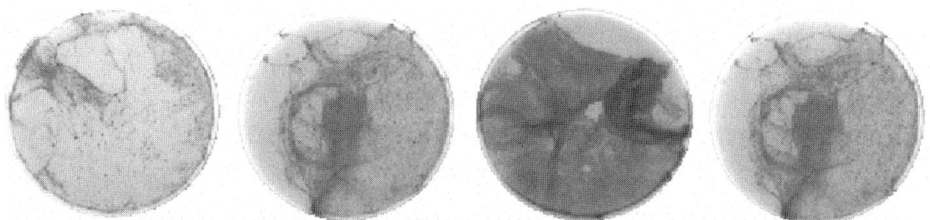

Fig. 6. Variations of high background staining and spot-looking artifacts, which can be caused by high number of dead cells added into the ELISPOT plate.

Frequent opening and closing of the incubator's door also may disturb cells in the plate. Nonspecific background is caused by the adsorption of detecting components (detection antibodies, enzyme conjugate, and precipitating substrate) onto the membrane. Both specific and nonspecific backgrounds hinder the detection and counting of spots. It is easier to troubleshoot one rather than both types of background. It is more difficult to identify the source of nonspecific background because of multiple factors contributing to it. We have found that one of the universal remedies against both specific and nonspecific backgrounds is aluminum foil. Wrapping ELISPOT plates into aluminum foil reduces background staining and improves contrast. It also produces a more uniform distribution of specific spots across the filter membrane *(34)*. In addition, application of foil appears to improve well-to-well reproducibility. The reason aluminum foil reduces the background staining is not known, but it is tempting to speculate that aluminum foil facilitates even distribution of heat over the bottom of ELISPOT plate during its incubation in CO_2 incubator.

7.2. Cells

The quality of staining also depends on the quality of cultured cells. It is of critical importance to determine the percent of dead cells because we have found that a high number of dead cells (30–50% and more) may be a reason for a high background staining and even lack of specific spots (**Fig. 6**). In some cases, even though the number of dead cells is low (e.g., approx 5%), there may be no spots formed at all because of apoptosis (**Fig. 7**). Intensity of staining also depends on the number of cells plated into the well—the addition of excessive number of cells per well may result in overstaining due to a specific background. Because the secretion capacity of cells is not known in advance, it is always recommended to test serial dilution of cells from each individual donor (e.g., 10^3, 10^4, 10^5, 10^6 cells per well) in the same ELISPOT plate. This ensures having enough data points to choose from in case over- or underdevelopment occurs. Using cells of unknown secretion capacity requires dedicating many

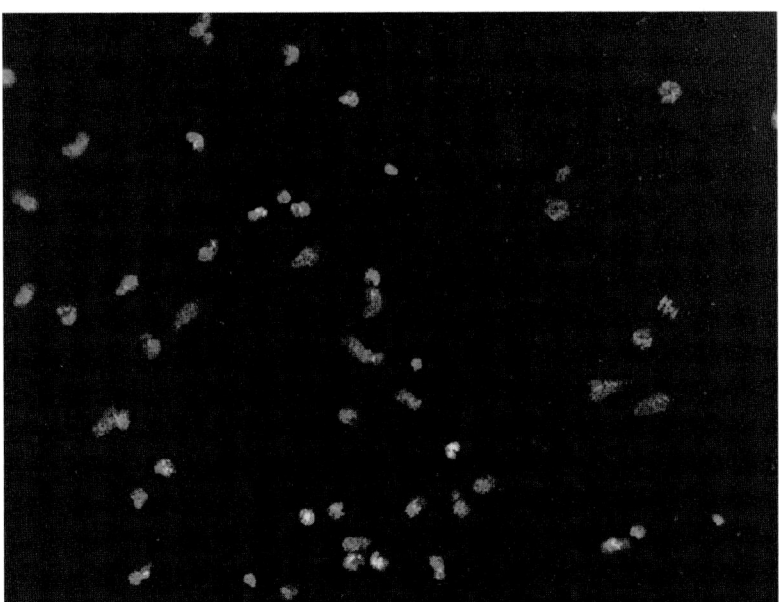

Fig. 7. High degree of programmed cell death (apoptosis) in cells plated into the ELISPOT plate may result in lack of spots. Cells attached to membranes (green fluorescence) were labeled immunocytochemically for an apoptosis marker active Caspase-3 using R&D Systems anticaspase-3 antibodies (red color). Note the high number of apoptotic cells. *See* **Color Plate 2** following page 50.

wells in the plate for cell optimization rather than for experimental groups. The solution to this problem is to preserve cell suspensions, freeze them, and store them in liquid nitrogen. It was reported that freezing of peripheral blood lymphocytes did not significantly affect their rosette formation *(35)* and that freezing of dendritic cells did not impair their ability to respond to maturation signals *(36)*. In the ELISPOT assay, cryopreserved peripheral blood mononuclear cells were shown to be similar to freshly isolated cells in their capacity to release IFN-γ *(37–39)* or even exceeded the latter *(27)*. A stock of cryopreserved cells with a known secretion capacity may be used in a single predetermined concentration in the entire ELISPOT plate. We have reported previously that the same cryopreserved peripheral blood lymphocytes can be used to study release of different cytokines *(40)*. Cryopreservation of cells for ELISPOT is advantageous for clinical trial studies because it helps to avoid variations in biological samples collected from the same donor but on different dates. Interestingly, cryopreserved cells are more active in secreting some cytokines *(41,42)*, which is thought to be caused by elimination of inhibitory platelets, which do not withstand freezing *(27)*.

7.3. Washing Procedures

The ultimate purpose of washing ELISPOT plates is to remove cultured cells and unbound reagents (detection antibodies, enzyme conjugate, enzyme substrate) form ELISPOT plates to minimize background staining. Plates can be washed, for example, with phosphate-buffered saline of various pH and molarity. It is necessary to remove as many cells as possible by washing since stained cells may be confused with specific spots and thus affect the accuracy of quantification. In some cases stimulated cells become very sticky and their complete removal may require incubation with enzymatic cell-detachment solutions (*see* Chapter 5).

8. ELISPOT Assay as a Tool for In Vitro Diagnostics

ELISPOT is widely used for research purpose but has a great potential as a diagnostic tool. For example, it was reported that the ESAT-6/CFP-10-based ELISPOT assay can be used to detect active tuberculosis in HIV-positive individuals with high sensitivity *(43)*. Authors of this study suggested that ELISPOT was more specific and more sensitive than PPD-based methods to detect latent *Mycobacterium tuberculosis* infection. ELISPOT may be also used for allergy diagnostics: it was reported that peripheral blood mononuclear cells from nickel-allergic individuals responded to Ni^{2+} with significantly greater production of IL-4, IL-5, IL-13, and IFN-γ, compared with the healthy controls *(44)*. It appears that the format of 96-well-based ELISPOT assay needs to be modified for diagnostic applications. First, the assay should be miniaturized to reduce the volume of samples needed for analysis: this is particularly important in pediatrics. Second, a fast and easy-to-operate turnkey ELISPOT reading system/scanner should be available to analyze staining and creating a report. Third, matrix with stained spots (e.g., membranes, plastics, etc.) should be both small enough for compact filing and have enough room for bar code labeling. Fourth, dyes used to stain spots should be stable to allow their re-evaluation after an extended period of time.

References

1. Sedgwick, J. D., and Holt, P. G. (1983) A solid-phase immunoenzymatic technique for the enumeration of specific antibody-secreting cells. *J. Immunol. Methods* **57**, 301–309.
2. Czerkinsky, C. C., Nilsson, L. A., Nygren, H., Ouchterlony, O., and Tarkowski, A. (2983) A solid-phase enzyme-linked immunospot (ELISPOT) assay for enumeration of specific antibody-secreting cells. *J. Immunol. Methods* **65**, 109–121.
3. Czerkinsky, C., Moldoveanu, Z., Mestecky, J., Nilsson, L. A., and Ouchterlony, O. (1988) A novel two colour ELISPOT assay. I. Simultaneous detection of distinct types of antibody-secreting cells. *J. Immunol. Methods* **115**, 31–37.

4. Tanguay, S. and Killion, J. J. (1994) Direct comparison of ELISPOT and ELISA-based assays for detection of individual cytokine-secreting cells. *Lymphokine Cytokine Res.* **13**, 259–263.
5. Bienvenu, J., Monneret, G., Fabien, N., and Revillard, J. P. (2000) The clinical usefulness of the measurement of cytokines. *Clin. Chem. Lab. Med.* **38**, 267–285.
6. Mashishi, T. and Gray, C. M. (2002) The ELISPOT assay: an easily transferable method for measuring cellular responses and identifying T-cell epitopes. *Clin. Chem. Lab. Med.* **40**, 903–910.
7. Pass, H. A., Schwarz, S. L., Wunderlich, J. R., and Rosenberg, S. A. (1998) Immunization of patients with melanoma peptide vaccines: immunologic assessment using the ELISPOT assay. *Cancer J. Sci. Am.* **4**, 316–323.
8. Asai, T., Storkus, W. J., and Whiteside, T. L. (200) Evaluation of the modified ELISPOT assay for γ interferon production in cancer patients receiving antitumor vaccines. *Clin. Diagn. Lab. Immunol.* **7**, 145–154.
9. Kamath, A. T., Groat, N. L., Bean, A. G., and Britton, W. J. (2000) Protective effect of DNA immunization against mycobacterial infection is associated with the early emergence of interferon-γ (IFN-γ)-secreting lymphocytes. *Clin. Exp. Immunol.* **120**, 476–482.
10. Eriksson, K., Nordstrom, I., Horal, P., Jeansson, S., Svennerholm, B., Vahlne, A., Holmgren, J., and Czerkinsky, C. (1992) Amplified ELISPOT assay for the detection of HIV-specific antibody-secreting cells in subhuman primates. *J. Immunol. Methods* **153**, 107–113.
11. Howell, D. M., Feldman, S. B,. Kloser, P., and Fitzgerald-Bocarsly, P. (1994) Decreased frequency of functional natural interferon-producing cells in peripheral blood of patients with the acquired immune deficiency syndrome. *Clin. Immunol. Immunopathol.* **71**, 223–230.
12. Schmittel, A., Keilholz, U., Thiel, E., and Scheibenbogen, C. (2000) Quantification of tumor-specific T lymphocytes with the ELISPOT assay. *J. Immunother.* **23**, 289–295.
13. Smith, S. M., Brookes, R., Klein, M. R., Malin, A. S., Lukey, P. T., King, A. S., et al. (2000) Human CD8+ CTL specific for the mycobacterial major secreted antigen 85A. *J. Immunol.* **165**, 7088–7095.
14. Pelfrey, C. M., Cotleur, A. C., Lee, J. C., and Rudick, R. A. (2002) Sex differences in cytokine responses to myelin peptides in multiple sclerosis. *J. Neuroimmunol.* **130**, 211–223.
15. Schmid-Grendelmeier, P., Altznauer, F., Fischer, B., Bizer, C., Straumann, A., Menz, G., Blaser, K., et al. (2002) . Eosinophils express functional IL-13 in eosinophilic inflammatory diseases. *J. Immunol.* **169**, 1021–1027.
16. Sho, M., Sandner, S. E., Najafian, N., Salama, A. D., Dong, V., Yamada, A., et al. (2002) Sayegh: new insights into the interactions between T-cell costimulatory blockade and conventional immunosuppressive drugs. *Ann. Surg.* **236**, 667–675.
17. Kemeny, D. M. (1997) Enzyme-linked immunoassays, in *Immunochemistry 1* (Johnstone, A. P., and Turner, M. W., eds)Oxford University Press, Oxford. p. 147–175.

18. Savage, M. D., Mattson, G., Desai, S., Nielander, G. W., Morgensen, S., and Conklin, E. J. (1992) *Avidin-Biotin Chemistry: A Handbook*. Pierce Chemical Co., Rockford, IL, p. 467.
19. Okamoto, Y., Abe, T., Niwa, T., Mizuhashi, S., and Nishida, M. (1998) Development of a dual color enzyme-linked immunospot assay for simultaneous detection of murine T helper type 1- and T helper type 2-cells. *Immunopharmacology* **39**, 107–116.
20. Okamoto, Y., Gotoh, Y., Tokui, H., Mizuno, A., Kobayashi, Y., and Nishida, M. (2000) Characterization of the cytokine network at a single cell level in mice with collagen-induced arthritis using a dual color ELISPOT assay. *J. Interferon. Cytokine Res.* **20**, 55–61.
21. Favre, N., Bordmann, G., and Rudin, W. (1997) Comparison of cytokine measurements using ELISA, ELISPOT and semi-quantitative RT-PCR. *J. Immunol. Methods* **204**, 57–66.
22. Herr, W., Schneider, J., Lohse, A. W., Meyer zum Buschenfelde, K. H., and Wolfel, T. (1996) Detection and quantification of blood-derived CD8+ T-lymphocytes secreting tumor necrosis factor alpha in response to HLA-A2.1-binding melanoma and viral peptide antigens. *J. Immunol. Methods* **191**, 131–142.
23. Arlen, P., Tsang, K. Y., Marshall, J. L., Chen, A., Steinberg, S. M., Poole, D., et al. (2000) The use of a rapid ELISPOT assay to analyze peptide-specific immune responses in carcinoma patients to peptide vs. recombinant poxvirus vaccines. *Cancer Immunol. Immunother.* **49**, 517–529.
24. Janetzki, S., Song, P., Gupta, V., Lewis, J. J., and Houghton, A. N. (2000) Insect cells as HLA-restricted antigen-presenting cells for the IFN-γ elispot assay. *J. Immunol. Methods* **234**, 1–12.
25. Ronnelid, J., and Klareskog, L. (1997) A comparison between ELISPOT methods for the detection of cytokine producing cells: greater sensitivity and specificity using ELISA plates as compared to nitrocellulose membranes. *J. Immunol. Methods* **200**, 17–26.
26. Schielen, P., van Rodijnen, W., Tekstra, J., Albers, R., and Seinen, W. (1995) Quantification of natural antibody producing B-cells in rats by an improved ELISPOT technique using the polyvinylidene diflouride membrane as the solid support. *J. Immunol. Methods* **188**, 33–41.
27. McCutcheon, M., Wehner, N., Wensky, A., Kushner, M., Doan, S., Hsiao, L., et al. (1997) A sensitive ELISPOT assay to detect low-frequency human T-lymphocytes. *J. Immunol. Methods* **210**, 149–166.
28. Taguchi, T., McGhee, J. R., Coffman, R. L., Beagley, K. W., Eldridge, J. H., Takatsu, K., et al. (1990) Detection of individual mouse splenic T-cells producing IFN-γ and IL-5 using the enzyme-linked immunospot (ELISPOT) assay. *J. Immunol. Methods* **128**, 65–73.
29. Klencke, B., Matijevic, M., Urban, R. G., Lathey, J. L., Hedley, M. L., Berry, M., et al. (2002) Encapsulated plasmid DNA treatment for human papillomavirus 16-associated anal dysplasia: a Phase I study of ZYC101. *Clin. Cancer Res.* **8**. 1028–1037.

30. McGhee, M. L., Ogawa, T., Pitts, A. M., Moldoveanu, Z., Mestecky, J., McGhee, J. R., and Kiyono, H. (1989) Cellular analysis of functional mononuclear cells from chronically inflamed gingival tissue. *Reg. Immunol.* **2**, 103–110.

31. Merville, P., Pouteil-Noble, C., Wijdenes, J., Potaux, L., Touraine, J. L., and Banchereau, J. (1993) Detection of single cells secreting IFN-γ, IL-6, and IL-10 in irreversibly rejected human kidney allografts, and their modulation by IL-2 and IL-4. *Transplantation* **55**, 639–646.

32. Herr, W., Linn, B., Leister, N., Wandel, E., Meyer zum Buschenfelde, K. H., and Wolfel, T., (1997) The use of computer-assisted video image analysis for the quantification of CD8+ T-lymphocytes producing tumor necrosis factor alpha spots in response to peptide antigens. *J. Immunol. Methods* **203**, 141–152.

33. Holst, G. C. (1998) *CCD Arrays, Cameras, and Displays.* 2nd ed., Society of Photo-optical Instrumentation Engineers, Wintrer Park, FL.

34. Kalyuzhny, A., and Stark, S. (2001) A simple method to reduce the background and improve well-to-well reproducibility of staining in ELISPOT assays. *J. Immunol. Methods* **257**, p. 93–97.

35. Merker, R., Check, I., and Hunter, R. L. (1979) Use of cryopreserved cells in quality control of human lymphocyte assays: analysis of variation and limits of reproducibility in long-term replicate studies. *Clin. Exp. Immunol.* **38**, p. 116–126.

36. Lewalle, P., Rouas, R., Lehmann, F., and Martiat, P. (2000) Freezing of dendritic cells, generated from cryopreserved leukaphereses, does not influence their ability to induce antigen-specific immune responses or functionally react to maturation stimuli. *J. Immunol. Methods* **240,** 69–78.

37. Keane, N. M., Price, P., Stone, S. F., John, M., Murray, R. J., and French, M. A. (2000) Assessment of immune function by lymphoproliferation underestimates lymphocyte functional capacity in HIV patients treated with highly active antiretroviral therapy. *AIDS Res. Hum. Retroviruses* **16**, p. 1991–1996.

38. Smith, J. G., Liu, X., Kaufhold, R. M., Clair, J., and Caulfield, M. J. (2001) Development and validation of a γ interferon ELISPOT assay for quantitation of cellular immune responses to varicella-zoster virus. *Clin. Diagn. Lab. Immunol.* **8**, p. 871–879.

39. Sobota, V., Bubenik, J., Indrova, M., Vlk, V., and Jakoubkova, J. (1997) Use of cryopreserved lymphocytes for assessment of the immunological effects of interferon therapy in renal cell carcinoma patients. *J. Immunol. Methods* **203**, p. 1–10.

40. Bailey, T., Stark, S., Grant, A., Hartnett, C., Tsang, M., and Kalyuzhny, A. (2002) A multidonor ELISPOT study of IL-1beta, IL-2, IL-4, IL-6, IL-13, IFN-γ and TNF-alpha release by cryopreserved human peripheral blood mononuclear cells. *J. Immunol. Methods* **270**, p. 171–182.

41. Venkataraman, M. (1994) Effects of cryopreservation on immune responses: VII. Freezing induced enhancement of IL-6 production in human peripheral blood mononuclear cells. *Cryobiology* **31**, 468–477.

42. Venkataraman, M. (1995) Effects of cryopreservation on immune responses. VIII. Enhanced secretion of interferon-γ by frozen human peripheral blood mononuclear cells. *Cryobiology* **32**, 528–534.

43. Chapman, A. L., Munkanta, M., Wilkinson, K. A., Pathan, A. A., Ewer, K., Ayles, H., et al. (2002) Rapid detection of active and latent tuberculosis infection in HIV-positive individuals by enumeration of Mycobacterium tuberculosis-specific T-cells. *Aids* **16**, p. 2285–2293.
44. Jakobson, E., Masjedi, K., Ahlborg, N., Lundeberg, L., Karlberg, A. T., and Scheynius, A. (2002) Cytokine production in nickel-sensitized individuals analysed with enzyme-linked immunospot assay: possible implication for diagnosis. *Br. J. Dermatol.* **147**, p. 442–449.

3

Membranes and Membrane Plates Used in ELISPOT

Alan J. Weiss

Summary

Membrane-bottomed, 96-well plates constitute the format in which the overwhelming majority of enzyme-linked immunospot (ELISPOT) assays are performed. The membranes in these plates are made from either nitrocellulose or polyvinylidene fluoride. These membranes are well suited for ELISPOT because they have high antibody binding capacities and because their white color provides an excellent backdrop for ELISPOT enumeration. These two membranes and, ultimately, the 96-well plates used in ELISPOT assays were commercialized for filtration applications and later optimized for deoxyribonucleic acid hybridization and protein chemistry applications. In this chapter, an overview of the development and biotechnology applications of nitrocellulose and polyvinylidene fluoride membrane is provided and characteristics and attributes of each of the membranes that are relevant to ELISPOT are summarized.

Key Words: ELISPOT; nitrocellulose; polyvinylidene fluoride; membrane; filter plate.

1. Introduction: A Chronology of Relevant Developments

There are two different types of membranes (in 96-well plate formats) that are used in enzyme-linked immunospot (ELISPOT) applications: nitrocellulose (NC) and polyvinylidene fluoride (PVDF). The reason that these are the predominant membrane types used in ELISPOT is based largely on their fortuitous suitability in non-ELISPOT applications rather than on the development of optimized assay substrates to address the specific needs of the ELISPOT assay. The chronology of relevant membrane and applications developments is outlined below.

Timeline of Membrane, Applications, and ELISPOT Developments

1954:   NC membranes become commercially available. Principal use is in removing bacteria from aqueous samples (sterile filtration)

1975:   0.45-μm NC membrane used in deoxyribonucleic acid (DNA) hybridization (Southern blotting) assay *(1)*

From: *Methods in Molecular Biology, vol. 302: Handbook of ELISPOT: Methods and Protocols*
Edited by: A. E. Kalyuzhny © Humana Press Inc., Totowa, NJ

1975:   PVDF membranes become commercially available. Initially, principal use is in removing bacteria from aqueous samples. PVDF is more durable and more solvent resistant than NC and as such is more suited to large-scale filtration applications.

1979:   0.45-µm NC membrane used in immunodetection of electroblotted proteins (Western blotting) *(2)*.

1983:   ELISPOT assay developed on plastic, 96-well plates *(3,4)*.

1985:   96-well plate with NC membrane commercially available. Intended application is dot blotting of nucleic acids using vacuum transfer (instead of capillary transfer).

1986:   0.45-µm PVDF membrane used in Western blotting assay *(5,6)*.

1988:   96-well plate with NC membrane used in ELISPOT *(7)*.

1992:   96-well plate with PVDF membrane commercially available. Intended application is dot blotting of proteins using vacuum transfer (instead of electro-blotting). Millipore Corporation (Bedford, MA) reports in its 1994/5 Catalog that this plate can be used in ELISPOT.

1995:   Articles citing improved ELISPOT results using PVDF plates are published *(8)*.

NC and then PVDF membranes were developed to serve the needs of sterile filtration applications. In ways that were never anticipated by membrane manufacturers, the porosity and binding properties of these two membrane types (NC and PVDF) enabled them to be used in two extremely important and burgeoning research applications. Finally, in response to the specific requirements of these molecular biology and protein chemistry applications, NC and PVDF membrane-bottomed, 96-well plates were developed and made commercially available. Independently and separately, ELISPOT assays were developed on 96-well plastic plates and took advantage of enzyme-linked immunosorbent assay (ELISA) techniques that had been perfected in that format. Because, fundamentally, the immunodetection component of ELISPOT assays and Western blotting is essentially identical, it was only a matter of time until the overwhelming majority of ELISPOT assays were performed on membrane-bottomed, 96-well plates.

2. NC Membrane Development and Applications

NC is made by boiling partially hydrolyzed cellulose in a solution of nitric and sulfuric acid. The cellulose is normally derived from tree pulp that is enzymatically and chemically pretreated and washed to remove lignin, reduce fiber length, and eliminate most of the noncellulose components. Cellulose nitration specifications (completeness, homogeneity, and level of impurities, mostly with regard to residual sulfates) are based on the intended final application of the polymer (e.g., explosive, lacquer or varnish component, membrane constituent.). Nitration of cellulose was first conducted in large-scale quantities

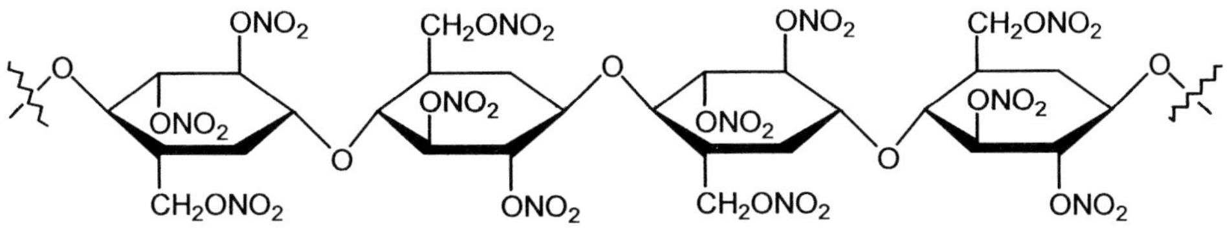

Fig. 1. Chemical structure of trinitrocellulose.

for purposes of producing gun-cotton, an explosive solid material that has many of the same uses as its chemical cousin, trinitroglycerin. The structure of NC (fully nitrated) is illustrated in **Fig. 1**. Chemical engineers in Germany discovered that NC polymers of varying molecular weights (polymer lengths) could be completely dissolved in solvents such as ethanol, methanol, and methylethylketone. If water or other nonsolvents (for NC) were subsequently added to these polymer solutions, it was possible to form membranes (homogeneous porous films) by controlling the evaporation of the (more volatile) solvents. Degree of cellulose nitration (*see* **Note 1**), the amount and average nominal length of the nitrocellulose polymers, solvent strength, ratio of solvent to nonsolvent, and the conditions used (temperature, air flow, relative humidity) to evaporate the solvents all affected the pore size, porosity, and structure of the NC membrane. Technology for making NC membranes was developed with the intent of producing filters that could remove bacteria from contaminated water.

After World War II, this technology was acquired by a number of different companies, including Millipore (Bedford, MA), Schleicher and Schuel (Darmstadt, Germany), and Sartorius (Groningen, Germany). The first commercially available products made by these companies were sterilizing filters (principally 0.45-μm nominal pore size membranes and devices). In 1975, E. M. Southern developed a protocol for detecting specific DNA sequences. This was accomplished by capillary transfer of DNA that had been separated by size in an agarose gel onto NC (*1*). DNA bound to the surface of the NC membrane can be detected more easily and more rapidly than DNA in the agarose gel. This is true for a number of reasons, including improved accessibility of the DNA on the membrane surface and, perhaps most importantly, the DNA distributed throughout the 1-mm thick gel becomes concentrated onto the two-dimensional membrane surface during capillary transfer. DNA binds to NC strongly (as do proteins) and instantaneously by virtue of electrostatic interactions. The nitrate ester, illustrated in **Fig. 2**, has an extremely strong dipole that interacts with the dipoles present in both nucleic acids and proteins. It has been reported erroneously in the literature that NC binds macromolecules, especial-

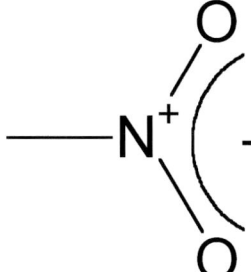

Fig. 2. Chemical structure of the nitrate ester illustrating its strong dipole.

ly proteins, by a hydrophobic mechanism. Whereas the surface energy of NC makes it intrinsically nonwetting (and a detergent or surfactant must be added to the membrane, normally during manufacture to make it compatible with aqueous solutions), the principal mode of protein and DNA binding to NC is electrostatic. In 1979, a method was developed by Towbin et al. *(2)* for identifying (using immunodetection) proteins that were electroblotted from an acrylamide gel onto NC. In many ways, Towbin's method is essentially a modification of the DNA blotting and detection technique adapted to protein chemistry. The name chosen for the protein method, "Western blotting," reflected and honored the similarities to the DNA methodology which by then was called "Southern blotting." Subsequently, a method for detecting ribonucleic acid (RNA) sequences *(9)*, called "Northern blotting" and for identifying proteins that bind to DNA *(10)*, called "Southwestern blotting" were developed on various filter media.

NC has a long-standing and deserved reputation for being difficult to handle. Although it is a strong membrane, its brittleness makes it susceptible to cracking and breaking. These handling issues were noted by both molecular biologists and protein chemists, and protocol "tricks" were developed in both applications areas to mitigate or eliminate serious problems. The brittleness of NC and lack of broad chemical compatibility were much more serious challenges in terms of its suitability in larger-scale filtration applications. It is difficult to seal NC to plastic housings (e.g., cartridges) without breakage, essentially impossible to pleat (to increase effective filtration surface area) and, in the end, NC is fairly susceptible to rupture (loss of filtration integrity) under high pressure unless it is very well supported. For these reasons, membrane manufacturers began to look for alternative membrane polymers once the need for higher-volume, higher-throughput filtration applications was established.

3. PVDF Membrane Development and Applications

PVDF overcame the majority of NC limitations, and many of the same companies that were able to produce NC also developed the ability to make PVDF membranes over a range of different pore sizes. The PVDF polymer itself is highly resistant to chemical degradation, except in the presence of strong alkali (pH greater than 12.0), and membranes made from PVDF are sufficiently elastic to withstand a broad range of fabrication conditions (including sonic welding and pleating) and high-pressure filtration applications. Millipore, an early provider of PVDF membranes, created the trademark, "Durapore®" to call attention to these attributes. PVDF, like NC, is intrinsically hydrophobic. Whereas NC is only marginally hydrophobic and can be made water wettable by adding a surfactant (e.g., glycerin) or detergent (e.g., Triton®-X 100) to the membrane, PVDF is extremely hydrophobic and requires significant surface modification to make it compatible with aqueous solutions (*see* **Note 2**). A considerable amount of chemistry and patented technology was subsequently developed to make PVDF membranes hydrophilic. The important point to note is that even though the polymer itself is very hydrophobic, membranes made from PVDF can be either hydrophilic or hydrophobic depending on whether the manufacturer has modified or covered over the polymer surface with a secondary chemical treatment that is water compatible.

The discussion of PVDF is relevant to ELISPOT in that the hydrophobic version of the 0.45-μm *(5)* membrane was found to be an excellent Western blotting substrate. PVDF binds proteins by hydrophobic interactions (van der Waal's forces). This applies, of course, only to hydrophobic PVDF membranes. Most types of hydrophilic PVDF will not bind proteins to any appreciable degree. Interestingly, hydrophobic PVDF will bind single-stranded DNA and RNA, but will not bind double-stranded DNA. The structure of PVDF is illustrated in **Fig. 3**.

The use of PVDF membranes in Western blotting type applications grew rapidly. Unlike NC membranes, PVDF membranes could stand up to automated protein-sequencing chemistries and were better able to retain low molecular weight proteins and peptides. PVDF was also better suited to a wider range of detection techniques, including fluorescence and chemiluminescence. One significant drawback of PVDF is the need to prewet the membrane in an alcohol solution prior to using it in Western blotting and most other applications that include immunodetection. The requirement to prewet with alcohol, which is completely necessary in Western blotting applications, is not universally applicable in ELISPOT. Differences in ELISPOT protocols with regard to the prewetting step are significant and are likely to have an impact on the performance of the assay. This topic will be discussed in greater detail later in this chapter and also in Chapter 4 (*see* **Note 3**).

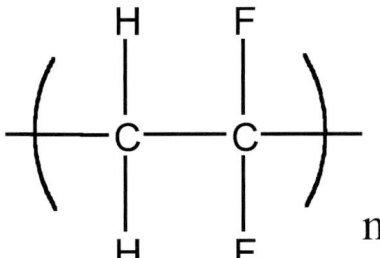

Fig. 3. Chemical structure of PVDF.

4. 96-Well Filter Plate (NC and PVDF) Development and Applications

The rationale behind the development and commercialization of NC and PVDF is clear from the perspective of filtration applications. As has been pointed out already, the use of each of these membrane types in molecular biology and protein chemistry applications was based on a fortuitous combination of membrane properties; high permeability (because of high porosity), and high DNA/RNA and protein binding. The development of 96-well PVDF and NC bottomed plates was primarily driven by the secondary (i.e., biochemistry) applications. One of the attributes of both Southern and Western blotting is that (DNA) hybridization and antibody binding are diffusion-limited reactions. In other words, the reactant in solution (complimentary DNA or antibody) must diffuse to the surface of the membrane before it can couple with the immobilized reactant (DNA or protein). In these types of solid-phase reactions, the times required to reach equilibrium binding are much longer as compared with reactions in which both reactants are in solution. Typically, DNA hybridization (Southern blotting) and immunodetection (Western blotting) require from 2 to 24 h. Membrane-bottomed, 96-well plates made it possible to reduce the time requirements associated with solid-phase binding (**Fig. 4**). (Earlier, simpler versions of these plates, called "Dot Blot" or "Slot Blot" apparatuses provided the first opportunity to exploit the benefits of filtration in these applications.) Filtration of the reactant in solution through the membrane brings it into intimate contact with the reactant immobilized on the membrane surface. So long as the filtration rate is kept low enough for hybridization or binding to take place, the time required to achieve efficient capture can be dramatically reduced. Additionally, all wash steps in between various reactions can be accomplished using filtration. The development of membrane-bottomed plates in conjunction with compatible vacuum manifolds made it possible to carry out from 1 to 96 different hybridization or

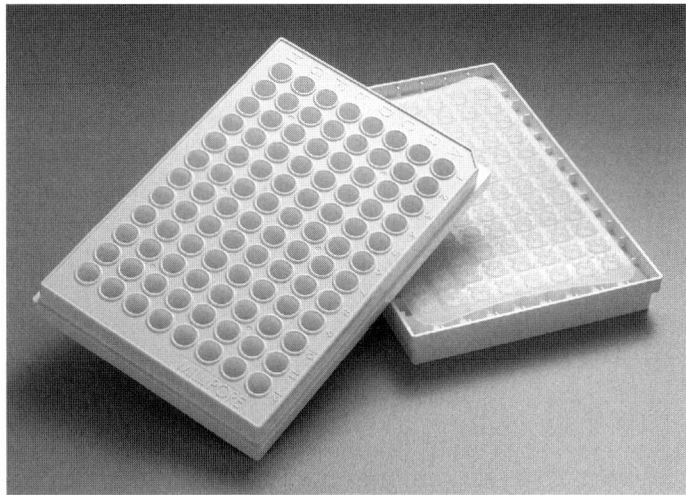

Fig. 4. Photograph of a 96-well filter-bottomed plate used for ELISPOT.

immunodetection assays using less reagents and requiring less time as compared to standard methodologies.

In addition to the membranes and other features that were useful for these applications, the plates were also designed to allow for discrete liquid transfer from the top (membrane-containing) plate to a (standard, plastic, 96-well) receiver plate. The filter-plate components that allow for this to occur have the potential to interfere with ELISPOT applications in at least three different ways.

1. If alcohol is used to prewet the membrane (this applies only to the use of PVDF membrane plates) and becomes trapped under the membrane, the alcohol can suppress or completely inhibit cytokine release (*see* **Note 3**).
2. If biotinylated antibody or avidin-enzyme conjugate becomes trapped under the membrane, high background will likely result (*see* **Note 3**).
3. If the membrane can not be removed because of these components, it may be difficult to analyze and essentially impossible to archive ELISPOT results

Preventing these types of problems during the ELISPOT assay and optimizing the different parts of the protocol will be reviewed in detail in Chapter 4.

There are other plate attributes that may have an impact on ELISPOT assay performance or analysis. The membrane inside each well needs to be planar within a fraction of a millimeter (e.g., ±0.1 mm) to provide the best results. Lack of membrane flatness may contribute to difficulties in imaging depending on the type of microscopy being used. Additionally, if the membrane is bowed, cells may tend to settle unevenly or roll to the relative low points (often either the center or the periphery) during incubation. Consequently, spots may become very unevenly distributed and difficult to enumerate accurately, especially if there is any spot confluence. The plate itself should also be flat (within a tolerance of perhaps 1 mm corner to corner) to assure compatibility with plate washers and most of the imaging software that supports automated image acquisition and analysis. There is one other feature of 96-well plates that has the potential to introduce variability into the ELISPOT assay. The 96-well plate is arrayed as 8 rows of 12 wells. Wells at the periphery (columns 1 and 12, rows A and F) of the plate are fundamentally different from "interior" wells insofar as they are in the most direct contact with the plate surroundings. Depending on incubation conditions and other protocol steps, this physical distinction may have some impact on one or more parts of the ELISPOT assay.

5. 96-Well Filter Plates in ELISPOT

ELISPOT assays were first developed on plastic, 96-well plates. Shortly after, NC-bottomed filter plates became available, the majority of ELISPOT assays were conducted in those plates. When PVDF filter plates were introduced, some investigators chose to use PVDF plates, and some continued to use NC. The reasons for choosing one plate (membrane) over the other are highly varied and will not be addressed in detail here. The fact that some laboratories and individual researchers feel strongly that one membrane is superior to the other runs contrary to the large body of Western blotting experience: Despite some clear-cut differences in how each of the membranes can be used, there is essentially no reported difference in terms of detection sensitivity or signal to noise on NC vs PVDF in the Western blotting application.

This having been said, it is clear is that the two membranes and their properties are quite different (*see* **Figs. 5** and **6**). The remainder of this chapter will be devoted to reviewing these properties which are summarized in **Table 1**, and highlighting the differences, especially as they might pertain to ELISPOT applications.

6. Properties of Nitrocellulose Membrane That Affect Its Performance in ELISPOT

The NC membrane in plates used for ELISPOT has a nominal pore size of 0.45 µm. Selection of this particular pore size is not based on the requirements of any of the major biochemical applications (Southern blotting, Western blotting, or ELISPOT) but rather on the pore size that best served these membranes' first application, namely sterile filtration (removal of bacteria from aqueous solutions). NC typically is 150 µm thick, a property that is governed mostly by manufacturing constraints. Whereas it would be most desirable in terms of filtration speed and raw material consumption to make the membrane as thin as possible, the ability to cast hundreds of meters of membrane and ensure filtra-

Fig. 5. A scanning electron micrograph of a 0.45-µm PVDF membrane.

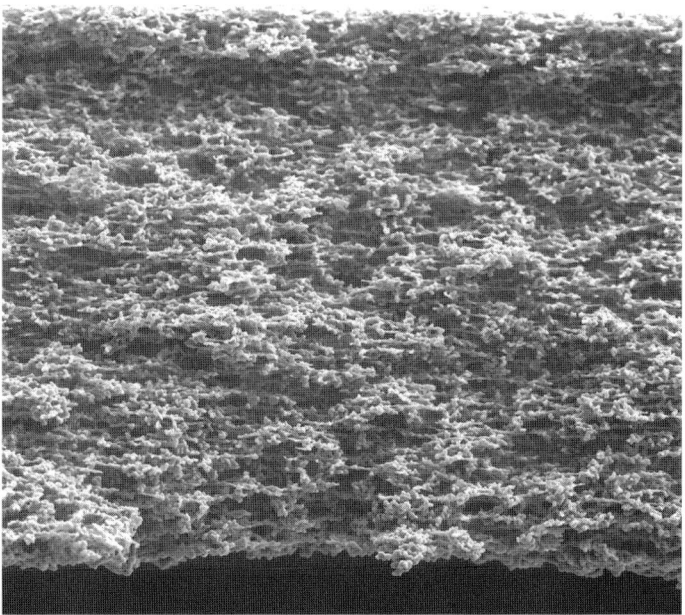

Fig. 6. A scanning electron micrograph of an NC membrane.

Table 1
Comparison of PVDF and NC Membranes

Attribute or characteristic	NC (used in ELISPOT) [nominal or average values]	PVDF (used in ELISPOT) [nominal or average values]
Pore size[a]	0.45 μm	0.45 μm
Porosity[b]	70– 5%	65–70%
Thickness	150 μm	135 μm
BET surface area (**6**)	6.5 m^2/g	6 m^2/g
Surface area ratio[c]	250	350
Saturation binding capacity (IgG)	250 μg/cm^2	350 μg/cm^2
(IgG) Binding capacity of top 1 μm	2 μg	3 μg
Wettability	Not wettable without prior addition of surfactants or detergents	Not directly wettable in water. Must be prewet with alcohol and then exchanged with water
Additives	Glycerin	None
Solvent compatibility	Not compatible with methanol or ethanol	Broadly compatible with a wide range of aqueous and organic solvents. Avoid prolonged exposure to strong alkali (e.g., pH >12.0)
Mechanism of binding	Electrostatic	Hydrophobic
Things that will interfere with or destabilize binding of anticytokine antibodies	Chaotropes (e.g., Tween-20, Triton-X 100). Water (if never dried), proteins, especially larger molecular weight proteins	Detergents (e.g., sodium dodecyl sulfate), low polarity solvents (e.g., dimethyl formamide)
Compatibility with different detection modes	✓ Colorimetric ✗ Fluorescence ✓ Chemiluminescence	✓ Colorimetric ✓ Fluorescence (marginal) ✓ Chemiluminescence

[a]Pore size is nominal and corresponds to the diameter of the largest particle that can pass through the membrane structure. A 0.45-μm pore size membrane is expected to retain 100% of particles whose diameter exceeds 0.45 μm.

[b]Porosity is the portion of the membrane volume that is occupied by air (not occupied by polymer). 1cm^2 of membrane whose thickness is 140μm (i.e., 0.014 cm) will have a volume of 0.014 cm^3 (14 μL). If the membrane is 70% porous, it will contain approx 10 μL of void space.

[c]Surface area ratio is the ratio of internal to frontal surface area. A surface area ratio of 250 means that in a 1cm^2 piece of membrane, the polymer surface area is 250 cm^2.

tion performance throughout dictates that membrane thickness be in the range of 150 μm. The porosity of NC is also governed by manufacturing considerations. Higher porosity leads to faster filtration and lower polymer consumption. The upper limit of achievable porosity is dictated by the need to be able to handle and manufacture the membrane. Typical porosity for NC is in the range of 70 to 75%. Porosity and thickness, along with polymer characteristics, combine to control the amount of polymer surface area per frontal area of membrane. As can be seen in **Fig. 6** (scanning electron micrograph of NC), the membrane looks like a microscopic sponge and there is a tremendous amount of polymer surface in contact with the porous structure. The surface area can be quantified (m^2/g of polymer) using the Brunauer-Emmett-Teller (BET) method *(6)*. Multiplying the BET surface area by the basis weight of the membrane (g/m^2) produces a ratio of total (internal + surface) polymer surface area per unit of frontal area. For NC, this ratio is in the range of 250, meaning that the amount of total surface area per square centimeter of frontal area is approx 250 cm^2. As the (frontal) surface area of membrane in a typical 96-well plate for ELISPOT is approx 0.3 cm^2, the polymer surface area per well is about 75 cm^2. This is an important characteristic because the binding capacity of most adsorptive polymers, including both NC and PVDF, is approx 1 $\mu g/cm^2$ for immunoglobulin (IgG). Consequently, the binding capacity of NC in a single well of a 96-well plate is on the order of 75 μg. In a typical ELISPOT application, the amount of IgG (anti-cytokine antibody) added to a well is in the range of 1 μg (e.g., 100 μL of a 10 μg/mL solution) so that most of membrane binding sites (surface area) remain unoccupied after antibody coating. Consequently, the need to block after antibody coating is required to reduce non-specific binding of subsequently added reagents (e.g., biotinylated antibody and enzyme-avidin conjugate, etc.; *see* **Note 4**). In most ELISPOT protocols, this blocking step (normally with serum containing tissue culture media) immediately follows antibody coating. In the event that it didn't, the sparse coating of antibody could result in a significant (<50%) and somewhat rapid (<5 d) loss of antibody activity. Antibodies absorbed onto NC (and PVDF) will continue to interact with the membrane surface if there are no other proteins or polymeric blocking agents to interfere. Multipoint attachment can lead to denaturation and ultimately to a loss of antibody activity. If plates are going to be stored for an extended period of time (more than a few hours) before use, it is highly recommended that they be blocked and then rinsed with deionized water before storage. Adsorption of antibody to the membrane surface is instantaneous once the protein comes into contact with the NC surface. Therefore, the reaction is limited by diffusion. Reducing the volume (e.g., from 100 μL to 50 μL) of antibody coating solution and increasing the temperature (e.g., from 4°C to 20°C) during coating will shorten the time required to achieve maximum antibody adsorption (e.g., from overnight to 4 h). Based on the kinetics (instantaneous), directionality of how the antibody comes in contact with the membrane (from the top surface), and the membrane's binding capacity, it is reasonable to consider whether antibody binding is uniform throughout the depth of the membrane. Essentially all of the antibody is immobilized on the top surface of the membrane with maximum penetration of immobilized antibody probably not exceeding one or two microns of depth. Consequently, all of the cytokine subsequently bound by the antibody is accessible to the secondary reagents (biotinylated antibody, avidin-enzyme conjugate and conjugate). This localization is ideal for the ELISPOT application in terms of maximizing sensitivity, limiting reagent costs and reducing incubation times.

NC is a marginally hydrophobic polymer and membranes made from it, including NC, would not wet out instantly in water if they did not contain some added surfactant, humectant, or detergent. Different membranes (depending on the manufacturer and the particular membrane) contain different wetting agents including Triton-X 100, glycerin, various hydrophilic polymers (e.g., polyvinyl alcohol fixative, polyvinylpyrrolidone), and detergents. The most biocompatible of these wetting agents is glycerin and this is the additive of choice, especially in applications like ELISPOT that involve contact with mammalian cells. One drawback to the use of glycerin is that glycerin is a liquid at room temperature. As a consequence, it can migrate (out of the membrane) or evaporate (despite its high boiling point) during prolonged storage. If the membrane in one or more wells do not wet out immediately (within a few seconds) after the addition of the antibody coating solution, it may suggest that those wells will perform poorly in the ELISPOT assay (*see* **Note 5**). Under no circumstances should NC be prewet with a solution containing any more than 20% methanol or ethanol (PVDF frequently is prewet using 15 µL to 50 µL of 70% methanol or ethanol). NC is not compatible with alcohol and will begin to dissolve after only a few minutes.

As was stated earlier in the chapter, NC binds proteins (e.g., antibodies) by electrostatic interactions. Binding is independent of pH and unaffected by the presence of even high levels of detergent (e.g., 1% sodium dodecyl sulfate). However, the antibody coated onto the membrane can be displaced by proteins present in the blocking solution (normally 10% serum in tissue culture media) especially if the membrane is not washed and dried prior to blocking. Consequently, it is common practice in both Western blotting and immunodiagnostic applications to dry the membrane (e.g., for 30 min or more at 37°C) after protein has been blotted or added prior to the blocking step.

NC is compatible with ELISA detection involving precipitated substrates (e.g., 5-5-bromo-4-chloro-3-indolyl phosphate/Nitroblue tetrazolium for alkaline phosphate, 3-amino-9-ethylcarbazole, for peroxidase, etc.) as well as chemiluminescent substrates (e.g., CDP-*Star*™; Tropix, Bedford, MA).

Because the surface of NC is extremely light-scattering, NC is a poorly suited for fluorescence detection due to very high background.

7. Properties of PVDF Membrane That Affect Its Performance in ELISPOT

Like NC, and for the same reasons, the PVDF membrane in plates used for ELISPOT has a nominal pore size of 0.45 μm. PVDF is nominally 135 μm thick and approx 65 to 70% porous. The BET surface area is about the same as the surface of NC, and its surface area ratio is somewhat higher (approx 350). Consequently, PVDF can bind upwards of 350 μg/cm² of IgG or in excess of 100 μg per well of a 96-well plate. As with NC, blocking of PVDF should occur within a few hours (or less) of antibody coating. Failure to do so may result in a rapid and significant loss of antibody activity. Once antibody has been coated and the membranes have been blocked (and washed using deionized water or very-low molarity buffer), plates can be stored (desiccated and at room temperature) for weeks or even months (*see* **Note 6**). PVDF that has been coated with protein (e.g., as a consequence of antibody coating and blocking) will rewet spontaneously upon the addition of aqueous media.

The major difference between NC and PVDF in ELISPOT applications is related to their mechanisms of binding and associated differences in handling or pretreatment. PVDF is very hydrophobic (its surface energy is approx 21 dynes/cm) and will not wet out in water. In Western blotting applications, PVDF is always prewet in alcohol (typically 50 to 100% methanol), then normally exchanged in water, and ultimately equilibrated in a (transfer) buffer solution before applying the membrane to the polyacrylamide gel for electrophoresis. The fact that hydrophobic PVDF membranes will not wet out spontaneously in water, unless coated by (blotted) proteins, is even exploited in a Western blotting application called "transillumination" *(11)*. The overwhelming majority of literature in Western blotting references the prewetting step, so it is not a surprise that many ELISPOT protocols also include an alcohol prewet step.

What is surprising is that some ELISPOT protocols do not include a prewet step. At Millipore (Danvers, MA), experiments were performed to determine the relative performance of PVDF 96-well plates (cat. no.: MSIPS4510) that were either prewet with 15 μL of 70% v/v methanol in water and rinsed, or not pretreated at all prior to antibody coating. Briefly, after the pretreatment with alcohol (or no pretreatment), plates were coated with 1 μg of antihuman interferon-gamma (Mabtech, Stockholm, Sweden), and blocked for 2 h in tissue culture media (RPMI, Invitrogen, Carlsbad, CA) containing 10% fetal bovine serum (Invitrogen). Fifty thousand peripheral blood mononuclear cells (*see* **Note 7**) were added per well to 16 wells per plate, stimulated with 0.5 μg phytohemag-

Table 2
The Effect of Alcohol Pretreatment of PVDF on ELISPOT Assay Performance

Experiment number	Prewet spot number mean ± SD[a]	Non-prewet spot number mean ± SD[a]	Non-prewet as a percent of prewet
1	606 ± 46	413 ± 37	68%
2	577 ± 37	416 ± 34	72%
3	604 ± 35	440 ± 42	73%
4	609 ± 40	391 ± 28	64%

[a]n = 12–16.

glutinin (PHA-L, Sigma, St. Louis, MO), and the plates were incubated overnight in a humidified, 37°C, 5% CO_2 tissue culture incubator. ELISPOTs were visualized using biotinylated anti-human interferon-γ (Mabtech), conjugated avidin-alkaline phosphatase (Mabtech) and 5-5-bromo-4-chloro-3-indolyl phosphate/Nitroblue tetrazolium Plus (Moss, Inc., Pasadena, CA), and then enumerated using an automated microscope (KS Elispot, Zeiss, Thornwood, NY) and its associated software. The results of four different experiments are summarized in **Table 2**.

In these experiments, the cells in the untreated (non-prewet) plates produced approx 30% fewer detectible spots. However, the consistency well-to-well and plate-to-plate was equivalent. Spot quality (intensity, uniformity, and size) and overall assay background were comparable in both plate types. Considering that half these results were obtained without prewetting, the comparable side-by-side performance is quite remarkable. It would appear that the determination to prewet with alcohol or not can be made by individual laboratories based on their reagent selections and particular assay requirements.

PVDF, like NC, is fully compatible with ELISA detection involving precipitating, color-forming substrates, and chemiluminescent substrates. Although the fluorescence background of PVDF is also high, as a result principally again of light scattering, ELISPOT assays have been developed on PVDF that are based on the use of fluorescently labeled antibodies *(12)*.

8. Final Remarks

Although with proper optimization, it is likely that comparable ELISPOT performance can be achieved using either type of membrane, it is unlikely that the same protocol will work equally well in both cases. Neither PVDF nor NC is a drop-in for the other (*see* **Note 8**). Fundamental differences in the two membranes, especially with regard to their mechanisms of binding and their ability to wet directly in water, will affect their behavior in ELISPOT. Modifications that have been made in the design of 96-well plates to make them compatible with automation, including stricter dimensional specifications and rigid side-walls (to allow handling by laboratory robotics and provide space for bar codes), also have benefited ELISPOT applications. These plates are now fully compatible with standard plate washers as well as with imaging equipment and image analysis software. It is reasonable to believe, based on the importance of ELISPOT and the impressive growth in the number of assays being performed, that membranes and membrane-based plates may some day be optimized specifically for this application.

9. Notes

1. The degree of nitration is critical. Theoretically, there is the potential to form nitrate esters on all three hydroxyls of each cellulose monomeric unit. In fact, because of the steric hindrance, the maximum level of nitration is approx 2.6 nitrate esters per residue. This type of polymer is highly explosive. Nitrocellulose used to make membranes is nitrated to a level of approx 2 nitrate esters per residue. This type of polymer is highly flammable, but will not explode.
2. PVDF in 96-well plates (for example, Millipore cat. no. MAIPS4510) used for ELISPOT is extremely hydrophobic. It will not wet out spontaneously upon the addition of water. The behavior of PVDF in this regard is significantly different from that of NC. 96-well plates containing hydrophilic PVDF (for example, Millipore cat. no. MAHVS4510) will not work at all in the ELISPOT application.
3. As required or desired, the hydrophobic PVDF membrane should be prewet by adding 15 µL of 70% methanol to each well. Within 1 or 2 min (or less) of adding the methanol solution, the membrane should be rinsed by adding 100 µL of water or coating buffer to the well and aspirating or decanting immediately thereafter. The rinse step may be repeated once more. The antibody solution should be added immediately after rinsing the membrane. Adding larger (e.g., 50 µL) or more con-centrated (e.g., 100%) volumes of methanol creates the risk of liquid collecting under the membrane. This liquid cannot be washed out effectively and may create serious problems later on in the ELISPOT assay.
4. The binding capacity of both NC and PVDF ($\geq$75 µg/well) far exceeds the amount of specific antibody (typically $\leq$5 µg/well) that is used to coat the membrane. Whereas this is beneficial insofar as it results in the localization of the antibody at or very near the membrane's top surface, it makes it absolutely necessary to block the membrane to prevent high levels of background owing to nonspecific binding. Membrane should be blocked within a few hours of antibody coating to prevent loss of antibody activity.
5. 96-well plates containing NC (for example, Millipore cat. no. MAHAS4510) may contain some wells, especially at the periphery of the plate, that will not wet out immediately upon addition of aqueous solution (e.g. antibody coating buffer). This may be the result of storage and handling conditions. In any event, wells that do not wet out immediately may produce spurious results and should not be used in the ELISPOT assay.

6. Once membranes have been coated and blocked, they may be rinsed in water or low molarity buffer (e.g., 10 mM phosphate, pH 7.4) and stored desiccated for weeks or months.

7. The membrane frontal surface area in a typical 96-well plate is approx 0.3 cm^2. If the responding T-cell is assumed to be round and estimated to have a nominal diameter of 10–15 µm, approx 150,000 cells would constitute a monolayer. Adding more than approx 100,000 cells creates the risk of some of these cells not being in intimate contact with the membrane. If the cell-secreting cytokine is not in direct contact with the antibody-coated membrane, it is possible that the shape and intensity of its corresponding ELISPOT may be so irregular as to disqualify it from being enumerated. In instances when the response rate to antigen is anticipated to be so low that it is advisable to stimulate 500,000 or 1,000,000 cells to get a significant response above background, it might be best to add 100,000 cells per well to 5 or 10 wells and determine the aggregate response ($\Sigma_{10 \text{ wells}}$), not the average response.

8. It is nearly certain that substituting a PVDF plate into a protocol that has been optimized using an NC plate will produce unsatisfactory results. The opposite is also true. The ELISPOT assay, especially the antibody coating, blocking, and color development steps should be optimized for whichever plate is going to be used.

References

1. Southern, E. M. (1975) Detection of specific sequences among DNA fragments separated by gel electrophoresis. *J. Mol. Biol.* **98**, 503–517.

2. Towbin, H., Staehelin, T., and Gordon, J. (1979) Electrophoretic transfer of proteins from polyacrylamide gels to nitrocellulose sheets: procedure and some applications. *Proc. Natl. Acad. Sci. USA* **76**, 4350–4354.

3. Czerkinsky, C. C., Nilsson, L. A., Nygren, H., Ouchterlony, O., and Tarkowski, A. (1983) A solid-phase enzyme-linked immunospot (ELISPOT) assay for enumeration of specific antibody-secreting cells. *J. Immunol. Methods* **65**, 109–121.

4. Sedgwick, J. D. and Holt, P. G. (1983) A solid-phase immunoenzymatic technique for the enumeration of specific antibody-secreting cells. *J. Immunol. Methods* **57**, 301–309.

5. Pluskal, M. F., Przekop, M. B., Kavonian, M. R., Vecoli, C., and Hicks, D. A. (1986) Immobilon PVDF transfer membrane: a new membrane substrate for Western blotting of proteins. *BioTechniques* **4**, 272 – 282 [Japanese reference to the use of 0.2-µm membrane (GVHP) in protein blotting].

6. Brunauer, S., Emmett, P. H., and Teller, E. (1938) Adsorption of gases in multimolecular layers. *J. Am. Chem. Soc.* **60**, 309–319.

7. Czerkinsky, C., Andersson, G., Ekre, H.-P., Nilsson, L.-A., Klareskog, L., and Ouchterlony, O. (1988) Reverse ELISPOT assay for clonal analysis of cytokine production. I. Enumeration of gamma-interferon-secreting cells. *J. Immunol. Methods* **110**, 29–36.

8. Schielen, P., van Rodijnen, W., Tekstra, J., Albers, R., and Seinen, W. (1995) Quantification of natural antibody producing B cells in rats by an improved

ELISPOT technique using the polyvinylidene difluoride membrane as the solid support. *J. Immunol. Methods* **188**, 33–41.

9. Alwine, J. C., Kemp, D. J., and Stark, G. R. (1977) Method for detection of specific RNAs in agarose gels by transfer to diazobenzyloxymethyl-paper and hybridization with DNA probes. *Proc. Natl. Acad. Sci. USA.* **74**, 5350–5354.

10. Miskimins, W. K., Roberts, M. P., McClelland, A. and Ruddle, F. H. (1985) Use of a protein-blotting procedure and a specific DNA probe to identify nuclear proteins that recognize the promoter region of the transferrin receptor gene. *Proc. Natl. Acad. Sci. USA.* **82**, 6741–6744.

11. Reig, J. A., and Klein, D. C. (1988) Submicron quantities of unstained proteins are visualized on polyvinylidene fluoride membranes by transillumination. *Appl. Theoret. Electrophoresis* **1**, 59–60.

12. Gazagne, A., Claret, E., Wijdenes, J., Yssel, H., Bousquet, F., Levy, E., et al. (2003) A fluorospot assay to detect single T lymphocytes simultaneously producing multiple cytokines. *J. Immunol. Methods* **283**, 91–98.

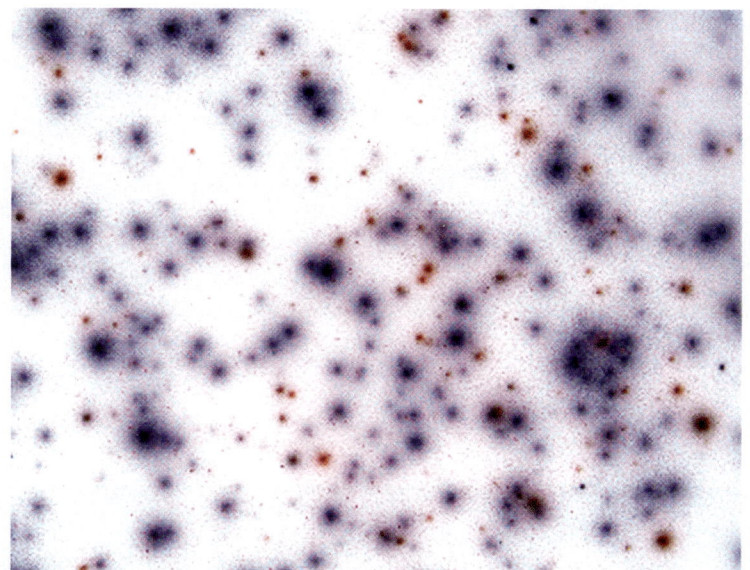

Color Plate 1, Fig. 3. (*See* full caption and discussion in Chapter 2, p. 19.) Two-cytokine ELISPOT assay (custom-made kit; R&D Systems).

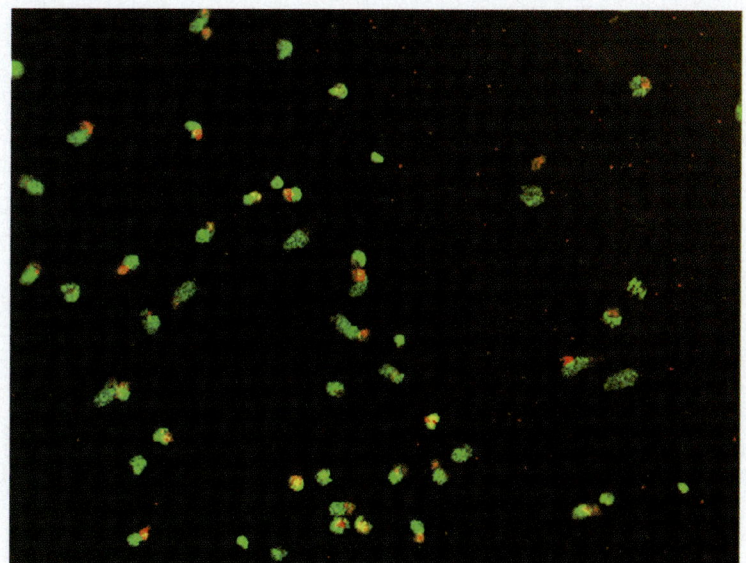

Color Plate 2, Fig. 7. (*See* full caption and discussion in Chapter 2, pp. 25–26.) High degree of programmed cell death (apoptosis) in cells plated into the ELISPOT plate may result in lack of spots.

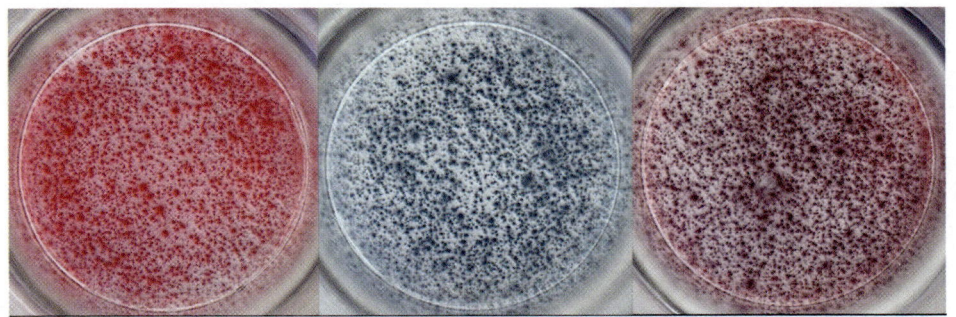

Color Plate 3, Fig. 10. (*See* full caption and discussion in Chapter 8, p. 148.) ELISPOT specimen with similar number of spots per well.

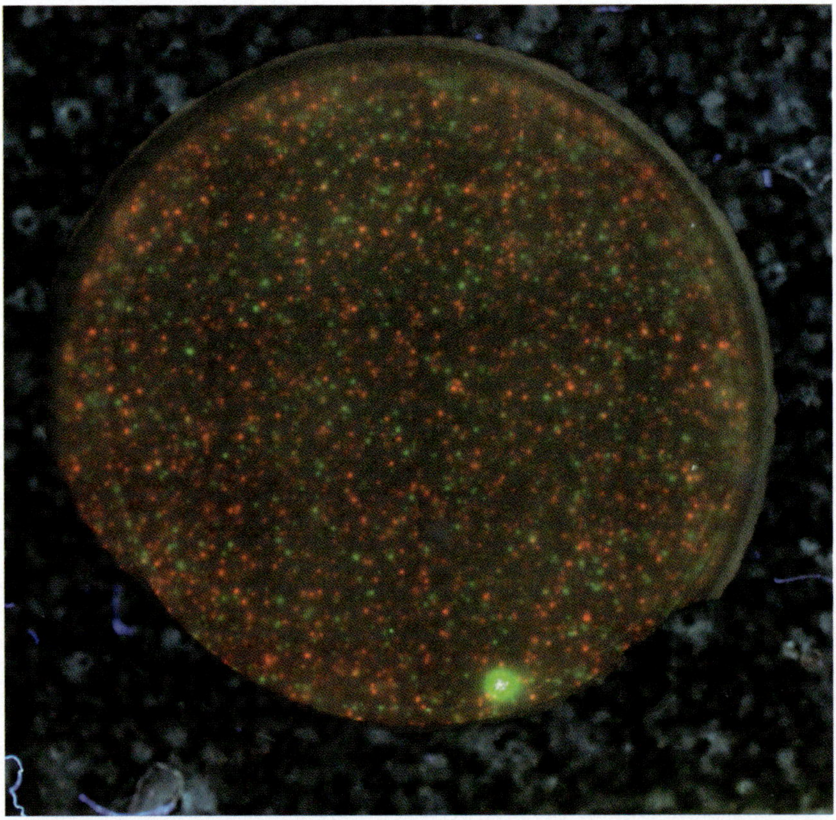

Color Plate 4, Fig. 11. (*See* full caption and discussion in Chapter 8, p. 149.) ELISPOT specimen labeled with fluorescence markers for IFN-γ (fluorescein isothiocyanate, green spots) and IL-5 (rhodamine, red spots).

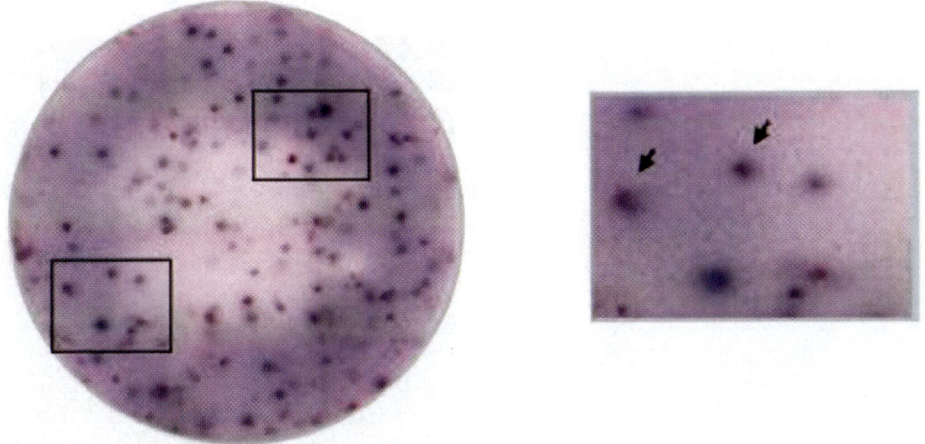

Color Plate 5, Fig. 1. (*See* full caption and discussion in Chapter 19, p. 290.) Dual-color immunoenzymatic ELISPOT for the detection of IFN-γ and IL-4 producing cells.

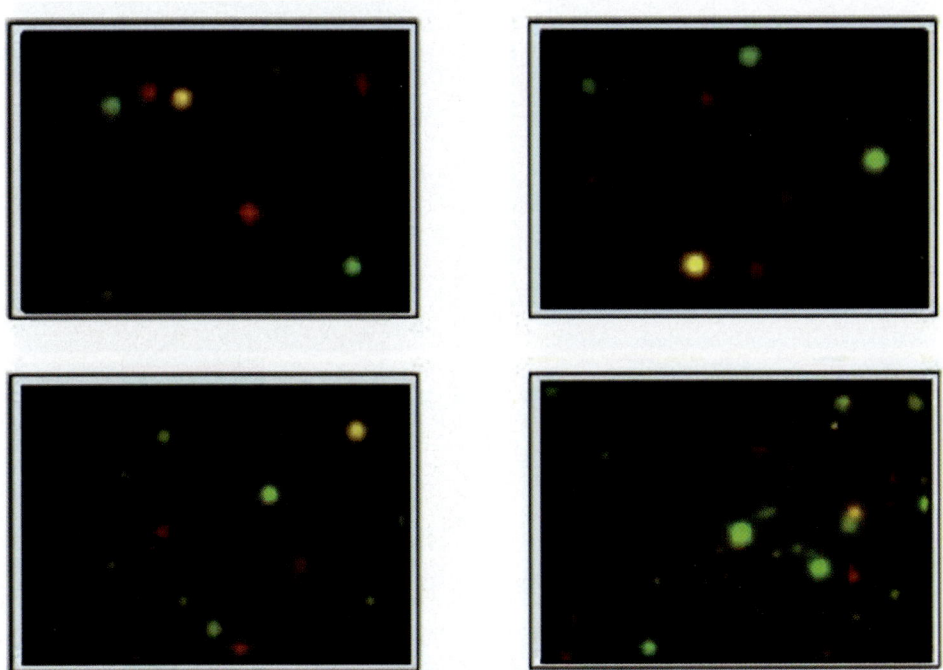

Color Plate 6, Fig. 2. (*See* full caption and discussion in Chapter 19, pp. 292,293.) Dual-color fluorospot for the detection of IL-2 and/or IFN-γ-producing cells.

A

B

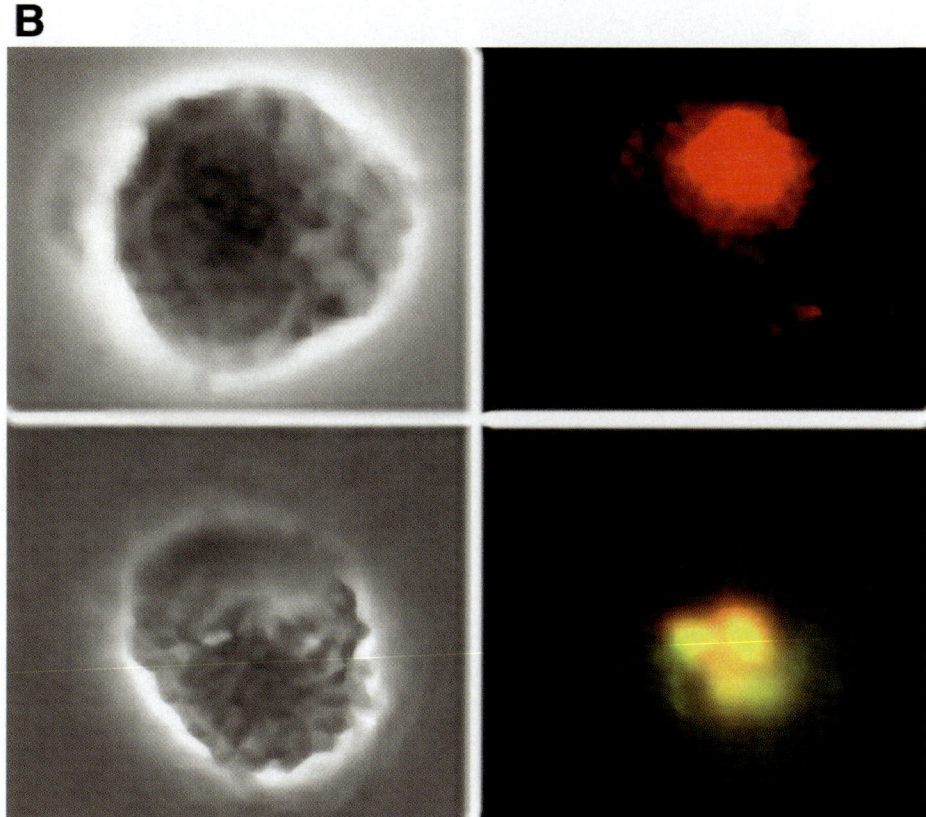

Color Plate 7, Fig. 3. (*See* full caption and discussion in Chapter 20, pp. 306,307,309.) Positive staining results using EliCell system.

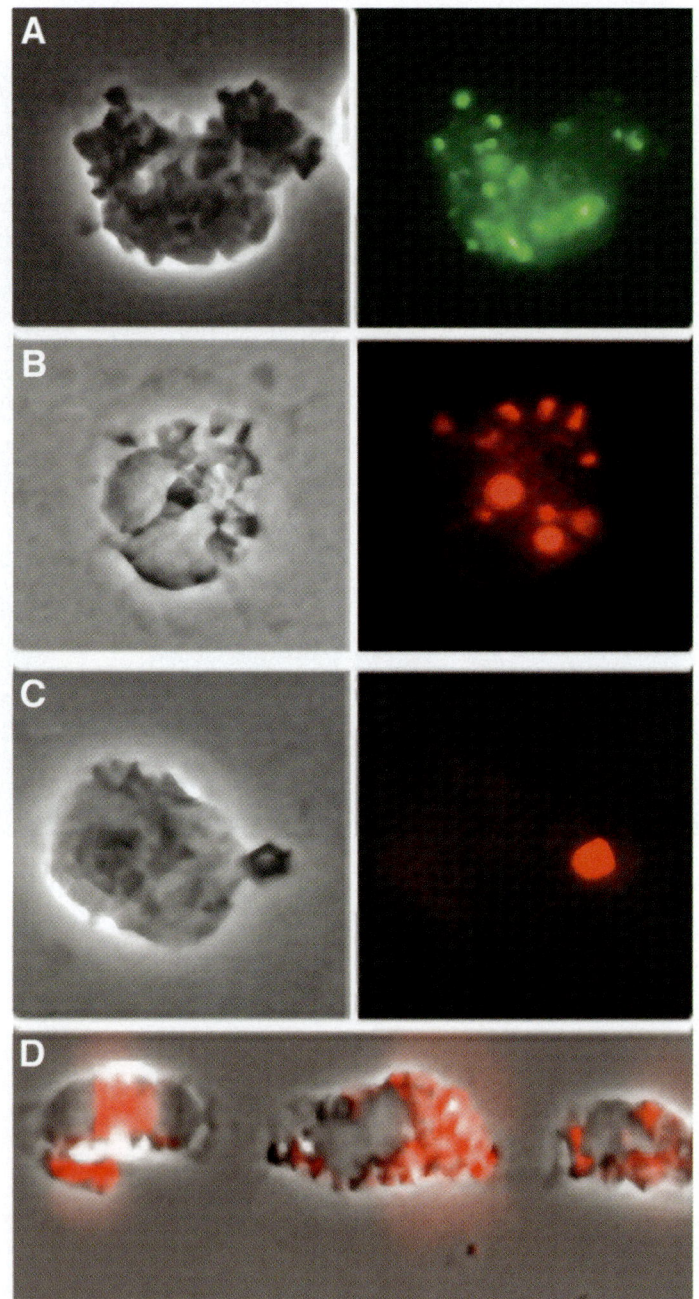

Color Plate 8, Fig. 4. (*See* full caption and discussion in Chapter 20, pp. 307–309.) Staining artifacts using EliCell system.

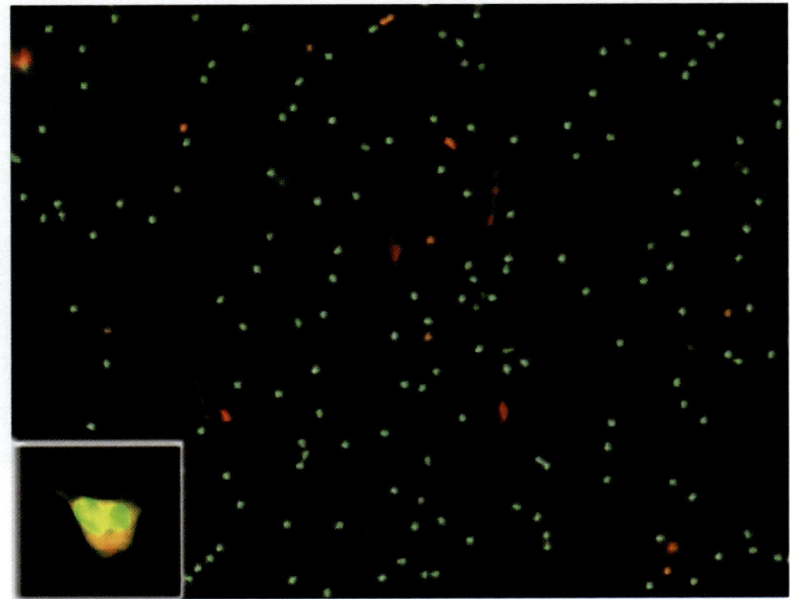

Color Plate 9, Fig. 5. (*See* full caption and discussion in Chapter 20, p. 309.) Viability of cells after EliCell assay.

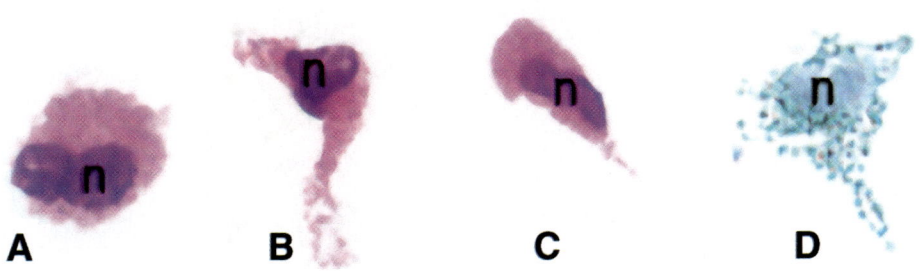

Color Plate 10, Fig. 6. (*See* full caption and discussion in Chapter 20, p. 310.) Morphology of eosinophils during EliCell assay.

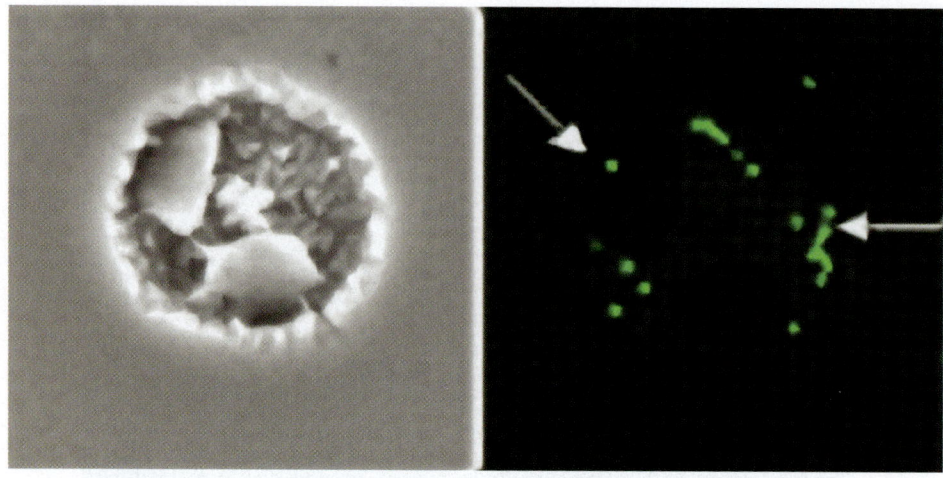

Color Plate 11, Fig. 7. (*See* full caption and discussion in Chapter 20, p. 312.) Intracellular lipid body staining of eosinophils using EliCell system.

4

Standardization and Validation Issues of the ELISPOT Assay

Sylvia Janetzki, Josephine H. Cox, Neal Oden, and Guido Ferrari

Summary

During the last 20 yr, the enzyme-linked immunospot (ELISPOT) assay has emerged as one of the most important and widely used assays to monitor immune responses in humans and a variety of other species. With the ELISPOT assay, immune cell frequencies can be measured at the single cell level without elaborate expansion or manipulation of cell populations. Its usefulness has led to its application in vaccine design and development and, most importantly, in monitoring vaccination efforts. The impact of results measured with this assay can be profound. In addition to ease of performance, repeatability and reliability are major features expected of an ELISPOT assay. The focus today is on standardization of the technique, validation strategies to comply with these required features, and accommodation of the growing demand of Good Laboratory Practice (GLP) compliance. This chapter will give the experienced scientists as well as newcomers to the field an overview over the major standardization issues for each step of the protocol. Guidelines are given on how to validate the ELISPOT performance.

Key Words: ELISPOT assay; standardization; validation; ELISPOT protocol; ELISPOT plates; ELISPOT antibodies; effector cells; antigen-presenting cells; ELISPOT evaluation; plate washers; cell counters; spot appearance.

1. Introduction

The enzyme-linked immunospot (ELISPOT) assay was first described for antibody-secreting B-cells in 1983 *(1,2)*, and in 1988, Czerkinsky reported the reverse ELISPOT technique protocol for the detection of T-cells secreting cytokine *(3)*. Since then, the general protocol has remained nearly unchanged. The introduction of the assay for monitoring vaccination responses in patients in immuntherapeutical trial settings in the late nineties, however, set forth strict demands to the assay *(4–7)*. Until then, one of the major concerns among scientists was the difficulty of obtaining repeatable results so that results could be compared not only among different laboratories but also

From: *Methods in Molecular Biology, vol. 302: Handbook of ELISPOT: Methods and Protocols*
Edited by: A. E. Kalyuzhny © Humana Press Inc., Totowa, NJ

among scientists in one group. Despite the straightforwardness and apparent ease of the general ELISPOT protocol, minor differences in the procedure and source of materials seem to be the source of often rather significant differences in the outcome of ELISPOT experiments *(8–11)*.

A study supported by the National Institutes of Health of the ELISPOT Collaborative group is the biggest study performed so far comparing the performance of 11 well-established ELISPOT laboratories performing HIV vaccine trials *(12)*. In this study, data concerning cell recovery, viability, and number of antigen-specific interferon (IFN)-γ-secreting cells in 11 peripheral blood mononuclear cell (PBMC) samples were obtained. Despite a good concordance in defining responder and nonresponder status in these samples, the differences in all three parameters tested were significant. The results of this study call for better standardization strategies of protocols and reagents to concur with reliability and reproducibility issues necessary to enhance all aspects of vaccine studies.

ELISPOT standardization can be defined as the imposition of a standard to techniques to remove as far as possible the effects of differences in confounding variables when comparing experiments performed under the same conditions. This chapter addresses all steps of the ELISPOT assay, including coating techniques, choices of materials, concentration of chemicals, important washing and incubation procedures, spot development, and evaluation of ELISPOT assays. Furthermore, because of the high impact of cell recovery and cell viability on the assay outcome, cell preparation, cryopreservation, thawing methods, as well as usefulness of automated cell counters will be discussed in detail. Each section will specifically highlight standardization strategies and recommend desirable outcomes. Many of the data presented were obtained from human IFN-γ ELISPOT assays. Our recommendations, however, apply for any ELISPOT system independent of species and cytokine. It needs to be stressed that, even if a standard operating procedure (SOP) for a specific cytokine has been validated in a laboratory, a new cytokine ELISPOT requires new standardization and validation procedures.

Validation is defined as making the technique effective and, importantly, producing the desired result. It should be stressed that "desired" result does not necessarily mean the best result, or highest spot count, but a result that is validated. Validation criteria are described in this chapter, and various statistical methods found to be most suitable for the purpose of establishing the ELISPOT assay are discussed.

2. Materials
1. HA plates (Millipore, Bedford, MA).
2. IP plates (Millipore, Bedford, MA).
3. ELISPOT HP and IP plates (Millipore, Bedford, MA).

4. 70% Ethanol.
5. Sterile phosphate-buffered saline (PBS).
6. Antibodies with high sensitivity and specificity (monoclonal), tested for ELISPOT and equal batch performance.
7. VACUTAINER® CPT™ tubes (Becton Dickinson, Franklin Lakes, NJ).
8. Accuspin tubes (Sigma-Aldrich Co, St. Louis, MO).
9. Leucosep tubes (Greiner Bio-One Inc., Longwood, FL).
10. Biotinylated antibodies with high sensitivity and specificity (preferably monoclonal), tested for ELISPOT and equal batch performance.
11. Horseradish peroxidase or alkaline phosphatase-coupled antibodies (Mabtech, Sweden).
12. Alkaline phosphatase.
13. Horseradish peroxidase.
14. 3-amino-9-ethylcarbazole, $C_{14}H_{14}N_2$ (chromogen).
15. Novared (Vector Laboratories, Burlingham, CA; chromogen).
16. 5-5-bromo-4-chloro-3-indolyl phosphate/Nitroblue tetrazolium (chromogen).
17. Fluorescent dye-coupled antibodies (Diaclone, Besacon, France).

3. Methods

3.1. ELISPOT Plates

One of the first choices a scientist has to make concerns 96-well microtiter plates. Historically, plates with a nitrocellulose membrane have mainly been used for ELISPOT assays. However, some limitations have been found in these plates, for example, inconsistency in the coating ability of wells in the periphery of the plate, and often small and faint appearing spots. This led to a clear trend of using plates with polyvinylidene fluoride (PVDF) membranes, which were first introduced to this technique by Shaw et al. *(13)* in 1993. Both membranes exhibit a high protein binding capacity of greater than 100 µg per well of immunoglobulin. The mechanism of binding and retaining antibody on the membrane surface, however, is favored in PVDF plates, which enhance the retention of antibodies during incubation. This can result in not only more, but also in better defined spots (S. Janetzki, unpublished data). A further improvement has been achieved with the introduction of ELISPOT IP (PVDF) and HP (nitrocellulose) plates, as well as HTS plates. These plates contain improved membrane features, a straightened membrane for preventing cells rolling to the well periphery, and a new underdrain for the prevention of leakage. HTS plates also exhibit improved frame features.

In this context, it also should be mentioned that some protocols refer to the use of plastic ELISA plates that appear more cost-effective. Their use, however, is not recommended at all for this technique because of their limited protein-binding capacity and surface area for protein-cell contact. The resulting spot quality is not desirable.

There are certain logistics to be followed when choosing the best-suited plate type for ELISPOT experiments. First of all, the same plates, possibly from the same batch, should be used throughout a study. If a protocol has been validated, a switch from nitrocellulose to PVDF can result in nondesirable results, for example, spots will be too large under the specific testing conditions, and spot numbers might decrease in the PVDF plates because of spot confluence. However, introducing PVDF plates to the protocol can result in a dramatic increase of spot numbers. Therefore, a fine balance has to be found for sensitivity requirements. Switching membranes requires a revalidation of the protocol, and possibly involves protocol changes. The same cells should be run in the same experiment on both plate types, possibly including wells located in the plate periphery. Choosing the right plate type should be based on a series of those experiments. Finally, it is important to check with the manufacturer for quality control procedures, batch performance, and meeting of standards set by the Society for Biomolecular Screening. Those standards are important for applying various automation procedures to the ELISPOT technique, including plate washers and ELISPOT readers (*see* **Subheadings 3.4.1.** and **3.5.**).

3.2. Coating Procedures

The coating procedure involves the addition of the capture antibody to the 96-well plate. In general a very simple and fast procedure, slight changes in the protocol, like changing of the antibody pair used, or decrease of antibody concentration, can have dramatic effects on ELISPOT results.

When using plates with PVDF membranes, a prewetting step should be included in the protocol (**Note 1**). It is recommended to add 15 µL per well of 70% ethanol to overcome the hydrophobicity of the membrane. Higher volumes can lead to leakage; lower volumes are not enough for prewetting the entire membrane. Prolonged incubation should be prevented because evaporation of the ethanol can lead to dry membranes with high hydrophobicity. The ethanol has to be washed out efficiently, for example, by three washes with sterile PBS. The washing procedure should not be neglected or cut short, and the protocol, once established, should not be changed. Remaining ethanol will interfere with the spot development (*see* **Fig. 1A**). Prewetting PVDF membranes can result in higher spot counts and it will prevent coating problems, which occur sporadically in approx 1% of the wells (*see* **Fig. 1B**).

The choice of an antibody pair for the ELISPOT assay should be based on ELISPOT testing performed by the scientist and manufacturer, taking into account quality controls, antibody sensitivity, and batch performance variability. The costs of antibodies can be considerable, thus making cost-effectiveness another important feature to consider. It is important to recognize that antibodies that perform well in ELISA might not perform well in an ELISPOT assay.

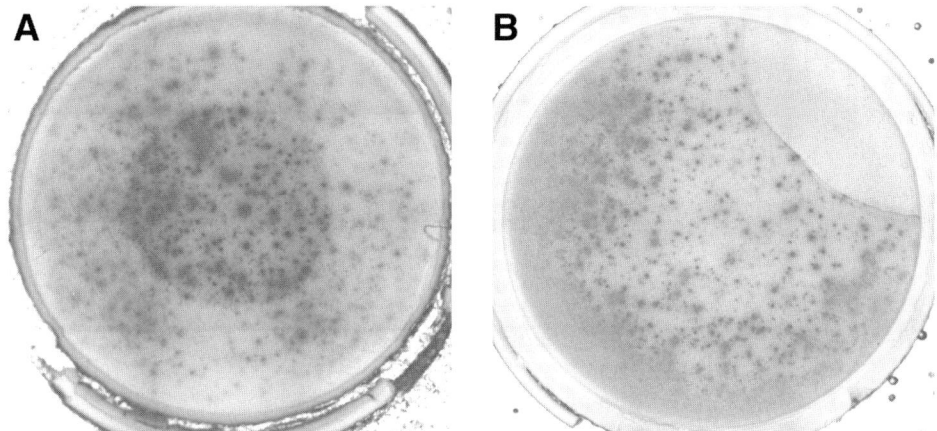

Fig. 1. Prewetting of PVDF plates. **(A)** A PVDF (MAIP plate, Millipore, Bedford MA) plate was prewetted with 50 µL of 70% ethanol and then washed as described. The high volume of ethanol saturated the membrane, and ethanol leaked into the underdrain. It could not be sufficiently removed by washing. The ethanol interfered with spot development, resulting in washed out spots in the periphery of the well. The center of the well has regularly developed spots since that part is located higher over the underdrain due to the concave shape of the membrane. **(B)** PHA control well of human PBMC in a PVDF plate that was not prewetted. Upper right part of well has no spots due to missing coating with antibody. Coating problems can affect a whole well, only the periphery or only parts of the well as shown. The pictures were taken with a KS ELISPOT reader (Carl Zeiss, Inc., Thornwood, NY).

Therefore, antibody pairs specifically designed for ELISPOT use should find priority consideration. Standardization and validation efforts become even more important in light of the fact that even commercially available ELISPOT kits for the detection of the same cytokine can exhibit differences in sensitivity as high as tenfold.

One of the most important parameters to standardize is the total amount of antibody used for coating. A general guideline is that approx 0.5–1 µg per well of coating antibody result in well-defined spots. Most importantly, the lower the total amount of antibody/well, the larger and fainter the spots become (*see* **Fig. 2**). This is attributable to the fact that with decreasing coating concentration, the antibody molecules bound on the membrane are further apart. Thus, the secreted cytokine has to diffuse further away from the secreting cell to be captured. Coating concentrations that are too low result in not only fainter spots but also fewer detectable spots. Therefore, the most efficient concentration needs to be established to concur with cost-effectiveness without compromising spot quality and ultimately spot count.

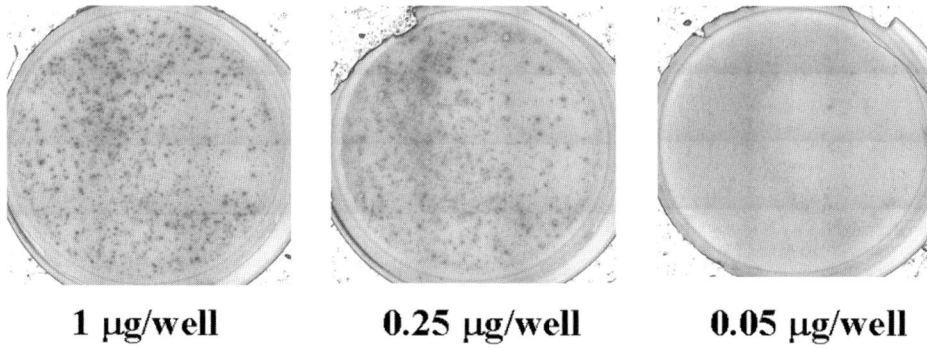

1 µg/well   0.25 µg/well   0.05 µg/well

Fig. 2. Influence of coating antibody concentration on spot outcome. A PVDF plate was coated with decreasing amount of coating antibody for detection of human IFN-γ as indicated. PBMCs were stimulated with 10 µg/mL PHA. Spots become fainter and more diffused with lower amount of coating antibody per well. The experiment was performed during the WHO-ELISPOT workshop at Duke University in April 2001. The pictures were obtained with a KS ELISPOT reader (Carl Zeiss, Inc., Thornwood, NY)

Various companies are now offering plates already precoated with an antibody and an immunological inert stabilizer. These plates clearly offer the advantage of availability whenever needed and the shortening of the protocol. However, it is important to validate the performance of those plates and to check with the manufacturer for quality controls and batch variability.

3.3. Cell Preparation

For conducting the ELISPOT assay, the integrity of the PBMC is critical for success. If performed correctly, the separation process yields a pure population of mononuclear cells consisting of lymphocytes and monocytes, with high viability, minimal red blood cell and platelet contamination and optimum functional capacity.

3.3.1. Preparation of Responding Cells

3.3.1.1. Cell Isolation

The most convenient source of T-cells for the ELISPOT assay is from peripheral blood derived from venipuncture (i.e., PBMC). Although it is possible to use whole blood for certain functional T-cell assays, for the ELISPOT assay, red cell hemolysis and secretion of inhibitors from platelets adversely affects T-cell function, and results of the assay are not interpretable. Even small amounts of red cells seem to adversely affect the results of the ELISPOT assay, resulting in high levels of artifacts. Thus, separation of mononuclear cells from peripheral blood is required. The technique for separating PBMCs from blood was first described in the 1960s and has changed very little since that time *(14)*. The original technique, consisting of layering blood onto a Ficoll-hypaque gradient, requires some technical expertise to yield sufficient PBMC. Layering Ficoll underneath the blood technically is easier to perform. In addition, two other improvements in the technique in the last few years have substantially reduced the technical burden of the procedure and improved yield and reproducibility of the technique. The Vacutainer® cell preparation (Becton Dickinson, Franklin Lakes, NJ), the Accuspin (Sigma-Aldrich Co, St. Louis, MO), and Leucosep (Greiner Bio-One Inc., Longwood, FL) tubes provide a convenient, single-tube system for the collection and separation of mononuclear cells from whole blood (**Note 2**). If red cells are visible in the cell pellet after processing and separation, one option is to remove these contaminating red cells by hypotonic lysis as described in **Note 3**.

3.3.1.2. Cryopreservation

Although PBMC processing may affect functional activity, we believe that cryopreservation is the most critical part of the procedure and the most likely to affect subsequent T-cell function if performed suboptimally. Poor recovery of PBMCs from stored specimens can be a source of immense frustration when conducting studies and therefore great attention should be paid to optimizing and standardizing cell freezing and thawing procedures. Real-time testing of T-cell responses on freshly isolated PBMCs without cryopreservation and storage is feasible for small phase I/II trials and for research protocols. There are clear advantages to being able to batch assays from multiple time points from a clinical trial, and this requires the cryopreservation of PBMC in a manner that maintains their functional capabilities. Under ideal conditions, fresh and cryopreserved PBMCs have been shown to have similar functional activity and preserve functional activity. Careful procedures used in PBMC cryopreservation and thawing result in minor losses in immune response as measured by the ELISPOT assay *(15–17)*.

Cryopreservation involves cooling cells at a rate of approx 1°C/min in the presence of a cryoprotectant such as dimethyl sulfoxide *(18–20)*. As with PBMC processing, there are several methods that can be used to cryopreserve PBMC. The use of automated controlled rate freezers provides the step-wise cooling that is necessary for optimal cryopreservation of cells and is most likely the procedure that will provide good standardization. However, the cost of an automated controlled rate freezer may be prohibitive for many laboratories, and a reliable supply of liquid nitrogen (LN) may not always be available. An easy and cheap option is the use of the Nalgene Mr. Frosty, these 18-vial plastic containers rely on the cooling rate of isopropanol which approximates 1°C/min in a –80°C freezer *(18,19)*.

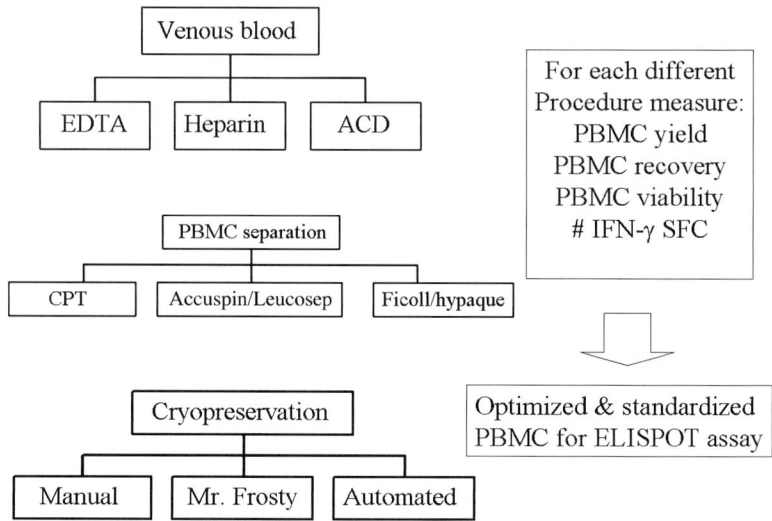

Fig. 3. Steps for optimizing and standardizing PBMC preparations for use in the ELISPOT assay.

The technical details of these separation procedures, the use of appropriate anticoagulants, and cryopreservation techniques are beyond the scope of this chapter. However, these procedures should be standardized within the laboratory setting prior to conducting ELISPOT assays as outlined in **Fig. 3**. Suggested reagents that can be used to standardize and perform quality assurance/quality control (QA/QC) on the ELISPOT assay are described in **Note 4**. These reagents can be used to directly quantify the integrity of the PBMCs and the number of IFN-γ or other cytokine-secreting cells when PBMCs are prepared under different conditions as shown in **Fig. 3**. Some examples of this type of approach to standardizing PBMC for use in the ELISPOT assay are shown below.

The optimal time frame between collection of blood sample to processing, separation and cryopreservation of PBMCs should be shorter than 8 h or on the same day as collection. It is not always feasible to process, separate and cryopreserve PBMCs within 8 h when samples are being shipped across the country to processing centers. Under these conditions, PBMCs left too long in the presence of anticoagulants or at noncompatible temperatures, adversely affect PBMC function and cause changes that affect the PBMC separation process *(8,20,21)*. We recently have shown that there is a dramatic effect on the number of IFN-γ-secreting cells if the PBMC are not processed in a timely fashion. Whole blood from 12 individuals was collected in tubes containing sodium–heparin anticoagulant. Processing of blood and ELISPOT assay set up was

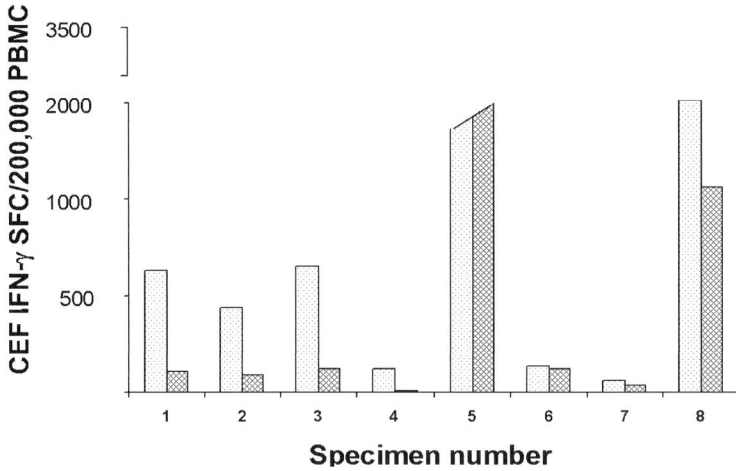

Fig. 4. Time to processing affects production of IFN-γ. Processing of eight blood samples and ELISPOT assay set up was either performed immediately (stippled bars) or after storage overnight of the blood at room temperature in a safety cabinet (hatched bars). PBMCs were stimulated with a pool of CMV, EBV, and influenza (CEF) peptides *(23)*. The average background SFC attributable to PBMCs only was subtracted. The scale has been readjusted for sample number 5. For this sample there were 2789 and 3056 CEF-specific SFCs/million PBMC, respectively, for blood processed immediately and overnight.

either performed immediately or after storage overnight of the blood at room temperature in a safety cabinet. Eight of the freshly isolated PBMC samples secreted IFN-γ in response to a pool of cytomegalovirus (CMV), Epstein–Barr virus (EBV), and influenza antigens (CEF pool, **ref. 23**). When the whole blood was left overnight before separation of PBMCs, seven of the eight PBMC samples had a reduction in the number of IFN-γ spot-forming cells (SFCs). Three samples exhibited a greater than fivefold reduction, and three had a one- to twofold reduction in the number of IFN-γ SFCs (**Fig. 4**). One of the responders changed to a nonresponder, with a 25-fold decrease from 123 to 5 CEF-specific SFC/10[6]. Aliquots of PBMCs from four of the fresh and stored blood samples were cryopreserved, and the ELISPOT assay was repeated. In two of the three samples, there was again a significant decrease in the SFCs in the stored compared with freshly processed samples. The responder with the highest CEF response (approx 3000 SFC/10[6]) appeared to be the least susceptible to the effects of delay in processing (sample ID #5, **Fig. 4**).

After optimizing the cell processing, separation, and cryopreservation techniques, it is essential that the cells are stored and shipped correctly so that deterioration of function does not occur. If cryopreserved and stored careful-ly, these PBMCs should maintain functional integrity for many years in stor-age before thawing. For the best functional T-cells, the best way to ship PBMCs is in specially constructed LN cryo-shippers. The LN is absorbed into a specially designed foam liner. LN is released slowly to ensure vapor phase LN is present inside the shipping container. Samples can be kept at a consis-tent temperature of approx –140°C for periods of 10–18 d depending on the size of the shipper. Many studies have shown that vapor LN is the optimal method for the long-term storage of cells. For shipping cells, LN cryo-ship-pers appear to be the only choice. One of the key parameters is keeping a con-stant temperature rather than fluctuating temperature. Changes in temperature per se appear to be detrimental to cells and subsequent function (Cox and Ferrari, unpublished observations). For this reason it is best to avoid shipping vials on dry ice.

For examination of functional T-cell responses, if samples have poor via-bility then they are unlikely to function correctly *(8,21,22)*. The reasons for the variability in cell recovery and viability can come from a number of sources; (1) errors in counting, (2) different methods of freezing and freeze media, (3) different methods of thawing cells, (4) number of centrifugation steps between thawing and counting, (5) length and type of storage condi-tions, and (6) the origin of the PBMCs. Because the goal of cryopreserving samples is so that batch testing of samples can be performed, it is critical to limit variability in the above parameters and thus functionality, viability and recovery of the cells will be optimal. A summary of the critical parameters for processing, collection, cryopreservation and storage of PBMCs is provided in **Note 5**.

3.3.1.3. THAWING OF CELLS

Differences in cell recovery may be the result of over- or underestimation of cell counts and different methods of thawing cells. The use of DNAses for thawing cells has been shown to improve cell yield and reduce clumping of cells *(17)*. For the ELISPOT assay in particular this is critical because of the need to be able to distinguish discrete spot forming cells in a well. A method that optimizes the recovery of PBMCs during the thawing procedure is described in **Note 6**. It also is recommended to allow the PBMC to "rest" overnight before plating in the ELISPOT assay. The rationale for resting cells overnight is to allow fragile apoptotic-prone cells to die, so that a true count of the number of viable cells can be assessed prior to plating. Overnight resting of the cells must be performed in media supplemented with 10–20% fetal bovine serum (FBS) or nonhuman serum (NHS) that has been tested using the QA/QC reagents as described in **Note 4**.

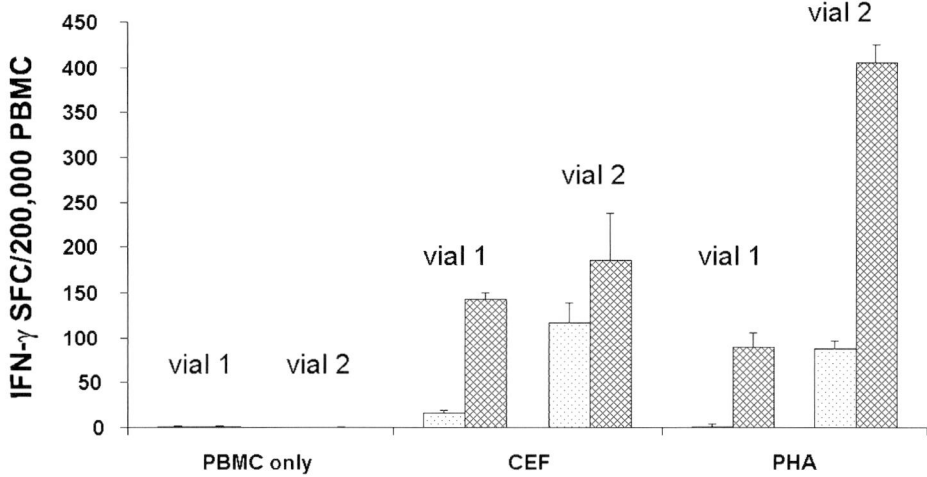

Fig. 5. Requirement for screening all reagents including media preparations for the ELISPOT assay. An ELISPOT assay was used to assess IFN-γ production in response to a pool of CMV, EBV, and influenza peptides (CEF pool, **ref.** *23*) and to PHA. Two vials of PBMC (vials 1 and 2) were assayed in two separate assays in "off-the-shelf" FBS supplemented media (stippled bars) or a new batch of FBS supplemented media (hatched bars). All other reagents for the assay were identical. Standard error bars are for three replicate wells.

A pertinent anecdotal experience provides an illustration of the importance of testing reagents prior to performing ELISPOT assays. At a training workshop organized by the World Health Organization and the authors (S. Janetzki, J. H. Cox, and G. Ferrari), all ELISPOT reagents except for FBS and RPMI media had been shipped to the laboratory in preparation for the workshop. An ELISPOT assay was set up with PBMC samples with known and very reproducible responses to the CEF pool *(23)*. Although the recovery and viability of the PBMCs was as expected, very few IFN-γ-secreting cells were detected either to the peptide pool or to PHA in three samples tested. Because PBMCs from one donor were still in culture, these were washed and resuspended either in the old culture medium or in freshly made medium that used a different source of FBS. A new vial of cells from the same donor was also thawed and prepared in the original media and the media with a different source of FBS. The results of the experiment are illustrated in **Fig. 5**. It is clear that the original medium was detrimental to elicitation of IFN-γ production in response to the peptide pool and PHA. ELISPOT reactivity was restored in the original aliquot of cells using the new medium. The second aliquot of PBMCs that was incubated in the old media only for the duration of the assay (approx 20 h) also had fewer IFN-γ SFCs compared with PBMCs incubated in the medium with new FBS.

3.3.1.4. Purification of Peripheral Blood Lymphocyte Cell Subsets

It is clear that the immune system can strongly influence the course of infectious diseases as well as of cancer. CD8[+] cytotoxic T-lymphocytes are involved in controlling infectious diseases and cancer by eliminating infected or tumor cells *(24–27)*, whereas CD4[+] are typically involved in sustaining the maturation of effector and memory CTL population as well as enhancing the CTL activity *(28–33)*. It is still not clear which of the two components should be preferentially elicited by candidate vaccine for infectious disease and cancer, if any. Because of the absence of correlates of immunoprotection in tumor and infectious disease models, it seems important to be able to quantify both CD4[+] and CD8[+] responses to candidate antigens to establish possible correlates of immunoprotection that may lead to successful immunotherapy interventions. In the ELISPOT assay, the identification and quantification of the T-cell effector subsets can be achieved by the utilization of macro- (Dynal, Upsala, Sweden; **ref. *34***), micromagnetic (Myltenyi Biotechnologies; **ref. *35***) beads, or magnetic colloid (StemCell Technologies; **ref. *36***) that are coated with an antibody or a tetrameric antibody complex directed to cell surface antigens. The combination of these beads or colloids can be used to remove from the PBMC suspension the population of interest (positive selection) or undesired cell subsets (negative selection). In case of positive selection using macro-beads or magnetic colloid, these reagents must be removed from the cell surface because they may interfere in cell-to-cell contact during antigen stimulation. The stimulation of purified T-cell subsets might also require supplementing the responding T-cells with appropriate antigen-presenting cells (APCs) to achieve an effective assay setup. It should be noted, that using purified T-cell populations might disrupt the physiological interaction between responding cells and accessory cells that may be important in conditioning the milieu to support antigen-specific responses.

Cell depletions can skew populations of T-cells in a way that estimated cell frequencies are no longer comparable among different individuals and should therefore not be used to quantify responses. In a recent study conducted at Duke University (Dr. Weinhold, unpublished observations), the removal of CD4 cells from 179 normal individuals created a new distribution of PBMCs where the frequency of CD8[+] T-cells ranged between 24.5 and 79.1% of the total CD4 depleted population (mean 50.8 ± 11.82%), as assessed by flow cytometry. Cell depletion as in the described approach would be only useful to determine the relative impact of the depleted population on the overall response, but not for estimation of cell frequencies.

The manufacturers have carefully calibrated the methodologies for positive and negative selection, so that cell populations are usually greater than 95% pure. The purity of these subsets should always be verified within the laboratory and validated within the ELISPOT assay SOP.

3.3.2. Antigens and APCs

3.3.2.1. ANTIGENS

Several platforms for antigenic stimulation of cells can be adopted as reported in **Table 1**. Advantages and limitations of these different antigens are also reported in the same table. Recombinant proteins can be used to detect CD4 mediated, and to a limited amount CD8 mediated, responses against the whole protein *(37)*. The use of recombinant proteins may be convenient for screening purposes to determine the frequencies of responders among a cohort of patients. Responses to recombinant proteins are dependent on the presence of APCs, which are decreased in number in cryopreserved samples. Replenishment of APCs may, therefore, be necessary. Epitopic regions cannot be identified with this method. Recombinant proteins may have solubility and stability problems at different storage temperatures. Similar advantages and limitations can be attributed to live recombinant viral vectors or virally infected cell lines *(38–40)*. The fact that some of them can express several proteins encoded by different genes within the same vectors is an advantage over recombinant proteins for screening of a large cohort of individuals. Nevertheless, the same limitations apply, and the possibility of increased background spots due to responses to the vector may occur.

Oligopeptides spanning from 14 to 22 amino acids (aa) length overlapping by 4-5 aa residues can be used successfully to screen responses against complex proteins. They can be used as pools of up to 122 individual peptides *(16,41,42)*. The possibility to use them in a matrix format for identifying optimal epitopes recognized by either CD4+ or CD8+ T-cells is their biggest advantage *(38, 43–48)*. The length of peptides may vary based on the type of effector population of interest. To design sets of overlapping peptides based on a protein sequence, it is recommended to use tools provided by the LANL website and related links *(49)*. This particular program can define sequences of peptides and apply filters that block "forbidden" aa residues that might affect the peptide binding to the HLA molecule. Twelve to eight aa peptides that represent well-defined epitopic sequences can also be used to identify CD4+ and CD8+ responses *(23,43)*. The same advantages and limitations of oligopeptides apply to this category of antigens.

When using oligopeptides, attention should be paid to their purity. Short contaminant sequences in peptide preparations may non-specifically activate cells and stimulate the production of cytokines. Moreover, low purity (<90%)

Table 1
Platforms for Antigenic Stimulation

Antigen	Advantage	Limitation
Recombinant protein	1. Large number of T-cell epitopes in one product; 2. Antigens are processed via the exogenous pathway; 3. Detection of CD4 responses.	1. Solubility of the protein; 2. Requirement of APC; 3. No epitope mapping. 4. Limited detection of CD8 responses
Live recombinant virus vectors	1. Several gene products or poliepitopic regions can be inserted in the same vector; 2. Antigens are processed via the endogenous pathway; 3. Detection of CD8 responses.	1. Possibility of increased background activity elicited by the viral vectors: 2. Possibility of cytopatic effects; 3. In some instances the presence of APC may be a pre-requisite; 4. No epitope mapping
Overlapping oligo-peptides	1. Overlapping peptides can be used to scan large proteins by using pools; 2. Overlapping peptides can be pooled together in a matrix system to identify new epitopes; 3. Detection of CD4 and CD8 responses.	1. High purity requirement of the oligo-peptides makes them an expensive reagent 2. Large number of cells required to cover multiple gene products or peptide pools
Epitopic peptides	1. Fine characterization of specific CD4 and CD8 responses.	1. Predicted epitopic sequences are not necessarily recognized by the immune system.

may create problems if different batches of the same peptides are used in clinical trials, whereas high purity allows the easy identification of contaminants.

It is important to understand that any stimulation strategy that will be used in the assay should always be validated in cohorts of individuals with detectable and nondetectable responses to the antigenic product. This will allow the determination of the level of background responses that might be elicited, and the specificity of such responses as part of validation of the reagents.

3.3.2.2. APCs

The preparation and use of APCs in the ELISPOT assay will be discussed in other chapters of this book (*see* Chapter 15).

3.3.3. Cell Counters

Because of its quantitative character as well as specific requirements for cell densities per well, the ELISPOT assay demands high accuracy in regards to cell numbers. It is also important to minimize the number of dead cells because the presence of dead cells can impact the integrity of the assay. Although no guidelines exist for optimum PBMC viability, a cut-off of >70% viability might be recommended as a minimum for specimens in an endpoint assay. Manual PBMC counting and viability assessment with a hemacytometer have been in use for decades. Although hemacytometer counts can be accurate, the procedure is entirely dependent on operator competency and hence prone to inconsistencies. Manual viability counts rely on the operator being able to distinguish overall size and shape of a lymphocyte as well as different hues of color. Thus, there is a need for accurate and reliable assessments of lymphocyte counts and viability, which do not depend on trypan blue or other dye exclusion and manual counting. Within the last year, several automated units that can determine cell counts and viability have appeared on the market *(50–53)*. Similar to other procedures used in the ELISPOT assay, cell-counting procedures should be validated in the laboratory by determining accuracy, linearity, precision, limit of detection, and range *(54–56)*.

3.4. ELISPOT Development

3.4.1. Plate Washers

The removal of cells from the ELISPOT plate before the addition of detection antibody is an important washing step because of its influence on spot appearance. Cells need to be carefully removed (*see* Chapter 5) to avoid nonspecific precipitation of reagents on the cell membrane that may create artifacts, such as those observed in **Figs. 6** and **7**. The washing procedure can be performed manually with a squirt bottle or pipet, or with an automated plate washer; and both procedures can generate equivalent results. Manual washes, however, may be the source of the following problems. First, removing cells considered infectious must be performed in the confined space of a safety cabinet. This usually creates problems for the disposal and treatment of supernatants and cells. Second, when there are a large number of plates to be washed, manual washing would take too long to achieve a consistent incubation time. Electronic multi-channel pipetors could be used for washing to increase the efficiency, but they cannot circumvent the difficulties related with the disposal of

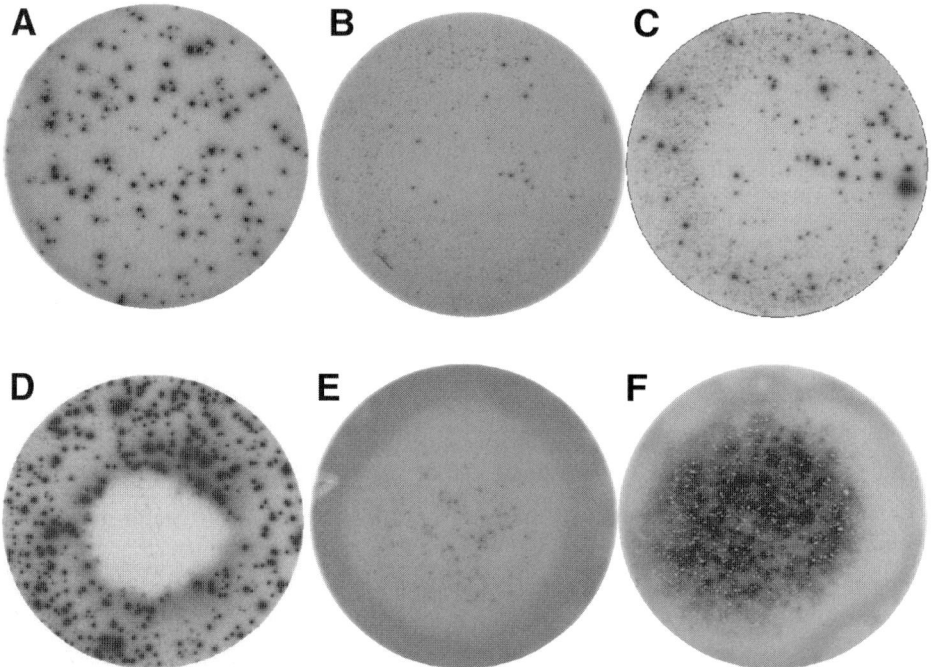

Fig. 6. Problems related to automated washing procedure. (**A**) well-distributed and developed spots; (**B**) artifactual spots caused by inefficient removal of cell debris; (**C**) target image with artifactual spots caused by inefficient washing caused by the high positioning of the probes and slow flow of washing buffer; (**D**) target image with absence of spots in the center due to inefficient washing caused by too low positioning of the probes and/or high flow of the washing buffer; (**E** and **F**) artifactual spots due to contamination and poor cleansing of the probes. The pictures were taken with an Immunospot Analyzer (Cellular Technology, Ltd., Cleveland, OH).

infectious waste. Finally, manual washes could fail to remove cells and reagents after each step if too little washing buffer and too low pressure is used, resulting in an increased number of artifacts on the membrane and accumulation of dye in the well periphery (**Fig. 6B**). Therefore, it is recommended that one invest in the purchase of an automated plate washer, when clinical trials with high numbers of assays have to be performed. At least two requirements should be observed during the purchase: the first is the ability to use at least two different washing buffers, and the second is the ability to adjust the height of the washing probes and the flow of the washing buffers. This will avoid possible confusion among different washing buffers. Automated washers that can use two bottles for different buffers at the same time can usually be programmed to

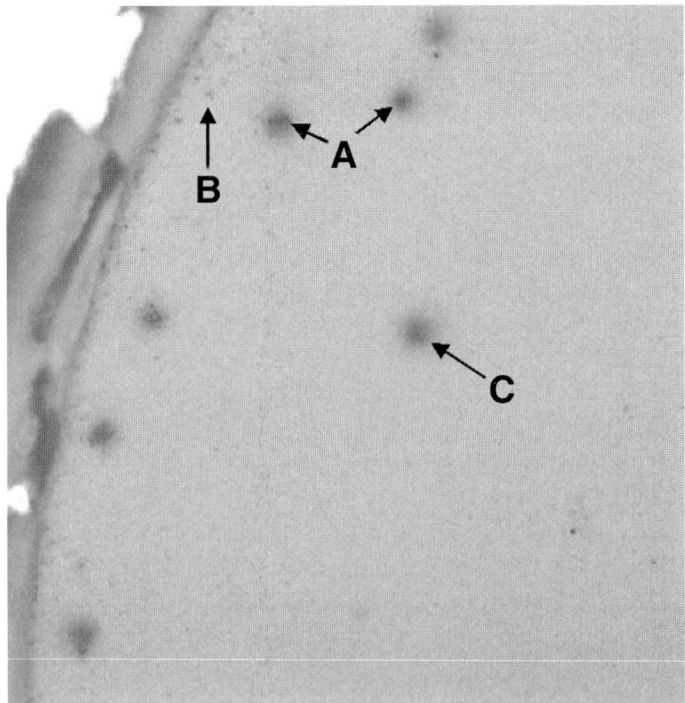

Fig. 7. False-positive spots. A partial well picture in high resolution with false positive spots in the well periphery is shown. Most spots are well defined, and only distinguishable from true spots by cell debris artifacts in the center (arrow **A**). There is also some accumulation of artifacts due to insufficient washing recognizable (arrow **B**). False-positive spots occurred due to a missing filter step of secondary biotinylated antibody what allows antibody aggregates to accumulate on the membrane, preferably around artifacts. False-positive spots also appear without cell debris artifacts when secondary antibody is not filtered, and are not distinguishable at all from true spots (arrow **C**). The picture was obtained with a KS ELISPOT reader (Carl Zeiss, Inc.; Thornwood, NY).

automatically switch from one source to the other. The feature of having adjustable settings for the height of the probes is important since not all plates have the same well depth. Furthermore, plate frames can be slightly tilted, resulting in uneven membrane heights. The probes may be placed too far or too close to the membrane generating a poor wash or damage of the membrane, respectively. Poor washing can also be caused by a slow flow of the washing buffer. Both conditions, location of probes too far from the membrane and slow flow, usually generate well-washed concentric areas in the center of the membrane with clearly defined spots, and a circle in the well periphery containing artifacts and an evident accumulation of substrate on the membrane. These effects are demonstrated in **Fig. 6C, 6D**.

It should be noted that automated washers significantly increase the number of plates that a single operator can handle in a given time. But automated washers require consistent decontamination and cleaning procedures; otherwise, accumulation of debris in the probes will generate the same type of problems observed in the event of probes too far from the membrane or of slow flow of the buffer (**Fig. 6E, 6F**). Because most of the decontamination procedures require the use of detergent, it is strongly recommended that several cycles of subsequent rinsing of the systems be performed to remove any residual detergent. Residual detergent could result in a loss of developed and readable spots in the subsequent assay. If the automated washer is also used to remove the coating antibody solution and the blocking solution preceding the plating of the cells for the incubation with the antigens, residual detergents may lyse the cell population. Therefore, though rinsing cycles are included in the decontamination programs provided by the manufacturer, they may not be sufficient. Despite the fact that automated washers are very useful for removal of cells after incubation, it is recommended to consider manual washes for steps preceding removal of cells. Since sterility is a necessity in ELISPOT assays until cell incubation is over, it is appropriate to perform all washes including the blocking procedure in a sterile cabinet using sterile pipet tips. Once again, washing procedures and setting of the plate washers should be addressed during the validation of the assay.

3.4.2. Spot Development

The spot development protocol is very similar to the ELISA protocol, except that it uses substrates that precipitate at enzyme sites on the membranes. Similar to the coating antibody, the correct antibody pair as well as concentration has to be chosen through validating experiments. As a general guideline, approx 0.1 μg of biotinylated detection antibody per well are sufficient for spot detection. Changes of antibody concentration do not have as dramatic effects as changes for the coating antibody. However, the lower the detection antibody concentration, the less enzyme will accumulate at a spot site, and the fainter the signal provided by the precipitating substrate. Therefore, the optimal amount of antibody, often meeting the manufacturer's recommendation, needs to be established and consistently followed. Before adding the diluted secondary antibody to the plate, it should be filtered in order to remove antibody aggregates that in turn are the cause of false positive (**Fig. 7**). Those spots have the same features as real spots, and are not distinguishable. A standardized and validated SOP should therefore always include necessary controls for false positive spots (e.g., medium control) and a filtering step for detection antibodies.

Similar attention needs to be given to the avidin–enzyme complex. It is important that the enzyme is stable over a certain time period, and does not require titration before use. This is a prerequisite for stable incubation times with the substrate, which in turn are necessary for spot conformity. Also batch variability of enzymes should be low to assure comparability of experiments. Optimally, a series of experiments should be started with a fresh avidin–enzyme complex. Recording of enzyme usage and storage is recommended.

A simplification of the development procedure has been achieved by the introduction of detection antibodies, which are directly coupled to enzymes. The challenge to create products with enzyme activities comparable to those reached by the biotin–avidin-amplification step has been met, and specific detection antibodies for the use in ELISPOT assays are available. However, to meet standardization guidelines it is recommended to use only one detection system throughout a series of experiments. Its performance has to be validated.

Substrate choices can influence the spot number detected. Various substrates for the same enzyme as well as the same substrate from different manufacturers can differ in sensitivity, resulting in considerable spot count differences. Furthermore, substrates with exceptional sensitivity need to be well titrated in amount used and incubation time to prevent overdevelopment, development of very large spots (which could become confluent), and pick-up of artifacts. The most commonly used substrates for peroxidase are 3-amino-9-ethylcarbazole (AEC), $C14H14N2$, 3,3¢-diaminobenzidine (DAB), Novared, and Tetramethylbenzidine (TMB); and for alkaline phosphatase, BCIP and BCIP/NBT.

A simplified procedure has been introduced by Gazagne et al. *(57)*. In their efforts they have been using detection antibodies directly coupled to fluorescent dyes. This method circumvents the use of enzymes and substrates. Even though there are so far only limited reports available, this technique appears promising especially for monitoring various cytokines simultaneously secreted by the same cell population.

3.5. ELISPOT Evaluation

In 1988, Czerkinsky suggested the use of a stereomicroscope for ELISPOT evaluation purposes *(3)*. Even though widely accepted, there are many drawbacks attached to that method: it is time- and labor-intensive, high spot numbers are difficult to evaluate, low resolution makes the differentiation of artifacts and real spots difficult, and the results are dependent on the experience and personal preferences of the scientist performing the evaluation. The bias and variability issues are the biggest problem attached to this method. The stereomicroscope has been used almost exclusively for ELISPOT evaluation, until the introduction of automated ELISPOT readers in 1997 *(58–60)*. Automated evaluation can circumvent many of the problems associated with manual counting. Today, multiple automated readers with various capabilities, resolution, and software packages can be found. They are discussed in Chapters 7 and 8.

When validating the ELISPOT protocol, the evaluation method has to be included in that process. Two main topics shall be discussed here: (1) the operator-dependent variability and (2) the use of assay-specific controls for defining reading parameters.

It can be easily recognized that manual evaluation with a stereomicroscope is dependent on the preferences of the operator. The high variability observed in manual evaluation can be drastically reduced when introducing an automated ELISPOT reader *(61,62)*. However, even when using an automated reader, there is still operator-dependent variability apparent *(62)*. This is attributable to the fact that each sample requires an adaptation of reading parameters, which are used by the software for automated recognition of spots and exclusion of artifacts. The better standardized the whole ELISPOT protocol and performance, and the better the materials used, the better the conformity of spots, and the lower the requirements for changes of reading parameters. However, it needs to be kept in mind that spot appearance is not only a function of assay performance, but also and especially a biological function *(63–65)*. This makes the use of predefined reading parameters for a series of experiments of limited usefulness, and it underlines the necessity of re-checking and re-adapting reading parameters in an automated setting. As discussed above, the more standardized all techniques involved in an ELISPOT assay, including cryopreservation, cell storage, the lower the possibility of varying spot definitions because of technical factors, artifacts, and nonspecific stimulation of cells.

The influence of personal preferences in establishing reading parameters can be partially overcome with assay-specific controls. Recently, Currier et al. introduced a suitable control for a majority of human IFN-γ ELISPOT assays *(23)*. This control contains a pool of 23 8-11mer peptides derived from the influenza virus, cytomegalovirus, and Epstein–Barr virus that are known to be recognized by $CD8^+$ T-cells and presented by class I HLA-A and HLA-B alleles. Because the produced spots are a result of a memory $CD8^+$ T-cell response, their appearance can provide a standard for spot definition, especially when using an automated reader system. In contrast, spots produced by other positive ELISPOT controls, like the mitogens such as phytohemagglutinin (PHA), ConA, or PMA/ionomycin should be used with caution for reading parameter setup. Mitogens induce cytokine release in different subsets of cells, and through different mechanisms than antigen recognition in memory responses. This leads to spots that can differ tremendously from the appearance of spots generated by antigen-specific memory cells. Furthermore, often the stimulation is so strong that the spots are too large and numerous to allow discrimination of single spots.

For evaluation purposes, the scientists need to be well trained, according to GLP guidelines. It is imperative that they have knowledge of the basic setup of the plate they are evaluating, especially of the location of assay-specific controls as discussed above. Furthermore, controls of effector cells alone, antigen-presenters alone, and similar controls should be used when defining the reading parameters. The performing scientist should be tested for variability by evaluating a plate multiple times. It might be advisable to limit the number of scientists performing the evaluation of experiments of one study to further decrease operator-dependent variability.

When using an automated reader, it is important to keep the machine well maintained and calibrated. Calibration cannot be performed on a scanned picture because the software will always give the same spot number when using the same reading parameters. Calibration should be performed on a newly scanned picture each time. It has to be kept in mind that spots will fade over time, even if kept protected from light. The rate of fading depends on the membrane and substrate used as well as the time of exposure to light. Finally, validation procedures should also include the establishment of the intrinsic variability of a reader due to lightening issues etc. Rescanning a well multiple times and evaluating the pictures with the same reading parameters can achieve this goal. The variability established here needs to be acknowledged when establishing recommendations for a positive response definition of ELISPOT results.

3.6. Assay Validation

The validation of an ELISPOT assay must be performed before initiating any clinical trial, but it is strongly recommended before initiating research projects that may lead to evaluation of therapeutic products in clinical trials. These studies should produce a "validated" SOP under GLP guidance *(55)* that can be used to submit data to the Food and Drug Administration for approval of clinical trials or licensing of vaccines, immunotherapeutic procedures, pharmaceutical products, etc. It should be noted that the use of a "validated" SOP established by a different laboratory does not automatically imply that the assay is "validated" for your laboratory.

The validation process should always start by identifying the variable within the assay that may create problems during the routine application of the assay (i.e., incubation times, purity of the reagents, concentration of reagents, purity of cells and their concentration), and subsequently verify that the whole procedure performs according to the guidelines described in the Federal Register (Part VIII; 60 FR 11260) based on the International Conference on Harmonisation documents ICH-Q2A (CDVS2.CDER.FDA.GOV; **ref.** *54*) and ICH-Q2B (http://www.eudra.org/emea.html). The seven characteristics that the

Table 2
Validation Criteria

Characteristics	Definition	Commonly referred to as:
Accuracy	Expresses the closeness of agreement between the found value and a reference value	

Precision
  Repeatability
  Intermediate
    precision
  Reproducibility
	a) Precision under the same operating conditions over a short interval of time;
	b) Precision within laboratory;
	c) Precision between laboratories.
		a) Intra-assay precision;
		b) Intra-laboratory variation;
		c) Inter-laboratory variation.

Specificity	To assess unequivocally the analyte in the presence of other components.	Absence of reactivity in unexposed negative individuals.
Detection limit	Lowest amount of analyte in a sample that can be detected but not necessarily quantitated.	
Quantitation limit	Lowest amount of analyte in a sample that can be quantitatively determined with suitable precision and accuracy.	
Linearity	Ability (within a given range) to obtain test results, which are directly proportional to the concentration of analyte in the sample.	To determine the linearity using different concentration of responding cells.
Range	The interval between the upper and lower concentration of analyte in the sample for which it has been demonstrated that the procedure has suitable level of precision, accuracy, and linearity.	To determine the expected ranges of responses to antigens based on the number of spots that can be counted without any approximation.

SOP should fulfill for a validated procedure were adapted from those documents and are listed in **Table 2**.

It should be noted that in those documents the term "analyte" corresponds, in fact, to the antigen-specific responding cells in the PBMC preparation.

Several articles have recently been published and address the "validation" of the ELISPOT assay *(9,10,16,17,66)*, and they may be helpful in addressing some of the requirements listed above for validation of the ELISPOT assay. However, validation of the assay for GLP compliance must always generate data that address all of the above criteria.

3.7. Statistical Considerations

We review below statistical considerations confronted when analyzing ELISPOT data in a large panel study. Scientific conclusions deriving from this study have been described elsewhere *(12)*. Data were drawn from 11 laboratories, each analyzing the same 11 donors. Each laboratory used six plates (one group of three rows per donor, two donors per plate). That is, there were three replicates for each donor–reagent–laboratory combination. Counting Diluent (media used for the assay), CEF, and PHA, there were nine reagents, each mapping to a different column on the plate. Additionally, in "batch 2" of the data, all plates were read a second time by a new reader.

Note that two missing-value codes are necessary for ELISPOT data; one to indicate experiments not done, and the other to indicate coalescent spots (too-numerous-to-count [TNTC] counts). TNTC counts can complicate analyses of means, but will usually not affect analyses of medians and ranks.

3.7.1. Numbers of Replicates

In the panel study, interest centered on determining which donors were responders to which reagents, in which laboratories. A planning issue not considered was that determination of responder status involves multiple comparisons of reagents to diluent. That is, the same diluent data were used in many separate comparisons to (other) reagents. Because the diluent data were more "heavily used", the study might have profited from a greater number of replicates for diluent than for the other reagents.

Dunnet's test *(67)* is a popular variant of the Analysis of Variance in which each of K experimental groups is in turn compared to a single control group. Hudgens et al. *(68)* cite Hochberg and Tamhane *(69)* to the effect that, if C is the number of replicates in the control group and E is the number of replicates in a single experimental group, then there is an optimal ratio of replicates: *optimal C/E = sqrt(K – 1)*. In this study, with $K = 8$ and $E = 3$, study power would probably have benefited from *8* replicates for each diluent in a batch–donor–laboratory instead of *3*.

3.7.2. Multiple Testing

A determination of responder status (Y/N) was required for each batch–laboratory–donor–reagent. If responder status is derived in each case from a statistical test (*see* below), many such statistical tests need to be performed simultaneously, engendering a multiple testing problem. To understand this, recall that when performing an ordinary statistical test at, say, the alpha = 0.05 level, one is carrying out a procedure that has a probability of 5% of being (incorrectly) significant *even if the null hypothesis is true.* That is, performing such a test when the null hypothesis is true is like flipping a coin that has a 5% probability of heads. If the coin comes up heads, one incorrectly rejects the null hypothesis, committing a "type I error."

The conclusions derived from the statistical test are believable in part because this probability of error is controlled to be fairly low, typically only 1 in 20. But suppose that, instead of performing one such test, one performs two independent tests, and both null hypotheses are true. If each test is carried out at $\alpha = 0.05$, the probability that at least one test will be (incorrectly) significant (i.e., the probability of at least one head in two flips) is about 10%. This probability increases with the number of tests. For five tests, the probability of at least one type I error is 23%. For 100 tests, it is 99%. This is the multiple testing problem, and the probability of making at least one type I error is called the "family-wise error rate." A classical way to deal with controlling the family-wise error rate is to adjust the 5% α-level for individual tests downward. If there are k such tests, it suffices to use an α-level of *0.05/k* for each individual test, whether or not the individual tests are statistically independent. This is the popular Bonferroni correction, known to be at worst conservative. Multiple testing is an area of active statistical research, and many other methods have been developed. We will describe one such, well-suited for ELISPOT analysis *(70)*.

3.7.3. Filters

Before analysis, it seemed prudent to filter out several types of data that were probably in error:

1. BADNEG: A few batch-lab-donor for whom any of the three diluent wells had SFCs exceeding $100/10^6$ PBMC were deleted.
2. BADEVEN: For each batch–laboratory–reagent–donor, the three replicate SFC were required to be comparable. This was determined by a chi-squared test for evenness (using SFC per 200,000) with an alpha of 0.05/1000. Cases failing this test were deleted. This method assumes the data are Poisson distributed, which may be unwarranted (*see* **Subheading 3.7.6.**). Another popular method, not used here, involves trimming the outer member of a triplet if its distance to the median exceeds 88.6% of the total range. BADEVEN drops an entire triplet, while trimming drops only the outlier. But trimming cannot be used of one member of the triplet is missing, while BADEVEN can. Simulations (not shown) suggest that, with random normal data, in which <u>no</u> data points are truly outliers, approx 20% of

triplets will trigger trimming. When applied to the panel data, the two methods have incomplete overlap, with trimming triggered more often.

3. PHALOW: Cases (excluding PHA) in which the batch–donor–laboratory was not a responder for a reagent, and also not a responder for PHA, were deleted. Note that, unlike the other two filters, PHALOW requires a pre-specified definition for "responder."

3.7.4. Poisson Model for SFCs

The Poisson distribution seems a natural model for the distribution of SFC counts because it requires only two (reasonable-sounding) assumptions: (1) there are a great number of cells (e.g., 200,000) in a well and (2) whether each cell forms a spot or not is governed by an independent coin toss with the same very low probability (e.g., p) for every cell in the well. Granting these two assumptions, it follows that the observed number of spots follows a Poisson distribution with expected value 200,000p.

Although the Poisson model seems natural, Lathey's data (*see* **Subheading 3.7.6.**; **ref. 9**) show considerable extra-Poisson variability, suggesting that another distribution, perhaps lognormal, may be more appropriate to model SFC.

3.7.5. Definitions of Responder

We explored several definitions of responder. We term two of the methods "arbitrary." Letting R be the reagent SFC/10^{-6} PBMC and D be the corresponding SFC for diluent, the conditions required for responder-positives according to the two arbitrary methods are given below:

1. ARB1: R > 3D and R-D > 20
2. ARB2: R > 4D and R > 55

An alternative ("statistical") definition involved comparison of reagent SFC to diluent SFC using the binomial distribution. Given that there are S total spots, if the case is not a responder, (i.e., if diluent and the other reagent do not differ), and assuming that spot counts are Poisson-distributed, the number of spots in, say, the Diluent group will follow a Binomial (1/2, S) distribution. If the numbers of replicates in the two groups differ, the probability parameter of the Binomial distribution changes. For example if there are three diluent replicate wells but only one "other reagent" replicate well, the number of Diluent spots will follow a Binomial (3/4, S) distribution. We reject (one-tailed) if the number of Diluent spots is much smaller than expected.

This test must be modified to incorporate TNTC counts. Typically, diluent will have no TNTC counts. So it seems reasonable that, if a reagent has a TNTC count, the donor should be counted as a responder. This means that the responder status of donor X to a given lab-reagent combination becomes:

Table 3
Data From Lathey's Table 1

	SFCs grouped by number formed					
Antigen	0–20	21–50	51–100	100–200	200–400	>400
Media	38	21	17			
	($n = 46$)	($n = 16$)	($n = 1$)			
Candida	30	33	18	22	5	
	($n = 21$)	($n = 16$)	($n = 12$)	($n = 12$)	($n = 4$)	
PHA				18	8	7
				($n = 2$)	($n = 20$)	($n = 44$)

1. Missing if there are no data for the reagent or for diluent.
2. Otherwise, nonresponder if both sums of counts (across repetitions) are 0.
3. Otherwise, responder if any reagent count is TNTC.
4. Otherwise, responder or not, according to the one-tailed binomial test.

A lenient application of the binomial test uses $\alpha = 0.05$. A Bonferroni adjustment for the panel study gave $\alpha = 0.05/1169$.

For the binomial test to work, the total number of spots in the two reagent groups must be large enough so that the p-value is capable of being less than 0.05 (or 0.05/1169). This means that, if the proportion of diluent replicates was D, we required the total spot count T to be at least log(a)/log(1 − D). For example, if the number of replicates was the same for both reagents, we required T to be at least 5 for a lenient test and at least 15 for a Bonferroni test. For total spot counts below these levels, we counted the donor as a nonresponder.

3.7.6. Poisson Variation and Extra-Poisson Variation

Lathey's Table 1 (abstracted in our **Table 3** with the kind permission of the author; **ref. 9**) gives SFC counts and coefficients of variation for media, Candida, and PHA, grouped into classes by number formed. Under the Poisson assumption, maximum likelihood methods allow estimation of the antigen-specific Poisson means of the SFC counts. The estimated means are 19, 61, and 400.

The mean of a Poisson distribution specifies the entire distribution. Thus, assuming SCF are Poisson-distributed, we can calculate the Coefficients of Variation expected in the SFC groups of Lathey's Table 1. These are given in **Table 4**. Rows 1–3 of **Table 4** are strikingly below Lathey's Table 1. Thus Lathey's data are far more variable than they would be if they were generated by three common underlying Poisson distributions (one for each antigen).

Another way to appreciate the size of the Lathey CVs is to compare them to the CVs one would expect in each SFC group if the values within the SFC

Table 4
Coefficients of Variation Implied by Poisson Distributions Matching the Data of Lathey's Table 1

Antigen	SFCs grouped by number formed					
	0–20	21–50	51–100	100–200	200–400	>400
Media	16.59	10.85	1.78	0.52	0.17	
Candida	4.04	6.01	10.91	1.74	0.39	
PHA	1.17	0.80	0.66	0.69	3.11	2.92

Table 5
CVs Expected Assuming SFCs Are Distributed Uniformly Within Each Interval

	SFCs grouped by number formed					
	0–20	21–50	51–100	100–200	200–400	>400
Uniform	58	24	19	19	19	

group were uniformly distributed across the interval of the group (for example, assuming that the SFC values in the 21–50 group were uniformly distributed on [21–50]). These are listed in **Table 5**. Lathey CVs often approximate the uniform CVs.

Is there a distribution that fits these data well? One possibility is that the counts are lognormally distributed, that is, that the logs of the counts are normally distributed. Ignoring the annoying possibility of a 0 count (requiring a log of minus infinity!), maximum likelihood estimates of the mean and variance of the log counts are: (2.64, 0.34), (3.62, 1.38), and (6.19, 0.22) for the three antigens, respectively. This implies mean counts of 17, 74, and 544. Simulations using sample sizes of 63, 65, and 66 (as in **ref. 9**) for the three antigens, respectively, generate mean sample sizes and CVs in the SFC groups as shown in **Table 6**. Judging by their similarity to Lathey's data, the lognormal distribution is moderately successful in simulating the observed distribution of counts.

3.7.7. Westfall/Young Analysis

Hudgens et al. **(68)** suggests that the method of Westfall and Young **(70–73)** is appropriate for determination of ELISPOT responder status. This method is conveniently implemented in the SAS MULTTEST procedure, and in the Bioconductor's Multtest Package in R.

As applied to the ELISPOT responder problem, the technique compares SFC for reagents to that of diluents by permutation (or alternatively resampling) tests

Table 6
Coefficients of Variation and Sample Sizes From the Simulation of Counts
Based on Lognormal Distributions (Results Rounded to the Nearest Integer)

	SFCs grouped by number formed					
Antigen	0–20	21–50	51–100	100–200	200–400	>400
Media	37	24	14			
	(n = 47)	(n = 15)	(n = 2)	(n = 1)		
Candida	46	25	20	19	18	32
	(n = 20)	(n = 19)	(n = 13)	(n = 8)	(n = 4)	(n = 2)
PHA				12	17	37
			(n = 1)	(n = 2)	(n = 20)	(n = 44)

based on *t*-tests. This approach is combined with theoretical advances in multiple testing to solve at one stroke the Poisson distribution and multiple testing problems described earlier. In the analysis discussed here, Hochberg's step-down method was used to adjust *p*-values. TNTC values still cause difficulties, and in the panel analysis, the TNTC values within each batch–laboratory–donor-reagent were mapped to a number slightly above the maximum of the non-TNTC counts. If all replicates were TNTC, they all were mapped to 2500. Note that, once having dealt with TNTC, future analyses should probably use the transformation SFC → log(SFC + 1) to make the data roughly normal and at the same time handle cases where SFC = 0.

3.7.8. Measures of Agreement

It was desired to investigate agreement between the responder status as calculated in the various ways (ARB1, ARB2, lenient Binomial, Bonferroni Binomial, Westfall/Young) described above. Statistics measuring pairwise agreement in responder status and in outcomes of the filters between the methods and batches generally are numerical summaries of a 2-×-2 table. For example, the rows may give responder status (Y/N) according to method A, whereas the columns give it according to method B. Two popular measures of agreement using these types of data presentation are: (1) the simple percent agreement (percent of cases falling in the diagonal cells), among all cases in which both methods had nonmissing data and (2) the κ statistic *(74)*. Even if two methods were completely unrelated, one might expect some agreement between them simply by accident. κ attempts to correct for this. The methods agreed well, with percent agreement ranging from 89 to 98%. κ values ranged from 0.74 to 0.97. Fleiss calls values above 0.75 "excellent," whereas Landis and Koch *(75)* call values above 0.6 "substantial" and values above 0.8 "almost perfect."

In addition to agreement, McNemar's test can be used to compare the off-diagonal cells of the table. This tests whether the positivity rates of the two methods are equal.

4. Notes

1. Prewetting of PVDF plates. Under hood, add 15 µL of 70% ethanol to wells. Tap plate slightly to assure even wetting of the entire membrane. Membrane will change color from white to dark gray-blue. Without further incubation, wash plate three times with 150 µL of sterile PBS. Continue with coating procedure.

2. The Vacutainer™ CPT™ cell preparation tube combines a blood-collection tube containing anticoagulant with a Ficoll-Hypaque or similar density separation fluid and a polyester gel barrier that separates the two liquids. The Accuspin and Leucospep tubes use a porous high-density polyethylene membrane frit that separates the blood from the lower chamber containing Ficoll. Anticoagulated whole blood is added to the top chamber, and upon centrifugation the whole blood descends through the gel barrier or frit to contact with the separation medium below giving a clear separation of the blood components. The erythrocytes aggregate and the granulocytes become slightly hypertonic, increasing their sedimentation rate, resulting in pelleting at the bottom of the tube. Lymphocytes and other mononuclear cells, that is, monocytes, remain at the plasma interface.

3. Hypotonic lysis of red cells. Resuspend the cell pellet in ammonium chloride lysis buffer (ACK lysing buffer, Quality Biological, Gaithersburg, MD), using 5 mL of solution per 10 mL of original blood volume. Allow to stand 5 min at room temperature, add 25 mL of PBS, mix, and centrifuge 15 min at 300g, room temperature. Remove supernatant, resuspend the cell pellet in media supplemented with ELISPOT-tested serum (*see* **Note 4**). Centrifuge again 15 min at 300g, room temperature. ACK lysing buffer can be made up as follows; to 800 mL of H_2O, add 8.29 g of NH_4Cl (0.15 M), 1 g of $KHCO_3$ (10 mM) and 37.2 mg of NA_2EDTA (0.1 mM). Adjust the pH to 7.2–7.4 with 1 N HCl. Add H_2O to make a total volume of 1 liter. Filter sterilize through a 0.2-µm filter and store at room temperature.

4. Testing reagents for use in the ELISPOT assay. It is absolutely critical to test all reagents prior to use to ensure that there is a low spontaneous background in unstimulated PBMCs and adequate and reproducible detection of spots in stimulated PBMCs. Ideally, all batches of media, batches of plates, serum, wash buffers as well as antibodies and detection reagents should be tested prior to conducting assays. An appropriate QA/QC reagent is one that can be used to consistently induce antigen-specific induction of IFN-γ in CD8⁺ T-cells or that induce antigen-specific induction of cytokine (s) of interest in T-cell subset (s) of interest. Examples of such QA/QC reagents are (1) a pool of CMV, EBV, and influenza peptides *(23)*, (2) commercially available pools of peptides from the CMV pp65 protein (BD Biosciences, San Diego CA), and (3) live virus preparations such as CMV (Advanced Biosciences Incorporated, Columbia, MD). Mitogens such as PHA, ionophores such as PMA/Ionomycin and staphylococcus enterotoxin also can be used to induce cytokine(s) of interest in T-cell subset(s) of interest. These types of

reagents can be obtained from a number of suppliers e.g., Aldrich-Sigma. However again it should be emphasized that the QA/QC reagents themselves should be test-ed to ensure reproducibility. PHA spots can vary widely in size, intensity and shape with a rather patchy occurrence. Because of the high spot number per well, they also can be confluent. Furthermore, PHA is known for losing its activity rather fast during storage, which can result in ineffective IFN-γ production and unclearly defined spots. The use of the CEF peptide pool described above circumvents these problems. The spots are well defined, comparable with spots in the remaining assay, and the spot number is typically below the number of spots resulting in con-fluence.

5. Critical parameters for processing, collection, cryopreservation and storage of PBMC: (1) Limit the time between blood draw and processing to as short a time as possible. For the **ELISPOT** assay if PBMC are not processed in a timely manner, the number of IFN-γ-secreting cells is dramatically reduced. (2) Use of freeze media and culture media that has been tested to ensure low spontaneous elicitation of cytokine secretion. (3) Avoid extremes of temperature during cell shipment and pro-cessing. (4) Minimize red blood cell and platelet contamination by use of good tech-nique and multiple washing steps. (5) Use correct centrifugation speed for PBMC separation. (6) Heparin, sodium citrate, and ethylenediamine tetraacetic acid anti-coagulants appear to be interchangeable for use in immunology assays, provided that the cells are processed within 8 h. (7) Slow freezing procedure to bring the tem-perature of the cells down to approx –140°C. (8) Store and ship cells in vapor phase LN at –140°C or lower. (9) Rapid thawing procedure to minimize cell disruption.

6. Thawing PBMCs with DNAse. When thawing PBMC, we recommend using DNAases such as Benzonase® nuclease, (CN Biosciences, Madison, WI). Benzonase nuclease is a genetically engineered endonuclease from Serratia marcescens. It degrades all forms of DNA and RNA (single stranded, double stranded, linear, and circular) while having no proteolytic activity. Add 50 U/mL of Benzonase to media to be used for cell thawing. Warm the media to at least room temperature (range, 25–37°C). Remove the cryovial containing the PBMC sample from the liquid nitrogen and hold in 37°C water bath. Do not shake vials; micro-crystals of ice may damage cell membranes and ultimately lyse PBMCs. As soon as the sample is just about thawed, pipet the sample into its previously labeled 50-mL conical. Add the media supplemented with Benzonase dropwise with gentle shaking, slowly bringing the volume in the conical up to 10 mL. Spin down the tube at 300*g* for 7–10 min at room temperature. Decant the supernatant and repeat the wash with the Benzonase media. Resuspend the cells in media without Benzonase for counting and subsequent procedures.

Acknowledgments

We gratefully acknowledge Ellen Kuta at the Henry Jackson Foundation, as well as Alicia Hedgepeth and Barbara Exley at the Duke University Medical Center for their technical support. We would like to thank Dr. Pat D'Souza for her support during the studies related to the NIH proficiency panel testing and

BBI for their technical support. We also gratefully acknowledge The EMMES Corporation, who supported Oden during the writing of this chapter. Finally we would like to thank Dr. Saladin Osmanov of the World Health Organization for helping disseminate the ELISPOT assay across four continents.

References

1. Czerkinsky, C. C., Nilsson, L. A., Nygren, H., Ouchterlony, O., and Tarkowski, A. (1983) A solid-phase enzyme-linked immunospot (ELISPOT) assay for enumeration of specific antibody-secreting cells. *J. Immunol. Methods* **65**, 109–121.
2. Sedgwick, J. D., and Holt, P. G. (1983) A solid-phase immunoenzymatic technique for the enumeration of specific antibody-secreting cells. *J. Immunol. Methods* **57**, 301–309.
3. Czerkinsky, C., Andersson, G., Ekre, H. P., Nilsson, L. A., Klareskog, L., and Ouchterlony, O. (1988) Reverse ELISPOT assay for clonal analysis of cytokine production. I. Enumeration of gamma-interferon-secreting cells. *J. Immunol. Methods* **110**, 29–36.
4. Janetzki, S., Palla, D., Rosenhauer, V., Lochs, H., Lewis, J. J., and Srivastava, P. K. (2000) Immunization of cancer patients with autologous cancer-derived heat shock protein gp96 preparations: a pilot study. *Int. J. Cancer* **88**, 232–238.
5. Lewis, J. J., Janetzki, S., Schaed, S., Panageas, K. S., Wang, S., Williams, L., et al. (2000) Evaluation of CD8(+) T-cell frequencies by the ELISPOT assay in healthy individuals and in patients with metastatic melanoma immunized with tyrosinase peptide. *Int. J. Cancer* **87**, 391–398.
6. Pass, H. A., Schwarz, S. L., Wunderlich, J. R., and Rosenberg, S. A. (1998) Immunization of patients with melanoma peptide vaccines: immunologic assessment using the ELISPOT assay. *Cancer J. Sci. Am.* **4**, 316–323.
7. Wang, F., Bade, E., Kuniyoshi, C., Spears, L., Jeffery, G., Marty, V., et al. (1999) Phase I trial of a MART-1 peptide vaccine with incomplete Freund's adjuvant for resected high-risk melanoma. *Clin. Cancer Res.* **5**, 2756–2765.
8. Cox, J., deSouza, M., Ratto-Kim, S., Ferrari, G., Weinhold, K., and Birx, D. (2002) Accomplishing cellular immune assays for evaluation of vaccine efficacy in developing countries., in *Manual of Clinical Laboratory Immunology* ASM Press, Washingon, DC, pp. 301–315.
9. Lathey, J. (2003) Preliminary steps toward validating a clinical bioassay. *BioPharm. Int.*, 42–50.
10. Mwau, M., McMichael, A., and Hanke, T. (2002) Design and validation of an enzyme-linked immunospot assay for use in clinical trials of candidate HIV vaccines. *AIDS Res. Hum. Retroviruses* **18**, 611–618.
11. Scheibenbogen, C., Romero, P., Rivoltini, L., Herr, W., Schmittel, A., Cerottini, J. C., et al. (2000) Quantitation of antigen-reactive T-cells in peripheral blood by IFNgamma-ELISPOT assay and chromium-release assay: a four-centre comparative trial. *J. Immunol. Methods* **244**, 81–89.
12. Cox, J., Ferrari, G., Kalams, S. A., Lopaczynski, W., Oden, N., D'Souza, P., and Group, a. t. E. C. S. (2005) Results of an ELISPOT proficiency panel conducted in

11 laboratories participating in international immunodeficiency virus type 1 vaccine trials. *AIDS Res. & Hu. Retroviruses* **21**, in press.

13. Shaw, R. D., Merchant, A. A., Groene, W. S., and Cheng, E. H. (1993) Persistence of intestinal antibody response to heterologous rotavirus infection in a murine model beyond 1 year. *J. Clin. Microbiol.* **31**, 188–191.

14. Boyum, A. (1968) isolation of mononuclear cells and granulocytes from human blood. *Scand. J. Clin. Lab. Invest.* **21**, 77–89.

15. Kreher, C. R., Dittrich, M. T., Guerkov, R., Boehm, B. O., and Tary-Lehmann, M. (2003) CD4+ and CD8+ cells in cryopreserved human PBMC maintain full functionality in cytokine ELISPOT assays. *J. Immunol. Methods* **278**, 79–93.

16. Russell, N. D., Hudgens, M. G., Ha, R., Havenar-Daughton, C., and McElrath, M. J. (2003) Moving to HIV-1 vaccine efficacy trials: defining T-cell responses as potential correlates of immunity. *J. Infect. Dis.* **187**, 226–242.

17. Smith, J. G., Liu, X., Kaufhold, R. M., Clair, J., and Caulfield, M. J. (2001) Development and validation of a gamma Interferon ELISPOT assay for quantitation of cellular immune responses to varicella-zoster virus. *Clin. Diagn. Lab. Immunol.* **8**, 871–879.

18. The Adult AIDS Clinical Trials Group (AACTG). Web site. Available at: http://aactg.s-3.com/pub/download/vir/freezingprotocol.pdf; Internet; accessed August 26, 2004.

19. Nalgene Cryopreservation Application Guide. Web site. Available at: http://www.nalgenelabware.com/techdata/Technical/manual.asp; Internet; accessed August 26, 2004.

20. Weinberg, A. (2002) Cryopreservation of peripheral blood mononuclear cells, in *Manual of Clinical Laboratory Immunology.* ASM Press, Washington, DC, pp. 220–223.

21. Betensky, R., Connick, E., Devers, J., Landay, A., Nokta, M., Plaeger, S., et al. (2000) Shipment impairs lymphocyte proliferative responses to microbial antigens. *Clin. Diagn. Lab. Immunol.* **7**, 759–763.

22. Weinberg, A., Betensky, R., Zhang, L., and Ray, G. (1998) Effect of shipment, storage, anticoagulant, and cell separation on lymphocyte proliferation assays for human immunodeficiency virus-infected patients. *Clin. Diagn. Lab. Immunol.* **5**, 804–807.

23. Currier, J., Kuta, E., Turk, E., Earhart, L., Loomis-Price, L., Janetzki, S., et al. (2002) A panel of MHC class I restricted viral peptides for use as a quality control for vaccine trial ELISPOT assays. *J. Immunol. Methods* **260**, 157–172.

24. Dudley, M. E., Wunderlich, J. R., Robbins, P. F., Yang, J. C., Hwu, P., Schwartzentruber, D. J., et al. (2002) Cancer regression and autoimmunity in patients after clonal repopulation with antitumor lymphocytes. *Science* **298**, 850–854.

25. Shankaran, V., Ikeda, H., Bruce, A. T., White, J. M., Swanson, P. E., Old, L. J., et al. (2001) IFNgamma and lymphocytes prevent primary tumour development and shape tumour immunogenicity. *Nature* **410**, 1107–1111.

26. Yee, C., Savage, P. A., Lee, P. P., Davis, M. M., and Greenberg, P. D. (1999) Isolation of high avidity melanoma-reactive CTL from heterogeneous populations using peptide-MHC tetramers. *J. Immunol.* **162**, 2227–2234.

27. Yee, C., Thompson, J. A., Byrd, D., Riddell, S. R., Roche, P., Celis, E., et al. (2002) Adoptive T-cell therapy using antigen-specific CD8+ T-cell clones for the treatment of patients with metastatic melanoma: in vivo persistence, migration, and antitumor effect of transferred T-cells. *Proc. Natl. Acad. Sci. USA* **99**, 16168–16173.

28. Camara, N. O., Sebille, F., and Lechler, R. I. (2003) Human CD4+CD25+ regulatory cells have marked and sustained effects on CD8+ T-cell activation. *Eur. J. Immunol.* **33**, 3473–3483.

29. Grakoui, A., Shoukry, N. H., Woollard, D. J., Han, J. H., Hanson, H. L., Ghrayeb, J., et al. (2003) HCV persistence and immune evasion in the absence of memory T-cell help. *Science* **302**, 659–662.

30. Matloubian, M., Concepcion, R. J., and Ahmed, R. (1994) CD4+ T-cells are required to sustain CD8+ cytotoxic T-cell responses during chronic viral infection. *J. Virol.* **68**, 8056–8063.

31. von Herrath, M. G., Yokoyama, M., Dockter, J., Oldstone, M. B., and Whitton, J. L. (1996) CD4-deficient mice have reduced levels of memory cytotoxic T-lymphocytes after immunization and show diminished resistance to subsequent virus challenge. *J. Virol.* **70**, 1072–1079.

32. Wei, W. Z., Morris, G. P., and Kong, Y. C. (2004) Anti-tumor immunity and autoimmunity: a balancing act of regulatory T-cells. *Cancer Immunol. Immunother.* **53**, 73–78.

33. Wherry, E. J., Blattman, J. N., Murali-Krishna, K., van der Most, R., and Ahmed, R. (2003) Viral persistence alters CD8 T-cell immunodominance and tissue distribution and results in distinct stages of functional impairment. *J. Virol.* **77**, 4911–4927.

34. Dynal Available at: http://www.dynalbiotech.com Internet: accessed November 8, 2004.

35. Miltenyi Biotec home page. Available at: http://www.miltenyibiotec.com/; Internet: accessed August 26, 2004.

36. StemCell Technologies home page. Available at: http://www.stemcell.com/; Internet: accessed August 26, 2004.

37. Rosenberg, E. S., LaRosa, L., Flynn, T., Robbins, G., and Walker, B. D. (1999) Characterization of HIV-1-specific T-helper cells in acute and chronic infection. *Immunol. Lett.* **66**, 89–93.

38. Currier, J. R., deSouza, M., Chanbancherd, P., Bernstein, W., Birx, D. L., and Cox, J. H. (2002) Comprehensive screening for human immunodeficiency virus type 1 subtype- specific CD8 cytotoxic T-lymphocytes and definition of degenerate epitopes restricted by HLA-A0207 and -C(W)0304 alleles. *J. Virol.* **76**, 4971–4986.

39. Larsson, M., Jin, X., Ramratnam, B., Ogg, G. S., Engelmayer, J., Demoitie, M. A., et al. (1999) A recombinant vaccinia virus based ELISPOT assay detects high frequencies of Pol-specific CD8 T-cells in HIV-1-positive individuals. *Aids* **13**, 767–777.

40. McAdam, S., Kaleebu, P., Krausa, P., Goulder, P., French, N., Collin, B., et al. (1998) Cross-clade recognition of p55 by cytotoxic T-lymphocytes in HIV-1 infection. *Aids* **12**, 571–579.

41. Pathan, A. A., Wilkinson, K. A., Wilkinson, R. J., Latif, M., McShane, H., Pasvol, G., et al. (2000) High frequencies of circulating IFN-gamma-secreting CD8 cytotoxic T-cells specific for a novel MHC class I-restricted Mycobacterium tuberculosis epitope in M. tuberculosis-infected subjects without disease. *Eur. J. Immunol.* **30**, 2713–2721.

42. Wilson, C. C., Palmer, B., Southwood, S., Sidney, J., Higashimoto, Y., Appella, E., et al. (2001) Identification and antigenicity of broadly cross-reactive and conserved human immunodeficiency virus type 1-derived helper T-lymphocyte epitopes. *J. Virol.* **75**, 4195–4207.

43. Draenert, R., Altfeld, M., Brander, C., Basgoz, N., Corcoran, C., Wurcel, A. G., et al. (2003) Comparison of overlapping peptide sets for detection of antiviral CD8 and CD4 T-cell responses. *J. Immunol. Methods* **275**, 19–29.

44. Edwards, B. H., Bansal, A., Sabbaj, S., Bakari, J., Mulligan, M. J., and Goepfert, P. A. (2002) Magnitude of functional CD8+ T-cell responses to the gag protein of human immunodeficiency virus type 1 correlates inversely with viral load in plasma. *J. Virol.* **76**, 2298–2305.

45. Ferrari, G., Neal, W., Jones, A., Olender, N., Ottinger, J., Ha, R., et al. (2001) CD8 CTL responses in vaccinees: emerging patterns of HLA restriction and epitope recognition. *Immunol. Lett.* **79**, 37–45.

46. Kern, F., Surel, I. P., Faulhaber, N., Frommel, C., Schneider-Mergener, J., Schonemann, C., et al. (1999) Target structures of the CD8(+)-T-cell response to human cytomegalovirus: the 72-kilodalton major immediate-early protein revisited. *J. Virol.* **73**, 8179–8184.

47. Masemola, A., Mashishi, T., Khoury, G., Mohube, P., Mokgotho, P., Vardas, E., et al. (2004) Hierarchical targeting of subtype C HIV-1 proteins by CD8+ T-cells: correlation with viral load. *J. Virol.* **78,** 3233–3243.

48. Novitsky, V., Rybak, N., McLane, M. F., Gilbert, P., Chigwedere, P., Klein, I., et al. (2001) Identification of human immunodeficiency virus type 1 subtype C Gag-, Tat-, Rev-, and Nef-specific ELISPOT-based cytotoxic T-lymphocyte responses for AIDS vaccine design. *J. Virol.* **75**, 9210–9228.

49. LANL HIV Molecular Immunology Database. Available at: http://www.hiv.lanl.gov/content/hiv-db/PEPTGEN/PeptGenSubmitForm.html; Internet; accessed August 26, 2004.

50. Beckman Coulter Vi-CELL™ Series Cell Viability Analyzer. Available at: http://www.beckman.com/products/instrument/partChar/pc_vicell.asp; Internet; accessed August 26, 2004.

51. Guava Technologies homepage. Available at: http://www.guavatechnologies.com/ ; Internet; accessed August 26, 2004.

52. Ioannides, C. (2003) Improving the accuracy and speed of mammalian cell counting. *Am. Biotechnology Lab.* **May,**, 10–12

53. Lem, L. (2003) Cell counting and viability assessments in the process. Development of cellular therapeutics. *BioProcessing J.* **July/August,** 57–60.

54. FDA Guideline for Industry. Text on validation of Analytical Procedures. Available at: www.fda.gov/cder/guidance/ichq2a.pdf ; Internet; accessed August 26, 2004.

55. FDA Code of Federal Regulations Part 58. Good Laboratory Practice for Nonclinical Laboratory Studies.Available at: http://www.access.gpo.gov/nara/cfr/waisidx_01/21cfr58_01.html; Internet; accessed August 26, 2004.
56. Feldkamp, C. S., and Carey, J. L. (2002) Standardization of Immunoassay methodologies, in *Manual of Clinical Laboratory Imunology* (Rose, N. R., Hamilton, R. G., Detrick, B., eds), ASM Press, Washington, DC pp. 1215–1226.
57. Gazagne, A., Claret, E., Wijdenes, J., Yssel, H., Bousquet, F., Levy, E., et al. (2003) A Fluorospot assay to detect single T-lymphocytes simultaneously producing multiple cytokines. *J. Immunol. Methods* **283**, 91–98.
58. Cui, Y., and Chang, L. J. (1997) Computer-assisted, quantitative cytokine enzyme-linked immunospot analysis of human immune effector cell function. *Biotechniques* **22**, 1146–1149.
59. Herr, W., Linn, B., Leister, N., Wandel, E., Meyer zum Buschenfelde, K. H., and Wolfel, T. (1997) The use of computer-assisted video image analysis for the quantification of CD8+ T-lymphocytes producing tumor necrosis factor alpha spots in response to peptide antigens. *J. Immunol., Methods* **203**, 141–152.
60. Vaquerano, J. E., Peng, M., Chang, J. W., Zhou, Y. M., and Leong, S. P. (1998) Digital quantification of the enzyme-linked immunospot (ELISPOT). *Biotechniques* **25**, 830–836.
61. Asai, T., Storkus, W. J., and Whiteside, T. L. (2000) Evaluation of the modified ELISPOT assay for gamma interferon production in cancer patients receiving anti-tumor vaccines. *Clin. Diagn. Lab. Immunol.* **7**, 145–154.
62. Janetzki, S., Schaed, S., Blachere, N. E. B., Ben-Porat, L., Houghton, A. N., and Panageas, K. S. (2004) Evaluation of ELISPOT assays: influence of method and operator on variability of results. *J. Immunol. Methods* **291**, 175–183.
63. Lewis, C. E., McCracken, D., Ling, R., Richards, P. S., McCarthy, S. P., and McGee, J. O. (1991) Cytokine release by single, immunophenotyped human cells: use of the reverse hemolytic plaque assay. *Immunol. Rev.* **119**, 23–39.
64. Hesse, M. D., Karulin, A. Y., Boehm, B. O., Lehmann, P. V., and Tary-Lehmann, M. (2001) A T-cell clone's avidity is a function of its activation state. *J. Immunol.* **167**, 1353–1361.
65. Karulin, A. Y., Hesse, M. D., Tary-Lehmann, M., and Lehmann, P. V. (2000) Single-cytokine-producing CD4 memory cells predominate in type 1 and type 2 immunity. *J. Immunol.* **164**, 1862–1872.
66. Bennouna, J., Hildesheim, A., Chikamatsu, K., Gooding, W., Storkus, W. J., and Whiteside, T. L. (2002) Application of IL-5 ELISPOT assays to quantification of antigen-specific T helper responses. *J. Immunol. Methods* **261,** 145–56.
67. Dunnet, C. W. (1964) New table for multiple comparison with a control. *Biometrics* **20**, 482–491.
68. Hudgens, M. G., Self, S. G., Chiu, Y., Russell, N. D., Horton, H., and McElrath, M. J. (2004) Statistical considerations for design and anlysis of the ELISPOT assay in HIV-1 vaccine trials. *J. Immunol. Methods* **288**, 19–34.
69. Hochberg, Y., and Tamhane, A. C. (1987) *Multiple Comparison Procedures*. Wiley, New York.

70. Westfall, P. H., and Young, S. S. (1993) *Resampling-Based Multiple Testing: Examples and Methods for P-Value Adjustment.* Wiley, New York

71. Ewens, W. J., and Grant, G.R. (2001) *Statistical Methods in Bioinformatics.* Springer Verlag, New York, pp. 356–360.

72. Westfall, P. H., and Tobias, R.D. (1999) *Multiple Comparisons and Multiple Tests Using the SAS System.* SAS Publishing, Cary, NC.

73. Westfall, P. H., and Wolfinger, R. D. Closed multiple testing procedures and PROC MULTTEST. Observations, The Technical Journal for SAS users. Available at: http://support.sas.com/documentation/periodicals/obs/obswww23/; Internet; accessed August 26, 2004.

74. Fleiss, J. L. (1973) *Statistical Methods for Rates and Proportions.* Wiley, New York.

75. Landis, J. R., and Koch, G. G. (1977) The measurement of observer agreement for categorical data. *Biometrics* **33**, 159–174.

5

A Cell-Detachment Solution Can Reduce Background Staining in the ELISPOT Assay

Angela Grant, Sarah Palzer, Chris Hartnett, Tanya Bailey, Monica Tsang, and Alexander E. Kalyuzhny

Summary

Enzyme-linked immunospot (ELISPOT) assays are widely used as a technique that allows determining the frequency of cytokine-releasing cells. Colored spots appear at the sites of cells releasing cytokines, with each individual spot representing a single cytokine-releasing cell. Porous membranes are used in ELISPOT plates to provide support for growing cells, thus making it difficult to remove them by washing. Cells that have adhered to the membrane may be stained nonspecifically, producing a background and then counted as specific spots. We have tested a cell detachment reagent, Accumax™, and found that it may be used to remove a large number of cells adhered to the microplate membranes. Accumax was tested in 16 different ELISPOT assays, including human interleukin (IL)-2, IL-4, IL-5, IL-6, IL-8, IL-13, IL-1β, interferon (IFN)-γ, and tumor necrosis factor (TNF)-α; mouse IL-4, IL-6, IFN-γ, and TNF-α; rat IL-2 and IFN-γ; and canine IFN-γ. Accumax was found to be compatible with human IL-13, IL-1β, IL-2, IL-4, IL-5, and IL-8 and mouse IL-4, IL-6, and TNF-α ELISPOT assays, allowing one to remove a large number of adhered cells without hindering ELISPOT assay performance. However, Accumax was incompatible with human IFN-γ, mouse IFN-γ, canine IFN-γ, and rat IFN-γ ELISPOT assays because Accumax reduced the intensity of staining and the number of spots formed.

Key Words: ELISPOT; cytokines; background; Accumax™.

1. Introduction

For the accurate quantification of spots produced in enzyme-linked immunospot (ELISPOT) assays, it is important to minimize or eliminate nonspecific staining, which may be caused for various reasons. Filter-bottom microplates are used predominantly in ELISPOT assays because, unlike plastic plates, membranes (i.e., nitrocellulose or polyvinylidene fluoride [PVDF]) can retain a larger quantity of capture antibodies, thus allowing one to achieve a

From: *Methods in Molecular Biology, vol. 302: Handbook of ELISPOT: Methods and Protocols*
Edited by: A. E. Kalyuzhny © Humana Press Inc., Totowa, NJ

strong staining color of spots that is required for their detection. Strong cell-adhesive properties of filter membranes have a negative impact on ELISPOT assays: adhered cells may be stained nonspecifically and either produce a background that covers the entire area of the membrane or stained like spots and thus confused with specific spots formed by cell-secreted cytokines. Cells may adhere to membranes for various reasons. For example, in ELISPOT assays such as human interleukin (IL)-4, IL-5, and IL-13, a large number of cells have to be plated to detect a quantifiable number of spots on the membrane *(1–3)*. Therefore, the chances of cell adherence increase with such assays. We also have observed that cells stimulated with Concanavalin A tend to adhere stronger to the membrane (data not shown). In our present study, we investigated the effect of a cell-detachment solution, Accumax™, in minimizing the adherence of peripheral blood mononuclear (PBMCs) to PVDF filter membranes in 96-well plates used in human IL-2, IL-4, IL-5, IL-6, IL-8, IL-13, IL-1β, interferon (IFN)-γ, and tumor necrosis factor (TNF)-α; mouse IL-4, IL-6, IFN-γ, and TNF-α; rat IL-2 and IFN-γ; and canine IFN-γ ELISPOT assays.

2. Materials

1. Sterile phosphate-buffered saline (PBS).
2. RPMI complete culture medium.
3. Calcium ionomycin ionophore (cat. no. C-7522, Sigma Chemical Co., St. Louis, MO).
4. Phytohemagglutinin (cat. no. L-3897, Sigma Chemical Co.).
5. Phorbol 12-myristate 13-acetate (cat. no. P-8139, Sigma Chemical Co.).
6. Anti-CD3ε antibodies (cat. no. MAB100, R&D Systems, Minneapolis, MN).
7. Mouse lyse and wash buffer (cat. no. WL2000, R&D Systems).
8. Human lyse solution: 155 mM NH$_4$Cl, 10 mM NaHCO$_3$, 0.1 mM ethylenediamine tetraacetic acid, pH 7.4.
9. 40 μm nylon cell strainer (Fisher Scientific Co.).
10. Concanavalin A (cat. no. C-7275, Sigma Chemical Co.).
11. Trypan Blue (cat. no. 15250-061, Gibco-BRL).
12. Fluoro Nissl Green (cat. no. N-21480, Molecular Probes, Inc., Eugene, OR).
13. Stereomicroscope or ELISPOT plate reader.
14. Human IL-1β ELISPOT kit (R&D Systems, EL201).
15. Human IL-2 ELISPOT kit (R&D Systems, EL202).
16. Human IL-4 ELISPOT kit (R&D Systems, EL204).
17. Human IL-5 ELISPOT kit (R&D Systems, EL205).
18. Human IL-6 ELISPOT kit (R&D Systems, EL206).
19. Human IL-8 ELISPOT kit (R&D Systems, EL208).
20. Human IL-13 ELISPOT kit (R&D Systems, EL213).
21. Human IFN-γ ELISPOT kit (R&D Systems, EL285).
22. Human TNF-α ELISPOT kit (R&D Systems, EL210).
23. Mouse IL-4 ELISPOT kit (R&D Systems, EL404).
24. Mouse IL-6 ELISPOT kit (R&D Systems, EL406).

25. Mouse IFN-γ ELISPOT kit (R&D Systems, EL485).
26. Mouse TNF-α ELISPOT kit (R&D Systems, EL410).
27. Rat IL-2 ELISPOT kit (R&D Systems, EL502).
28. Rat IFN-γ ELISPOT kit (R&D Systems, EL585).
29. Canine IFN-γ ELISPOT kit (R&D Systems, EL781).
30. Accumax™ (Sigma Chemical Co., cat. no. A-7089). This is a solution of prote-olytic, collagenolytic, and DNAse enzymes in Dulbecco's PBS. The solution does not contain mammalian or bacterial derived products. After thawing, Accumax can be stored at 4°C for up to 2 mo in sterile tubes.

3. Methods

3.1. Preparation of Rat and Mouse Splenocytes

1. Aseptically remove spleens from 4- to 6-wk-old rats and mice and place spleens into RPMI complete culture medium
2. Press spleens through a 40-µm nylon cell strainer.
3. Collect splenocytes by centrifugation (500g for 5 min).
4. Discard supernatant, break clumps of cells, and resuspend them in lyse solution (10 mL for mouse and 20 mL for rat) and incubate for 10 min at room temperature.
5. Remove clumps with a pipetor and add 40 mL of wash buffer.
6. Centrifuge cells for 5 min at 500g, discard supernatant, and resuspend cells in 10 mL of RPMI complete medium.

3.2. Preparation of Human PBMCs

1. Separate PBMCs from each donor by layering 25 mL of blood onto 20 mL of 1.077 g/mL Ficoll-Paque Plus at 25°C and centrifuge 500g for 30 min at room temperature.
2. Discard upper plasma layer after centrifugation and transfer PBMCs into two ster-ile 50 mL tubes.
3. Resuspend PBMCs in 45 mL of sterile PBS and centrifuged them for 5 min at 500g at room temperature.
4. Discard supernatant, break the pellet and remove the remaining red blood cells by adding 10 mL of human cell-lyse solution and incubate for 5 min at room temper-ature. After lysing is completed add sterile PBS (to reach 50-mL graduation mark on the tube) to resuspend PBMCs
5. Centrifuge PBMCs for 5 min at 500g at room temperature. Discard supernatants and add 30–40 mL of RPMI complete medium. PBMCs should be counted by Trypan Blue exclusion.

3.3. Counting Live Cells

1. Mix cells 1:2 with a Trypan blue solution.
2. Pipet about 10 µL of the mixture into each side of a hemacytometer.
3. Place hemacytometer under the microscope and count cells using a 10X or 20X lens and phase contrast illumination as described in hemacytometer's insert. Live

cells are those ones that are not stained with Trypan blue. Once the cell concentration is determined, cell dilutions can be made as needed.

3.4. ELISPOT Assays

Commercially available ready-to-use ELISPOT kits (R&D Systems, Inc.) were used in this study to measure release of human IL-1β, IL-4, IL-5 , IL-6, IL-8, IL-13, IFN-γ, and TNF-α; mouse IL-4, IL-6, IFN-γ, and TNF-α; rat IL-2 and FN-γ; and canine IFN-γ. Each kit included a dry 96-well PVDF-backed plate precoated with corresponding capture antibodies, a concentrated solution of detection antibodies, a concentrated solution of streptavidin-conjugated alkaline phosphatase, 5-5-bromo-4-chloro-3-indolyl phosphate/Nitroblue tetrazolium substrate, and wash and dilution buffers. Assays were performed according to the protocols included with each ELISPOT kit (*see* **Note 1**). PBMCs and splenocytes were incubated and stimulated at 37°C and 5% CO_2 for 16 to 24 h directly in ELISPOT plates wrapped in aluminum foil to maintain well-to-well reproducibility and to minimize background staining *(4)*. To stimulate the release of human IL-1β, IL-2, IL-5, IL-8, IL-13, and IFN-γ; mouse IFN-γ; rat IL-2 and IFN-γ; and canine IFN-γ, cells were incubated with 0.5 µg/mL of Calcium ionomycin ionophore plus 50 ng/mL of phorbol 12-myristate 13-acetate. To stimulate release of human IL-4 and IL-6, cells were incubated in the presence of 3 µg/mL of phytohemagglutinin. To stimulate the release of human and mouse TNF-α, cells were incubated in the presence of anti-CD3ε antibodies. Mouse IL-4 and IL-6 were stimulated with 4 µg/mL Concanavalin A.

3.5. Quantification of Cells Adhering to ELISPOT Membranes

The membranes were punched out, stained with Fluoro Nissl Green *(5)*, and counted using a manual tally counter under the fluorescence microscope (AX70 Provis, Olympus America, Lake Success, NY).

3.6. Quantification of ELISPOT Data

The cytokine-releasing activity of PBMCs were evaluated by quantifying spot-forming cells. Spot-forming cells, in turn, were determined by counting colored spots distributed over the entire area of the membrane using ImageHub image acquisition and spot-counting software (designed by MedBioComp, Inc. and distributed by MVS Pacific, LLC) assuming that one cell will produce one spot.

3.7. Effects of Accumax on Cell Removal and Spot Formation in the ELISPOT Assay

In three preliminary experiments, we examined conditions under which Accumax removes cells from the membranes. Cells were added to the plates according to the protocol. After cell incubation, 100 µL per well of Accumax was added directly to half of the wells containing cells and culture media. The other half of the plate was washed four times with the kit's wash buffer, and then 100 µL per well of Accumax was added. The incubation time with Accumax was either 5, 15, or 30 min. Two incubation temperatures, 37°C stationary and room temperature (approx 25°C) on the micro titer plate shaker (Lab-Line, cat. no. 4625) at 500 rpm, were examined to determine the effect of temperature on Accumax efficacy. After finishing the incubation, Accumax was washed off three times with the kit's wash buffer. Then plates were developed according to the kit's protocol. Incubation with Accumax at room temperature for 15 min on the micro titer plate shaker was optimal for removal of adherent cells (data not shown).

In the second experiment, we determined the effect of the number of washes and the volume of Accumax. One group of wells was washed seven times with wash buffer without the addition of Accumax. We determined that the extra washes were not responsible for removing adhered cells (data not shown). Various volumes of Accumax (50, 75, and 100 µL per well) were added to the remaining wells and incubated according to the optimal conditions described above. It was found that 100 µl per well removed the largest percentage of adhered cells (data not shown).

In the third experiment, we investigated the optimal concentration of Accumax. Dilutions of 1:2 and 1:3 in PBS, and 100% Accumax were tested at 50, 75, and 100 µL per well. Wells that contained 100 µl of 100% Accumax removed the highest number of adhered cells (data not shown). The optimal conditions found in the three preliminary experiments were applied to human IL-2, IL-4, IL-5, IL-6, IL-8, IL-13, IL-1β, IFN-γ, and TNF-α; mouse IL-4, IL-6, IFN-γ, and TNF-α; rat IL-2 and IFN-γ; and canine IFN-γ ELISPOT assays. Accumax did not affect spot formation in human IL-2, IL-4, IL-5, IL-8, IL-13, and IL-1β or mouse IL-4, IL-6, and TNF-α ELISPOT assays (**Table 1**). Spots appeared to be similar in shape and size when compared with the control wells that did not contain Accumax. For these assays, Accumax removed a large number of PBMCs that were adhered to the ELISPOT membrane (**Table 2** and **Fig. 1**). Accumax did not hinder spot formation in human TNF-α and IL-6 and in rat IL-2 ELISPOT assays. In these assays, few cells adhered to the membrane so Accumax did not prove to be advantageous. However, Accumax hinders the development of human IFN-γ, mouse IFN-γ, canine IFN-γ, and rat IFN-γ by reducing spot formation (**Table 1** and **Fig. 2**; *see* **Note 2** for simplified protocol).

The results of our studies demonstrate that Accumax can be used as an inexpensive tool that allows the removal of adhered PMBCs and thus minimizes the

Table 1
Effect of Accumax on Spot Formation in ELISPOT Assays

	Number of spots per well					
	With Accumax			Without Accumax		
Assay	Mean	±	SD	Mean	±	SD
Human IL-2	460.0	±	11.5	443.7	±	25.1
Human IL-5	48.0	±	5.2	51.0	±	5.3
Human IL-6	577.3	±	16.3	530.3	±	26.3
Human IFN-γ	9.7	±	1.5	24.0	±	8.2
Human IL-13	68.7	±	12.9	64.0	±	11.1
Human IL-1β	155.7	±	18.6	194.7	±	22.5
Human IL-8	298.0	±	22.5	298.0	±	13.1
Human IL-4	54.0	±	2.6	61.3	±	2.1
Human TNF-α	194.7	±	11.9	205.3	±	1.2
Mouse IL-4	77.3	±	4.0	88.3	±	14.6
Mouse IL-6	219.7	±	6.1	255.7	±	19.9
Mouse IFN-γ	31.0	±	1.7	474.0	±	16.5
Mouse TNF-α	354.7	±	27.2	386.3	±	17.8
Rat IFN-γ	44.3	±	4.5	70.7	±	10.7
Rat IL-2	52.7	±	7.5	49.0	±	7.9
Canine IFN-γ	193.3	±	7.6	336.0	±	20.8

formation of the nonspecific background staining, which hinders the quantification of spots formed by released cytokines.

4. Notes

1. We recommend that one wraps aluminum foil around the bottom of the ELISPOT plate to minimize nonspecific staining and maintain well-to-well reproducibility (*see* **ref. 4** for detailed protocol).
2. The suggested simplified protocol for cell removal with Accumax is used after finishing the incubation of cells in ELISPOT plates and before adding detection antibodies: (1) wash the plate four times with wash buffer; (2) add 100 μL of 100% Accumax into each well in the ELISPOT plate; (3) incubate plate with added Accumax on a titer plate shaker at 500 rpm for 15 min at room temperature; (4) wash the plate three times with wash buffer before loading detection antibodies; and (5) add detection antibodies and continue developing ELISPOT plate as usual.

Table 2
Effect of Accumax on the Removal of Cells Adhering to Filter Bottom Membranes in 96-Well PVDF ELISPOT Plates

	Number of residual cells per well					
	With Accumax			Without Accumax		
Assay	Mean	±	SD	Mean	±	SD
Human IL-2	1.7	±	1.2	36.0	±	7.2
Human IL-5	26.0	±	8.7	416.0	±	36.2
Human IL-6	3.7	±	1.2	4.7	±	1.2
Human IFN-γ	2.3	±	0.6	26.0	±	7.2
Human IL-13	13.0	±	4.4	334.7	±	51.2
Human IL-1β	5.7	±	1.5	90.0	±	20.4
Human IL-8	1.3	±	0.6	33.7	±	2.1
Human IL-4	1.3	±	1.5	18.0	±	4.4
Human TNF-α	1.3	±	0.6	3.7	±	2.1
Mouse IL-4	5.7	±	1.5	108.7	±	8.0
Mouse IL-6	6.3	±	3.1	127.0	±	14.9
Mouse IFN-γ	4.3	±	3.2	41.0	±	13.1
Mouse TNF-α	1.7	±	1.2	69.3	±	20.2
Rat IFN-γ	2.3	±	1.5	8.3	±	1.2
Rat IL-2	1.0	±	0.0	18.0	±	4.4
Canine IFN-γ	5.3	±	3.2	51.3	±	5.7
Human IL-2	460.0	±	11.5	443.7	±	25.1

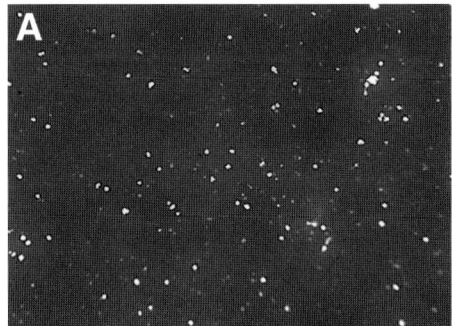

Fig. 1. Effect of Accumax on cell removal from PVDF membranes in human IL-13 ELISPOT assay. (**A**) Adhered cells as seen in control group without adding Accumax; (**B**) number of cells adhered to the membrane after treatment with Accumax.

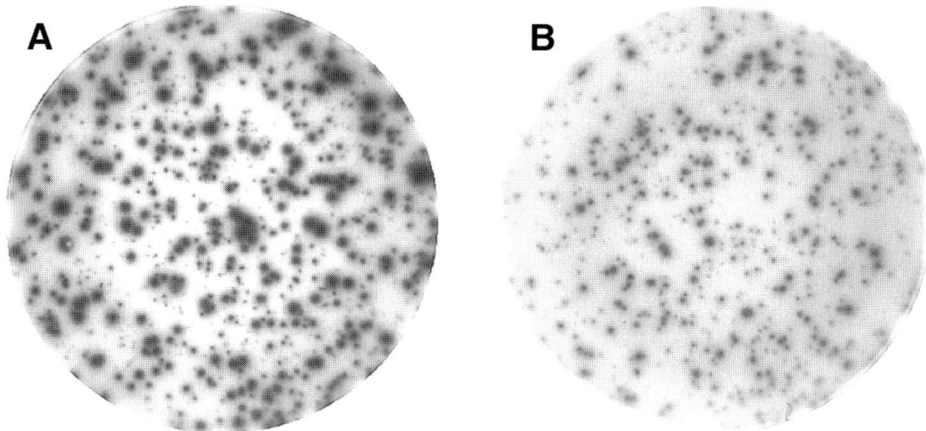

Fig. 2. Effect of Accumax on spot formation in a canine IFN-γ ELISPOT assay. **(A)** Spots in control well without treatment with Accumax; **(B)** Quality of formed spot after addition of Accumax. Images were captured using ELISPOT reading system SpotoGraphics (MBC Corporation, LA).

References

1. Bailey, T., Stark, S., Grant, A., Hartnett, C., Tsang, M., and Kalyuzhny, A. (2002) A multidonor ELISPOT study of IL-1beta, IL-2, IL-4, IL-6, IL-13, IFN-gamma and TNF-alpha release by cryopreserved human peripheral blood mononuclear cells. *J. Immunol. Methods.* **2,** 171–182.
2. Hauer, A. C., Bajaj-Elliott, M., Williams, C. B., Walker-Smith, J. A., and MacDonald, T. T. (1998) An analysis of interferon gamma, IL-4, IL-5 and IL-10 production by ELISPOT and quantitative reverse transcriptase-PCR in human Peyer's patches. *Cytokine* **8,** 627–634.
3. Schmid-Grendelmeier, P., Altznauer, F., Fischer, B., Bizer, C., Straumann, A., Menz, G., et al. (2002) Eosinophils express functional IL-13 in eosinophilic inflammatory diseases. *J. Immunol.* **2,** 1021–1027.
4. Kalyuzhny, A., and Stark, S. (2001) A simple method to reduce the background and improve well to well reproducibility of staining in Elispot assays. *J. Immunol. Methods* **257,** 93–97.
5. Quinn, B., Toga, A. W., Motamed, S., and Merlic, C. A. (1995) Fluoro Nissl green: a novel fluorescent counterstain for neuroanatomy. *Neurosci. Lett.* **184,** 169–172.

6

Isolation of Subsets of Immune Cells

Carrie E. Peters, Steven M. Woodside, and Allen C. Eaves

Summary

Subsets of immune cells can be isolated before analysis by the enzyme-linked immunospot (ELISPOT) assay with various cell separation techniques. This chapter describes techniques to select desired cells or deplete unwanted cells by crosslinking cells to dense or magnetic particles for subsequent separation. The RosetteSep™ method can be used to isolate specific cell types directly from human whole blood, using the red blood cells (RBCs) present in the sample as dense particles. Unwanted cells are crosslinked to multiple RBCs, forming "rosettes." The rosettes, free RBCs, and granulocytes pellet when the sample is centrifuged over a buoyant density medium. The unlabeled, desired cells are simply collected from the interface between the plasma and the buoyant density medium. The SpinSep™ method for isolation of mouse spleen or bone marrow cells is similar to RosetteSep, except that the unwanted cells are bound to dense particles rather than RBCs. The EasySep™ immunomagnetic system can be used with cell suspensions from a variety of species. Cells are crosslinked to nanometer-sized paramagnetic particles. Magnetically labeled cells are separated from unlabeled cells by placing the sample in a high gradient magnetic field. Both the labeled and the unlabeled fractions can be recovered for further use.

Key Words: Cell isolation; cell separation; cell enrichment; cell subsets; density separation; magnetic cell separation; RosetteSep; EasySep.

1. Introduction

Cell separation techniques can be used to isolate subsets of immune cells before analysis using the enzyme-linked immunospot (ELISPOT) assay. When performed on selected cells, the ELISPOT assay detects the response of single cells within the selected cell subset. For example, this combination of techniques can be used to evaluate the heterogeneity of cytokine release within defined cell subsets or to determine the frequency of effector cells responding to activation by antigen-presenting cells. The most selective cell separation

From: *Methods in Molecular Biology, vol. 302: Handbook of ELISPOT: Methods and Protocols*
Edited by: A. E. Kalyuzhny © Humana Press Inc., Totowa, NJ

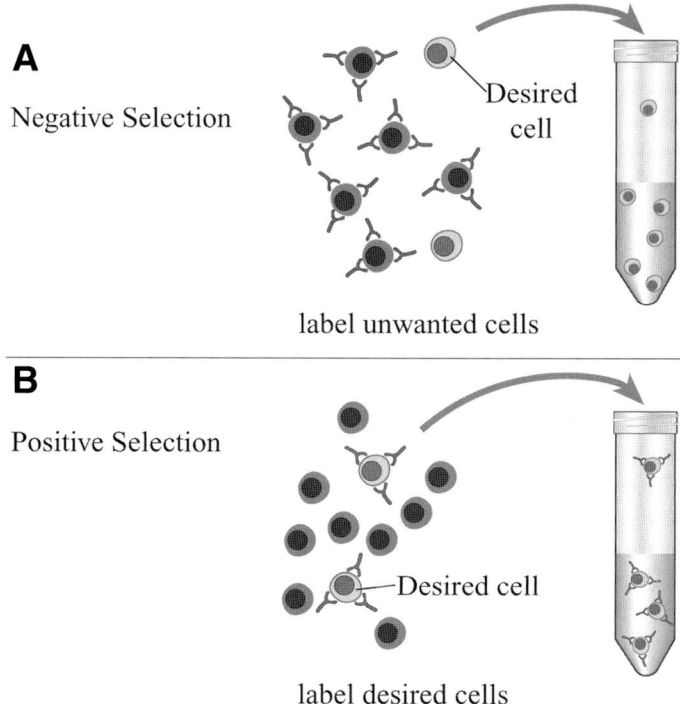

Fig. 1. Negative and positive selection approaches. (**A**) negative cell selection. Unwanted (dark) cells are labeled with antibody and removed. The desired, unlabeled cells are recovered. (**B**) positive cell selection. Desired (light) cells are labeled with antibody and recovered.

techniques that are currently available use the high specificity of monoclonal antibodies to cell surface antigens to purify subsets of cells.

1.1. Positive and Negative Cell Selection

There are two basic approaches to the antibody-mediated isolation of a specific cell type from a heterogeneous cell suspension: positive and negative selection. The difference between the two approaches lies in whether the desired or undesired cells are selectively targeted with antibody (**Fig. 1**). With positive selection, an antibody to a marker that defines the desired cell type (e.g., CD3) is used to target and selectively recover marker-positive cells. Thus, the desired cells are labeled with antibody. With negative selection, the unwanted cells are labeled with antibodies and subsequently removed, leaving the desired cells ready for use. In most primary cell samples, including blood, the "unwanted cells" are a mixture of cell types, each characterized by specific cell

Table 1
Advantages and Disadvantages of Negative and Positive Cell Selection

	Negative selection	Positive selection
Advantages	Desired cells are not labeled with antibody Dead cells, which tend to become non-specifically labeled, are removed The removal of desired cells from a separation matrix (e.g., column or beads) is not required	Requires only one antibody (against the desired cell) Can obtain very high purity of rare cells
Disadvantages	Requires antibodies against all unwanted cells Must know what cells are in the starting cell suspension.	Desired cells may be activated or altered in some way by antibody binding Desired cell must express a unique marker May also select dead cells

surface markers. A "cocktail" of different antibodies must be used to deplete all the unwanted cell types. The advantage of negative selection is that the recovered, desired cells have not been labeled with antibodies. This is often important for assays of cellular response such as the ELISPOT assay because antibody binding to cell surface antigens may induce cellular responses that interfere with the intended purpose of the assay. The advantages and disadvantages of positive and negative selection are listed in **Table 1**.

1.2. Immunodensity and Immunomagnetic Cell Selection

To separate antibody-targeted cells from nontargeted cells, the antibodies must bind to a surface (matrix) that can by physically removed. Matrices commonly used to remove selected cells include magnetic particles, dense particles, affinity columns, and panning flasks. However, particle-based separations are the most efficient because they offer a much larger surface area for binding cells and more rapid kinetics of contacting cells.

Cells that are labeled with dense particles are separated from unlabeled cells by gravity or centrifugal force. Optimal particle size and density is the result of a balance between two conflicting requirements: efficient cell separation and efficient cell labeling. Particles with rapid settling rates are the most efficient at separating cells, but inefficient at labeling cells, settling out of suspension before they contact the cells. Particles with settling rates only slightly faster

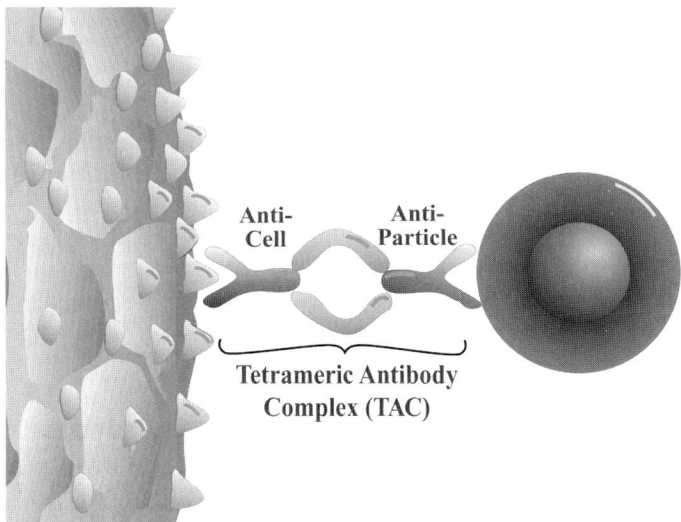

Fig. 2. Tetrameric Antibody Complexes. TACs are comprised of two mouse IgG$_1$ antibodies held in tetrameric array by two rat anti-mouse antibodies. One mouse antibody in the TAC recognizes a cell surface marker, and the other recognizes a particle.

than the settling rates of cells label cells efficiently and will rapidly separate labeled from unlabeled cells when centrifuged over a buoyant density medium. The labeled cells pellet and the unlabeled cells collect at the interface between the density medium and the sample above.

With immunomagnetic cell separation, cells are labeled with antibodies and magnetic particles and then physically separated in a magnetic field. Factors affecting separation efficiency include the delivery of antibody and magnetic particles to the cells, the magnetic susceptibility of the particles, and the strength of the magnetic field gradient. Immunomagnetic cell separation techniques can achieve higher cell purity than immunodensity separation as a result of the higher contrast between the separation forces on labeled and unlabeled cells. Nevertheless, immunodensity cell separation is particularly well suited to processing multiple samples in that it is fast, simple and requires no specialized equipment.

1.3. The Use of Tetrameric Antibody Complexes as Crosslinkers

Bispecific tetrameric antibody complexes (TACs) provide a flexible method of crosslinking target cells to dense or magnetic particles (**Fig. 2; refs. *1* and *2***). In positive selection, the TACs crosslink a single desired cell type to the particles, whereas in negative selection multiple unwanted cell types are crosslinked to the particles by a cocktail of different TACs. TACs also can be used to indirectly

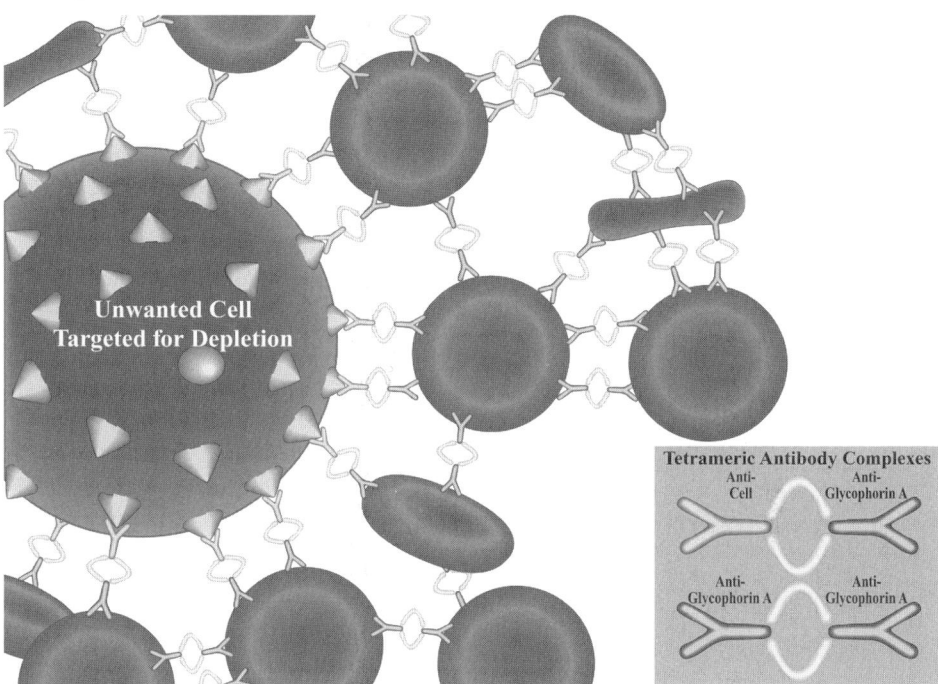

Fig. 3. Rosette of unwanted nucleated cell and RBCs formed by RosetteSep TACs. Multiple RBCs bind to an unwanted cell via TACs, forming a complex "rosette" with a density similar to that of the RBCs.

crosslink cells to particles via the use of hapten-conjugated (e.g., phycoerythrin [PE], fluorescein isothiocyanate [FITC], or biotin) anti-cell antibodies and TACs that recognize both the hapten and a particle. The ability to use TACs either directly or indirectly makes this crosslinking approach almost infinitely flexible.

The RosetteSep™ immunodensity method uses autologous red blood cells (RBCs) that are already present in the sample as dense particles to pellet unwanted white cells, thereby purifying specific cell subsets by negative selection. A cocktail of TACs targeting multiple cell types crosslink unwanted cells in a sample of whole blood to many RBCs. Bispecific TACs bind RBCs to the target cells, and monospecific anti-RBC × anti-RBC TACs bind additional RBCs, thus forming a complex ("rosette") with a density similar to RBCs (**Fig. 3**). The sample is then layered over a buoyant density medium and centrifuged. The rosettes pellet, along with any free RBCs and granulocytes. The desired cells, which have not been labeled with TACs and are not linked to RBCs, do not pellet and are recovered at the interface between the plasma and the buoyant density medium.

Table 2
Effect of Specific Manipulations on the Purity and Recovery
of Desired Cells Using Negative or Positive Selection

Manipulation	Negative selection	Positive selection
↑ labeling reagents / longer incubation times	↑ Purity ↓ Recovery	↓ Purity ↑ Recovery
↓ labeling reagents / shorter incubation times	↓ Purity ↑ Recovery	↑ Purity ↓ Recovery

The EasySep™ immunomagnetic cell selection system uses TACs to crosslink the target cells to dextran-coated paramagnetic particles. Either the desired cells (positive selection) or the unwanted cells (negative selection) are magnetically labeled (**Fig. 2**). The cell suspension is then placed in a magnet, and the magnetically labeled cells migrate to the sides of the tube. Finally, the unlabeled cells are simply poured off.

1.4. Choosing the Best Cell Selection Method

With any cell selection method, one must find the appropriate balance between purity and recovery because improvement in one parameter typically is at the expense of the other. In negative selection, for example, adding more labeling reagent or labeling for a longer period of time is likely to improve cell targeting and thus depletion of unwanted cells, increasing the purity of desired cells. However, higher concentrations of labeling reagents and longer labeling times are likely to increase non-specific labeling of desired cells, thereby decreasing recovery. The expected effects of various manipulations on the purity and recovery of desired cells using negative or positive selection approaches are outlined in **Table 2**.

The most appropriate cell selection method for a given application is determined by answering the following questions:

Is it important that the desired cells do not have antibody on their surface? (If yes, a negative selection approach should be used.)

Is the desired cell rare? (If yes, and high purity is required, a positive selection approach should be used.)

What is most important: purity, recovery, or a rapid processing time? (If purity, use positive selection; if recovery and/or processing time, minimize processing steps and use negative selection, for example, RosetteSep.)

What is the starting sample (e.g., whole blood or a mononuclear cell preparation)? (For human whole blood, use RosetteSep.)

Are multiple samples to be processed in parallel, or is there a single sample? (Immunodensity separations can be performed more readily in parallel than immunomagnetic separations.)

What species is being used? (Human and mouse—many options, see questions above; other species, use hapten-conjugated antibodies, *see* **Subheading 3.2.5.**)

Note that in some cases it may be necessary to use a combination of negative and positive selection methods. For example, CD4+ CD25+ T regulatory cells are isolated by first negatively selecting the CD4+ T-cells and then positively selecting the CD25+ cells.

2. Materials

2.1. Immunodensity Cell Selection

2.1.1. RosetteSep Negative Selection for Human Whole Blood

1. Ficoll-Paque® (Amersham Biosciences or other supplier) or RosetteSep DM-L (StemCell Technologies Inc, Vancouver, Canada; *see* **Note 1**).
2. RosetteSep Cell Enrichment Cocktail (StemCell Technologies).
3. Phosphate-buffered saline (PBS) without Ca^{2+} or Mg^{2+}.
4. Recommended medium: PBS without Ca^{2+} or Mg^{2+} containing 2% fetal bovine serum (FBS) or another protein.

2.1.2. SpinSep Negative Selection for Mouse Cells

1. Recommended medium: PBS without Ca^{2+} or Mg^{2+} containing 2% FBS **OR** Hank's balanced salt solution (HBSS) containing 2% FBS **OR** RPMI containing 2% FBS (*see* **Note 2**).
2. 15-mL Polypropylene conical centrifuge tubes (*see* **Note 3**).
3. SpinSep Mouse Cell Enrichment Cocktail (StemCell Technologies), comprising (1) antibody cocktail, (2) SpinSep dense particles, and (3) SpinSep Density Medium-Murine.

2.2. Immunomagnetic Cell Selection

2.2.1. EasySep Positive Selection: Human Cells

1. Recommended Medium: PBS without Ca^{2+} or Mg^{2+} and containing 2% FBS and 1 mM ethylenediaminetetraacetic acid (EDTA)
2. 5-mL Polystyrene round bottom tubes (Becton Dickinson, cat. no. 2058)
3. EasySep human positive selection Kit (StemCell Technologies), comprising (1) EasySep human positive selection cocktail and (2) EasySep magnetic nanoparticles
4. EasySep Magnet (StemCell Technologies, cat. no. 18000).

2.2.2. EasySep Negative Selection: Human Cells

1. Recommended Medium: PBS without Ca^{2+} or Mg^{2+} and containing 2% FBS **OR** HBSS without Ca^{2+} or Mg^{2+} and containing 2% FBS.
2. 5-mL Polystyrene round bottom tubes (Becton Dickinson, cat. no. 2058).
3. EasySep negative selection human cell enrichment kit (StemCell Technologies), comprising (1) EasySep negative selection human cell enrichment cocktail; and (2) EasySep magnetic nanoparticles.

4. EasySep magnet (StemCell Technologies, cat. no. 18000).

2.2.3. EasySep Positive Selection: Mouse Cells

1. Recommended Medium: PBS without Ca^{2+} or Mg^{2+} and containing 2% FBS and 1 mM EDTA.
2. 5-mL Polystyrene round bottom tubes (Becton Dickinson, cat. no. 2058).
3. EasySep mouse positive selection kit (StemCell Technologies), comprising: (1) mouse FcR blocker; (2) anticell labeling reagent; (3) EasySep PE selection cocktail; and (4) EasySep magnetic nanoparticles.
4. EasySep magnet (StemCell Technologies, cat. no. 18000).

2.2.4. EasySep Negative Selection: Mouse Cells

1. Recommended Medium: PBS without Ca^{2+} or Mg^{2+} and containing 2% FBS and 5% normal rat serum **OR**
   HBSS without Ca^{2+} or Mg^{2+} and containing 2% FBS and 5% normal rat serum. The addition of 2 mM EDTA is recommended for progenitor selection procedures.
2. 5-mL Polystyrene round bottom tubes (Becton Dickinson, cat. no. 2058).
3. EasySep negative selection mouse cell enrichment kit (StemCell Technologies), comprising (1) EasySep negative selection mouse cell enrichment cocktail; (2) EasySep biotin selection cocktail; (3) EasySep magnetic nanoparticles; and (4) normal rat serum.
4. Optional: FcR blocker (recommended for B-cell enrichment).
5. EasySep magnet (StemCell Technologies, cat. no. 18000).

2.2.5. EasySep Selection of Cells Labeled With PE-conjugated Antibodies (Any Species)

1. Recommended Medium: PBS without Ca^{2+} or Mg^{2+} and containing 2% FBS and 1 mM EDTA
2. 5-mL polystyrene round bottom tubes (Becton Dickinson, cat. no. 2058).
3. EasySep PE Selection Kit for Human Cells (StemCell Technologies), comprising (1) human FcR blocker; (2) EasySep PE selection cocktail; (3) EasySep magnetic nanoparticles **OR** EasySep PE selection kit for mouse cells (StemCell Technologies), comprising (1) mouse FcR blocker; (2) EasySep PE selection cocktail; and (3) EasySep magnetic nanoparticles **OR** EasySep PE selection kit (StemCell Technologies) comprising (1) EasySep PE selection cocktail and (2) EasySep magnetic nanoparticles.
4. Optional: When selecting cells from species other than human or mouse, an appropriate species-specific FcR blocking antibody may be required to achieve desired cell purity.
5. EasySep magnet (StemCell Technologies, cat. no. 18000).

2.3. Support Protocols

2.3.1. Freezing Cells

1. Sterile cryovials appropriate to the volume of cells to be frozen.
2. Indelible marker.
3. 100% FBS.
4. Dimethyl Sulfoxide (DMSO)l
5. Icel
6. 70% Isopropanol.
7. Freezing container (e.g., "Mr. Frosty," Nalgene, cat. no. 5100-0001).

2.3.2. Thawing Frozen Human Mononuclear Cells (MNCs)

1. 70% Isopropanol or ethanol.
2. 50-mL Centrifuge tubes.
3. PBS without Ca^{2+} or Mg^{2+}.
4. Recommended medium: PBS without Ca^{2+} or Mg^{2+} containing 2% FBS or another protein.
5. Optional: 1 mg/mL Deoxyribonuclease (DNAse; StemCell Technologies or other supplier of DNAse that is not toxic to cells; *see* **Note 4**).

2.3.3. Washing Cells

1. 50-mL Centrifuge tubes.
2. Recommended medium: PBS without Ca^{2+} or Mg^{2+} containing 2% FBS or another protein.

2.3.4. Density Gradient Separation

1. Ficoll-Paque® (Amersham Biosciences or other supplier).
2. PBS without Ca^{2+} or Mg^{2+}.
3. Recommended medium: PBS without Ca^{2+} or Mg^{2+} containing 2% FBS or another protein.

2.3.5. Preparing Mouse Spleen or Bone Marrow Samples

1. Recommended medium: PBS containing 2% FBS **OR**
   HBSS containing 2% FBS **OR**
   RPMI containing 2% FBS.
2. 15-mL Polypropylene centrifuge tubes (*see* **Note 3**).
3. 70-µm Mesh nylon strainer.

2.3.6. Preparing Mouse Spleen Cells for CD11c⁺ Selection

1. Recommended medium: PBS containing 2% FBS **OR**
   HBSS containing 2% FBS **OR**
   RPMI containing 2% FBS.

Table 3
Recommended Reagent Volumes and Tubes Sizes for Use With RosetteSep

Volume whole blood (mL)	Volume PBS+2% FBS (mL)	Volume density medium (mL)	Tube size (mL)
1	1	1.5	5
2	2	3	14
3	3	3	14
4	4	4	14
5	5	15	50
10	10	15	50
15	15	15	50

The use of less than 15 mL of density medium in 50-mL tubes is not recommended because it makes it difficult to recover the purified cell layer.

2. 15-mL Polypropylene centrifuge tubes (*see* **Note 3**).
3. Collagenase Type IV (Worthington Biochemical) 250 µg/mL in RPMI 1640.
4. 16-Gage needle.
5. Syringe.
6. EDTA in solution.
7. 70-µm Mesh nylon strainer.

2.3.7. RBC Lysis After RosetteSep

1. Ammonium chloride solution (StemCell Technologies, or other supplier).
2. Recommended Medium : PBS without Ca^{2+} or Mg^{2+} containing 2% FBS or another protein.

3. Methods

3.1. Immunodensity Cell Selection

3.1.1. RosetteSep Negative Selection for Human Whole Blood

This procedure can be used to enrich specific cell types (e.g., CD4 T-cells) directly from human whole blood. The autologous RBCs present in the sample are used as "dense particles" and crosslinked to unwanted cells using TACs. *See* **Note 5** for additional information regarding the selection of monocytes; *see* **Table 3** for recommended volumes of reagents and recommended tube sizes; and *see* **Fig. 4** for an example of results using RosetteSep CD4 T-cell enrichment.

1. Add 50 µL of RosetteSep cocktail per mL of blood and mix well.
2. Incubate for 20 min at room temperature.
3. Dilute sample with an equal volume of Recommended Medium (*see* **Subheading 2.1.1.**) and mix gently (*see* **Note 6**).

Start: 18% CD4⁺ T Cells

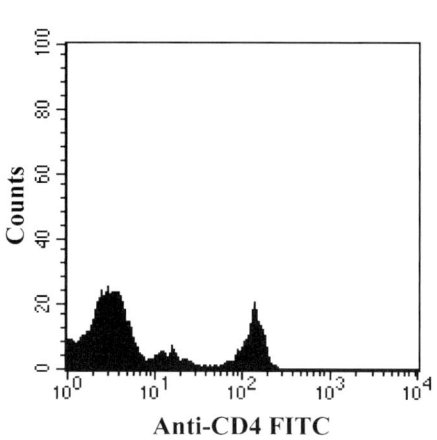

Anti-CD4 FITC

Enriched: 90% CD4⁺ T Cells
Recovery: 85% CD4⁺ T Cells

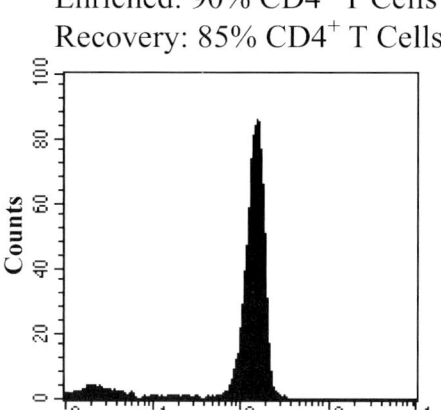

Anti-CD4 FITC

Fig. 4. Isolation of human CD4+ T-cells using RosetteSep. Note that negative selection of CD4+ T-cells removes the CD4+ monocytes.

4. Layer the diluted sample on top of density medium **OR** layer the density medium underneath the diluted sample. Be careful to minimize mixing of density medium and sample.
5. Centrifuge for 20 min at 1200g at room temperature, with the brake off (*see* **Note 7**).
6. Remove the purified cells from the interface between the density medium and the plasma above (*see* **Note 8**).
7. Wash enriched cells with recommended medium. Repeat.
8. Use purified cells as desired. Purified samples should be lysed with ammonium chloride (*see* **Subheading 3.3.7.**) to remove residual red blood cells prior to flow cytometric analysis (this can be done as one of the wash steps) or if residual red blood cells will interfere with subsequent assays.

3.1.2. SpinSep Negative Selection of Mouse Cells

This procedure can be used to enrich specific cell types directly from suspensions of mouse spleen cells or bone marrow. The procedure is optimized for use with 5×10^7 nucleated bone marrow cells or 10^8 nucleated spleen cells per separation in 15-mL polypropylene conical centrifuge tubes. Using less than 2 $\times 10^7$ cells is not recommended.

1. Prepare spleen cell or bone marrow sample as described in **Subheading 3.3.5.** and resuspend in recommended medium (*see* **Subheading 2.1.2.**) at 5×10^7 nucleated cells/mL (a range of 2–8 $\times 10^7$ cells/mL is acceptable).

Table 4
Recommended Volumes for Different Sample Sizes When Performing SpinSep

Total nucleated cells in start	Step 5 Volume in which to resuspend cells (mL)	Step 8 Volume of SpinSep density medium (mL)	Step 9 Volume of recommended medium (mL)
2×10^7 to 3×10^7	1	3	1
$>3 \times 10^7$ to 1×10^8	2	4	6

2. Remove bottle of SpinSep density medium from refrigerator, mix well by inversion, and let equilibrate to room temperature for a minimum of 30 min before use. While medium is equilibrating, perform **steps 3** to **7**.

3. Add 10 µL SpinSep antibody cocktail per 1 mL of cells. Mix well, and then incubate in the refrigerator (4–8°C) for 15 min or on ice for 30 min. Note that longer periods of incubation during this step will decrease desired cell purity and recovery.

4. Remove cells from refrigerator or ice and wash once by filling tube with recommended medium and centrifuging at 200g (*see* **Note 7**). Ensure that the cell suspension is well mixed during this wash step.

5. Resuspend washed cell pellet in recommended medium (*see* **Table 4** for recommended volumes).

6. Gently vortex tube of SpinSep Dense Particles for approx 30 s until the particles are well suspended and no clumps are observed. The dense particles settle to the bottom of the tube during storage and must be resuspended before use.

7. Add 500 µL of SpinSep dense particles per 10^8 total nucleated cells. Mix well and incubate on ice 20 min. Mix the tube occasionally during the incubation period to prevent the dense particles from settling; continuous mixing is not necessary. Note that incubating for longer than 20 min during this step will decrease desired cell purity and recovery.

8. While cells are incubating with the dense particles, aliquot SpinSep density medium (room temperature) into a separate tube (*see* **Table 4** for recommended volumes).

9. After incubation (**step 7**) is complete, dilute suspension of cells and particles with recommended medium (*see* **Table 4** for recommended volumes).

10. Layer the diluted cell suspension on top of the 3 or 4 mL of SpinSep density medium and centrifuge for 10 min at 1200g (*see* **Note 7**) at room temperature with the brake off.

11. Remove purified cells from the interface between the underlying SpinSep density medium and the recommended medium above. Transfer into a separate polypropylene tube and wash by filling tube with recommended medium (or other appropriate medium used for subsequent applications). Centrifuge at 200g (*see* **Note 7**).

12. Pour off supernatant, resuspend cell pellet in appropriate medium, and use purified cells as desired.

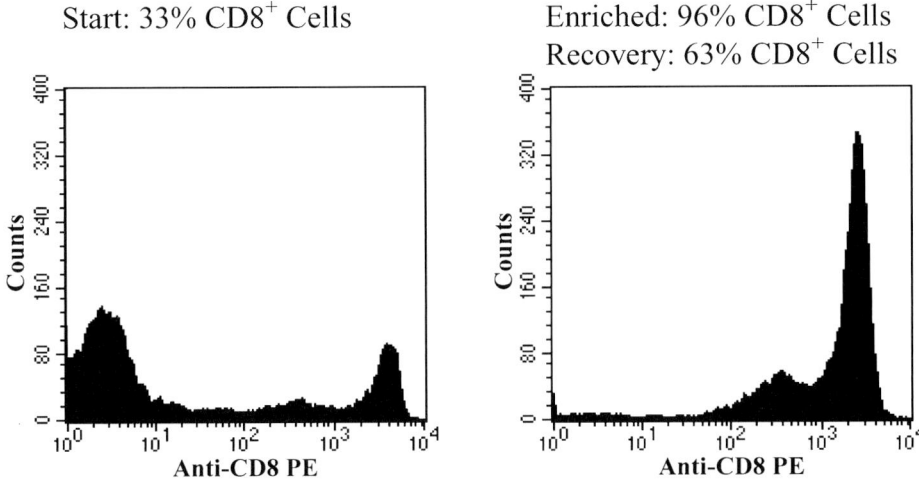

Fig. 5. Isolation of human CD8+ T-cells using EasySep positive selection.

3.2. Immunomagnetic Cell Selection

3.2.1. EasySep Positive Selection: Human Cells

(When isolating CD4+ T-cells, *see* **Note 9** first.) *See* **Fig. 5** for an example of results using EasySep Human CD8 positive selection.

1. Prepare nucleated cell suspension at a concentration of 1×10^8 cells/mL in recommended medium (*see* **Subheading 2.2.1.** and **Note 10**). Cells must be placed in a 12×75-mm polystyrene tube to properly fit into the EasySep magnet. Do not exceed a volume of 2.5 mL (i.e., 2.5×10^8 cells) per tube. For samples containing $<10^7$ cells, resuspend in 100 µL.
2. Add 100 µL of EasySep positive selection cocktail per milliliter of cells (e.g., for 2 mL of cells, add 200 µL of cocktail). Mix well and incubate at room temperature for 15 min.
3. Mix EasySep magnetic nanoparticles to ensure that they are in a uniform suspension by vigorously pipetting up and down more than 5 times (*see* **Note 11**). Vortexing is not recommended. Add 50 µL of the particles per mL cells (e.g., for 2 mL of cells, add 100 µL of nanoparticles). Mix well and incubate at room temperature for 10 min.
4. Bring the cell suspension to a **total volume** of 2.5 mL by adding recommended medium. Mix the cells in the tube by gently pipetting up and down two to three times. Place the tube (without cap) into the magnet. Set aside for 5 min.
5. Pick up the magnet, and in one continuous motion invert the magnet and tube, pouring off the supernatant fraction. The magnetically labeled cells will remain inside

the tube, held by the magnetic field of the EasySep magnet. Leave the magnet and tube in inverted position for 2–3 s, and then return to upright position. Do not shake or blot off any drops that may remain hanging from the mouth of the tube.

6. Remove the tube from the magnet and add 2.5 mL of recommended medium. Mix the cell suspension by gently pipetting up and down two to three times. Place the tube back in the magnet and set aside for 5 min.

7. Repeat **steps 5** and **6**, and then **step 5** once more, for a total of three 5-min separations in the magnet (*see* **Note 12**). Remove tube from magnet and resuspend cells in an appropriate amount of desired medium. The positively selected cells are now ready for use (*see* **Note 13**).

3.2.2. EasySep Negative Selection: Human Cells

1. Prepare nucleated single-cell suspension at a concentration of 5×10^7 cells/mL in recommended medium (*see* **Subheading 2.2.2.**; *see* **Note 14**). Cells must be placed in a 12×75-mm polystyrene tube to properly fit into the EasySep magnet. Do not exceed a volume of 2.0 mL (i.e. a total of 1×10^8 cells) per tube (when isolating monocytes, *see* **Note 15**).

2. Add 50 µL of EasySep negative selection enrichment cocktail per milliliter of cells. Mix well and incubate at room temperature for 10 min.

3. Mix EasySep magnetic nanoparticles to ensure that they are in a uniform suspension. Add 100 µL of the particles per mL cells. Mix well and incubate at room temperature for 10 min.

4. Bring the cell suspension to a total volume of 2.5 mL by adding the appropriate medium. Mix the cells in the tube by gently pipetting up and down two to three times. Remove the cap from the tube and place the tube into the magnet. Set aside for 5 min.

5. Pick up the magnet, and in one continuous motion, invert the magnet and tube, pouring off the supernatant fraction into a new tube (*see* **Note 16**). The magnetically labeled cells will remain inside the tube, held by the magnetic field of the magnet. Leave the magnet and the tube in inverted position for 2–3 s, and then return to upright position. Do not shake or blot off any drops that may remain hanging from the mouth of the tube. The negatively selected cells are now ready for use (*see* **Notes 17** and **18**).

3.2.3. EasySep Positive Selection: Mouse Cells

1. Prepare a single cell suspension from spleen or bone marrow (*see* **Subheadings 3.3.5.** and **3.3.6.**). Resuspend cells at a concentration of 1×10^8 cells/mL in recommended medium (*see* **Subheading 2.2.3.**). For samples containing 10^7 cells or fewer, resuspend in 100 µL (when isolating CD11c$^+$ cells from spleen, *see* **Note 19**).

2. Add Murine FcR Blocker at 10 µL/mL of cells. Mix well (*see* **Note 20**).

3. Add 15 µL of EasySep murine cell labeling reagent per milliliter of cell suspension. Mix well and incubate at room temperature for 15 min (*see* **Note 21**).

4. Wash once with 10-fold excess of recommended medium and resuspend to original volume (i.e., at a concentration of 10^8 cells /mL) in a 12×75-mm polystyrene tube.

5. Add 100 µL of EasySep PE selection cocktail per milliliter of cell suspension. Mix well and incubate at room temperature for 15 min.
6. Mix EasySep magnetic nanoparticles to ensure that they are in a uniform suspension by pipetting vigorously up and down five times. Add 50 µL of particles per milliliter of cell suspension. Mix well and incubate at room temperature for 10 min.
7. Bring the cell suspension to a total volume of 2.5 mL by adding recommended medium. Mix the cells in the tube by gently pipetting up and down two to three times. Place the tube (without cap) into the magnet. Set aside for 5 min.
8. Pick up the magnet, and in one continuous motion invert the magnet and tube, pouring off the supernatant fraction. The magnetically labeled cells will remain inside the tube, held by the magnetic field of the EasySep magnet. Hold the magnet and tube in inverted position for 2-3 s, and then return to upright position. Do not shake or blot off any drops that may remain hanging from the mouth of the tube.
9. Remove the tube from the magnet and add 2.5 mL of recommended medium. Mix the cell suspension by gently pipetting up and down two to three times. Place the tube back in the magnet and set aside for 5 min.
10. Repeat **steps 8** and **9**, and then **step 8** once more, for a total of three 5-min separations in the magnet. Remove tube from magnet and resuspend cells in an appropriate amount of desired medium. These positively selected cells are now ready for use (*see* **Note 12**). The purity of the desired cells can be verified directly by flow cytometry based on PE fluorescence.

3.2.4. EasySep Negative Selection: Mouse Cells

1. Prepare a single cell suspension from spleen or bone marrow (*see* **Subheading 3.3.5.**). Resuspend cells at a concentration of 1×10^8 cells /mL in recommended medium (*see* **Subheading 2.2.4.**). Cells must be placed in a 12×75-mm polystyrene tube to properly fit into the EasySep Magnet. Do not exceed a volume of 2.0 mL (i.e., 2.0×10^8 cells) per tube (when isolating B-cells or hematopoietic progenitors, *see* **Note 22**).
2. Centrifuge tube with EasySep negative selection enrichment cocktail before use to ensure recovery of entire contents. Add 20 µL of the cocktail per mL cells. Mix well and incubate in refrigerator (4 to 8°C) for 15 min.
3. Wash, resuspend at 1×10^8 cells/mL in recommended medium.
4. Add 100 µL of EasySep biotin selection cocktail per milliliter of cells. Mix well and incubate in refrigerator (4 to 8°C) for 15 min.
5. Mix EasySep magnetic nanoparticles to ensure that they are in a uniform suspension. Add 100 µL of the particles per mL of cells. Mix well and incubate in refrigerator (4 to 8°C) for 15 min.
6. Bring the cell suspension to a total volume of 2.5 mL by adding recommended medium. Mix the cells in the tube by gently pipetting up and down two to three-times. Remove the cap from the tube and place the tube into the magnet. Set aside for 5 min.
7. Pick up the magnet, and in one continuous motion invert the magnet and tube, pouring off the supernatant fraction into a new tube. The magnetically labeled cells will

remain inside the tube, held by the magnetic field of the magnet. Leave the magnet and the tube in inverted position for 2-3 s, and then return to upright position. Do not shake or blot off any drops that may remain hanging from the mouth of the tube.

8. Place the tube with the supernatant fraction inside the magnet to perform a second round of magnetic separation (*see* **Note 23**). Set aside for 5 min and repeat **step 7**. The negatively selected cells are now ready for use (*see* **Note 18**).

3.2.5. EasySep Selection of Cells Labeled With PE-Conjugated Antibodies (Any Species)

This procedure is designed to select cells labeled with any PE-conjugated antibody. It can therefore be used to positively or negatively select a wide variety of cell types from many different species. It only requires a PE-conjugated antibody that selectively labels the target cells (*see* **Note 24**) and possibly a species-specific FcR blocker to boost purities by minimizing nonspecific binding of antibody to cells such as monocytes and macrophages.

1. This procedure is used for processing up to 2.5×10^8 cells per separation. Prepare cell suspension at a concentration of 1×10^8 nucleated cells/mL in recommended medium (*see* **Subheading 2.2.5.**). For rare cells, start with a cell concentration of 2×10^8/mL (*see* **Note 10**). For samples containing 10^7 cells or fewer, resuspend in 100 µL.

2. Add species-specific FcR blocking antibody at 100 µL/mL for human cells or 10 µL/mL for murine cells and mix. When selecting cells from other species, an appropriate species-specific FcR blocking antibody may be required to achieve the desired cell purity. A final concentration of 0.5 to 3.0 µg/mL is recommended for the blocking antibody.

3. Add PE-conjugated antibody at a final concentration of 0.3 to 3.0 µg/mL. Mix well and incubate at room temperature for 15 min (*see* **Note 25**).

4. Wash once with 10-fold excess of recommended medium and resuspend to original volume in a 12×75-mm polystyrene tube. Do not exceed an initial volume of 2.5 mL (i.e., 2.5×10^8 total cells) per tube.

5. Add EasySep PE selection cocktail at 100 µL/mL cells. Mix well and incubate at room temperature for 15 min.

6. Mix EasySep magnetic nanoparticles to ensure that they are in a uniform suspension by vigorously pipetting up and down more than five times. Vortexing is not recommended. Add the particles at 50 µL/mL cells. Mix well and incubate at room temperature for 10 min.

7. Bring the cell suspension to a total volume of 2.5 mL by adding recommended medium. Mix the cells in the tube by gently pipetting up and down two to three times. Place the tube (without cap) into the EasySep magnet. Set aside for 5 min.

8. Pick up the magnet, and in one continuous motion invert the magnet and tube, pouring off the supernatant fraction. The magnetically labeled cells will remain inside the tube, held by the magnetic field of the EasySep magnet. Leave the magnet and tube in inverted position for 2-3 s, and then return to upright position. Do not shake or blot off any drops that may remain hanging from the mouth of the tube.

9. Remove the tube from the magnet and add 2.5 mL of recommended medium. Mix the cell suspension by gently pipetting up and down two to three times. Place the tube back in the magnet and set aside for 5 min.

10. Repeat **steps 8** and **9**, and then **step 8** once more, for a total of three 5-min separations in the magnet. (For mouse separations or low frequency cell types, additional separations may improve purity; *see* **Note 26**.) Remove tube from magnet and resuspend cells in an appropriate amount of desired medium. The positively selected cells are now ready for use.

3.3. Support Protocols

3.3.1. Freezing Cells

Rapid freezing damages cells. Freezing rates can be controlled using a rate-controlled freezer or a freezer "box" such as "Mr. Frosty" (*see* Subheading **2.3.1.**) that contains 70% isopropanol. Alternately, one may place the vials in a Styrofoam container and then into the freezer.

1. Make up 20% DMSO in FBS. Keep on ice. (Do not put 100% DMSO on ice or it will form crystals.)
2. Label cryovials.
3. Resuspend cells at 2×10^7 cells / mL in ice-cold FBS. Keep on ice.
4. Mix cells gently with 20% DMSO in FBS at a ratio of 1:1. (The final cell suspension will be 90% FBS / 10% DMSO.) Rapidly transfer cells in freezing medium to each cryovial.
5. Place cryovials immediately into freezing container (e.g., "Mr. Frosty," Nalgene, containing 70% isopropanol). Place container in –135°C or –152°C freezer immediately. (Do not let cells sit in freezing medium at room temperature. Keep on ice and transfer rapidly.)

3.3.2. Thawing Frozen Human MNCs

1. Thaw cells quickly in a 37°C water bath or beaker of warm water. In tissue culture hood, wipe cryovial with 70% ethanol. Do not vortex cells at any time.
2. Gently transfer cells into a 50-mL tube (0.5 to 5.0 mL of cells per 50-mL tube). If cells are expected to be clumpy, add 0.25–0.50 mL of 1 mg/mL DNAse dropwise, while gently swirling the tube.
3. Slowly add 15 mL PBS without Ca^{2+} or Mg^{2+} dropwise while holding the tube and gently swirling.
4. Fill tube to 50 mL with PBS. Gently invert tube to mix.
5. Spin down cells at 300*g* (*see* **Note 7**) for 8 min.
6. Pour off supernatant and flick tube gently to resuspend the pellet. Resuspend cells at desired concentration in appropriate medium.
7. **If cells are clumpy:** Add DNAse at a final concentration of 0.1 mg/mL. Clumpy cell suspensions may also be filtered through a 70-μm mesh nylon filter to obtain a single cell suspension.

3.3.3. Washing Cells

1. Add recommended medium (*see* **Subheading 2.3.3.**) to fill tube containing cells.
2. Centrifuge at 300g (*see* **Note 7**) for 8 min.
3. Pour off supernatant and flick tube gently to resuspend the pellet.
4. Resuspend cells at desired concentration in appropriate medium.

3.3.4. Density Gradient Separation

This procedure will enrich total MNCs from human whole blood. It is often used as a preparatory step before another antibody-mediated method of specific cell selection.

1. Dilute whole blood sample with an equal volume of PBS.
2. Layer the diluted sample over a volume of Ficoll-Paque® equivalent to the initial blood volume. Be careful to minimize mixing of the sample and the Ficoll-Paque.
3. Centrifuge for 20 min at 1200g or 30 min at 300g at room temperature with the brake off (*see* **Note 7**).
4. Remove the MNCs from the interface between the Ficoll-Paque and the plasma layer.
5. Wash enriched cells with PBS + 2% FBS.

3.3.5. Preparing Mouse Spleen or Bone Marrow Samples

1. For spleen, disperse into 5 mL of recommended medium (*see* **Subheading 2.3.5.**). Further disperse clumps by gently pipetting up and down several times. Remove remaining clumps of cells and debris by passing cell suspension through a 70-µm mesh nylon strainer. For bone marrow, flush bone marrow cells from femur and tibia into recommended medium using a syringe equipped with a 23-gage needle. Disperse clumps by gently passing the cell suspension through the syringe several times to obtain a single-cell suspension.
2. Centrifuge the cells for 6 min at 400g (*see* **Note 7**), resuspend cell pellet, and adjust cell concentration in medium recommended for the following procedure.

3.3.6. Preparing Mouse Spleen Cells for CD11c$^+$ Selection

1. Cut spleen into small pieces in a Petri dish. Add 2–5 mL of collagenase and incubate 20 min at 37°C.
2. Disrupt tissue by gently passing several times through a 16-gage needle using syringe.
3. Add stock EDTA to the cell suspension to a final concentration of 1 mM EDTA. Disrupt tissue further by gently passing several times through a 16-gage needle using syringe.
4. Incubate for 5 min at 37°C.
5. Filter through a 70-µm mesh nylon strainer into a 15-mL tube. Rinse strainer with recommended buffer and top up tube with buffer.
6. Centrifuge the cells for 6 min at 400g (*see* **Note 7**), discard supernatant completely, and resuspend cells at 2×10^8 cells / mL in medium recommended for the following procedure.

3.3.7. RBC Lysis After RosetteSep

1. Resuspend cells after the first wash (**step 7**) in 3 mL of ammonium chloride.
2. Incubate for 5 min.
3. Fill tube with recommended medium (*see* **Subheading 2.3.7.**).
4. Centrifuge at 300*g* (*see* **Note 7**) for 8 min.
5. Pour off supernatant and flick tube gently to resuspend the pellet.
6. Resuspend cells at desired concentration in recommended medium.

4. Notes

1. The buoyant density medium Ficoll-Paque can be used with all the RosetteSep mononuclear cell enrichment cocktails. The density medium RosetteSep DM-L can be used for T, B, and NK cell enrichment. Both are used in exactly the same manner. RosetteSep DM-L is slightly denser than Ficoll-Paque; therefore, more cells are "buoyant" on this medium and the recovery of desired cells is higher.
2. The use of a medium containing phenol red will aid in detecting the interface between the medium and the underlying SpinSep Density Medium, which is colorless.
3. To optimize the recovery of desired cells from mouse tissues, use 15-mL polypropylene conical centrifuge tubes during all steps in the protocol. Use of centrifuge tubes made from polystyrene or modified polystyrene adversely affects cell recoveries.
4. To ensure that the DNAse is not toxic to cells, purchase from a supplier who indicates that the product can be used with cells. DNAse may be aliquotted and stored at –20°C; storage at 4°C is not recommended.
5. When enriching monocytes, add 1 m*M* EDTA to the whole blood before adding the RosetteSep cocktail; also add 1 m*M* EDTA to the recommended medium used in **step 3**.
6. Diluting the whole blood prior to density gradient centrifugation may improve the recovery of MNCs (*3,4*).
7. To convert g to rpm, use the following formula:
   $g = 118 \times 10^{-7} \times r \times n^2$, where g = relative centrifugal force; r = rotating radius in cm; and n = rpm (revolutions per minute).
8. When enriching rare cells, it may be difficult to see the cells at the interface. It is advisable to remove some of the density medium along with the enriched cells in order to ensure their complete recovery.
9. To isolate pure CD4+ T-cells by positive selection, it is necessary to first deplete monocytes because they also express the CD4 antigen. A monocyte-depleted MNC suspension can be prepared directly from whole blood using the RosetteSep monocyte depletion cocktail (StemCell Technologies). Monocytes can also be depleted from MNC suspensions using the EasySep Human CD14 Positive Selection Cocktail (StemCell Technologies).
10. For very rare cells (i.e., cells representing <5% of the initial sample), increasing the cell concentration to 2×10^8 from 1×10^8 in **step 1** will likely improve purity.
11. When selecting cells based on markers with relatively low cell-surface expression levels (e.g., CD25 antigen), recovery will likely be improved by using EasySep SA

(Special Application) magnetic nanoparticles (cat. no. 18250) instead of the standard EasySep magnetic nanoparticles. Recovery may also be improved by increasing the magnetic separation time (**steps 4** and **6**) from 5 to 10 min.

12. When the desired cell concentration in the starting sample is less than 10–15%, additional rounds of magnetic separation will likely improve purity. Each round of separation progressively enriches the labeled cells by washing away unlabeled cells. However, recovery of the desired cells will decrease with each additional round of separation.

13. The anti-cell antibody in the positive selection cocktail can block some clones of fluorescently labeled antibodies used for assessing purity of the enriched cell sample by flow cytometry. In cases where a suitable antibody clone is not available one of the following methods may be used instead: (1) Add fluorescently labeled antibodies at the same time as the selection cocktail at a concentration of between 0.15 and 0.4 μg/mL. This method labels the cells in the entire sample. (2) Use alternate markers (e.g., for CD8$^+$ T-cells, assess CD3$^+$/CD4$^-$ population). (3) Use a secondary fluorochrome-conjugated antibody, such as FITC-labeled sheep anti-mouse IgG.

14. Separation from samples with <5×10^7 total cells can result in lower cell purity and recovery.

15. For monocyte selection, the addition of EasySep FcR blocker is recommended prior to labeling in **step 2** to improve recovery. For optimal recovery of monocytes, perform the labeling steps (**steps 2** and **3**) at 4°C, and allow 10 min for magnetic separation (**step 4**).

16. It is easier to pour off the supernatant into a 14-mL tube than a 5-mL tube (**step 5**).

17. The purity of desired cells may be improved for some applications by performing a second round of magnetic separation. Place the tube containing the supernatant from the first separation back into the magnet and set aside for 5 min. Pour off the supernatant fraction into a new tube. Some of the desired cells may be lost in this second separation.

18. Some of the desired cells may be left behind in the original tube after pouring off supernatant. These cells may be recovered by re-suspending the nanoparticles in 2.5 mL of medium and performing a third magnetic separation. The supernatant of this separation step can either be combined with the primary supernatant or kept separately to assess purity and yield (recommended for cell populations with lower start percentages).

19. When selecting CD11c$^+$ cells from spleen, resuspend the cells at 2×10^8 cells/mL in recommended medium in **steps 1** and **4**.

20. This agent inhibits nonspecific binding of antibodies to Fc receptors on monocytes, macrophages and other FcR$^+$ cells.

21. When selecting c-Kit$^+$ or Sca1$^+$ cells, incubate at 4°C in **steps 3** and **5**. In **step 4**, resuspend the cells at 2×10^8 cells/mL, e.g., twice the initial concentration.

22. For B-cell selection, addition of an FcR blocker is recommended prior to the addition of the enrichment cocktail in **step 4**. When isolating hematopoietic progenitors, 2 mM EDTA should be included in the recommended medium.

23. To achieve high recovery it may be desirable to perform only a single round of magnetic separation (i.e., stop the procedure after **step 7**). This may reduce the purity of the desired cells.

24. Recovery of positively selected PE-labeled cells is dependent on the quality of the PE-conjugated antibody used. Antibodies that have expired or that have been stored improperly may show lower affinity for the surface marker on the target cell, resulting in lower recovery. It is important to add sufficient PE-conjugated antibody to ensure a significant fluorescence intensity of the target cells, as there is a strong correlation between fluorescence intensity and cell recovery. The fluorescence intensity of the positively selected cells should be at least 100-fold (2 logarithms) greater than that of the negative control for adequate recovery. It is also possible to select cells using FITC- or biotin-conjugated labeling reagents in combination with FITC or Biotin selection cocktails. When using biotin-conjugated labeling reagents, the target surface antigen may block fluorescently conjugated reagents used for assessing purity of the enriched cell sample by flow cytometry. In cases where suitable reagents are not available one of the following methods may be used instead: (1) Add fluorescently conjugated reagents at the same time as the biotin-conjugated labeling reagents. For antibodies the recommended concentration is between 0.15 and 0.4 µg/mL. This method labels the cells in the entire sample. (2) Use alternate markers that define the desired population. (3) Use a secondary fluorochrome-conjugated antibody such as FITC-labeled sheep anti-mouse IgG.

25. Titrate PE-conjugated antibodies for optimal purity and recovery. Cell recovery increases with increasing fluorescence intensity of the PE-labeled cells. However, the use of excess labeling antibody can reduce purity.

26. For samples with a desired cell starting frequency of less than 10 to 15%, additional separation rounds will likely improve purity. If desired, repeat **steps 8** and **9** an additional one to three times. Please note that recovery will decrease with each additional round of separation.

References

1. Lansdorp, P. M., Aalberse, R. C., Bos, R., Schutter, W. G. and Van Bruggen, E. F. J. (1986) Cyclic tetramolecular complexes of monoclonal antibodies: a new type of crosslinking reagent. *Eur J. Immunol.* **16**, 679–683.
2. Lansdorp, P. M., and Thomas, T. E. (1990) Purification and analysis of bispecific tetrameric antibody complexes. *Mol. Immunol.* **27**, 659–666.
3. Boyum, A. (1968) Isolation of mononuclear cells and granulocytes from human blood. Isolation of mononuclear cells by one centrifugation, and of granulocytes by combining centrifugation and sedimentation at 1 g. *Scand. J. Clin. Lab. Invest. Suppl.* **97**, 77–89.
4. Boyum, A. (1968) Isolation of leucocytes from human blood. Further observations. Methylcellulose, dextran, and ficoll as erythrocyte-aggregating agents. *Scand. J. Clin. Lab. Invest. Suppl.* **97**, 31–50.

7

Image Analysis and Data Management of ELISPOT Assay Results

Paul Viktor Lehmann

Summary

The recent renaissance of enzyme-linked immunospot (ELISPOT) assays largely is the result of advances in image analysis. Information on the frequency of antigen-specific T-cells and also on the secretion rate of the individual cells is captured in spots generated using this technique. Although the overall assessment of ELISPOT results can be conducted visually, this is inevitably subjective, inaccurate, and cumbersome. In contrast, objective, and accurate measurements are fundamental to good science. Validated image analysis algorithms and procedures, therefore, have become critical for elevating the quality of ELISPOT assays results. As cytokine and granzyme B ELISPOT assays become the gold standard for monitoring antigen-specific T-cell immunity in clinical trials, the pressure increases to make ELISPOT analysis transparent, reproducible and tamperproof, complying with Good Laboratory Practice and Code for Federal Regulations Part 11 guidelines. In addition, ELISPOT assays in clinical and basic science settings frequently require high degrees of throughput, thus further raising the need for advanced data management and statistical analysis. The ImmunoSpot software portfolio has been specifically designed to meet all these needs, using the techniques described in this chapter.

Key Words: T-cells; ELISPOT; image analysis; cytokine productivity; spot morphology; single cell resolution; spot size gating.

1. Introduction

Each spot within an enzyme-linked immunospot (ELISPOT) assay is not "just a dot," but the footprint of a single cells' secretory activity—one that contains detailed information about the secretory process itself. Understanding the basics of spot formation is critical for performing image analysis, which can withstand rigorous scientific scrutiny.

From: *Methods in Molecular Biology, vol. 302: Handbook of ELISPOT: Methods and Protocols*
Edited by: A. E. Kalyuzhny © Humana Press Inc., Totowa, NJ

1.1. ELISPOT Size and Morphology

In an interferon (IFN)-γ ELISPOT assay, for example, IFN-γ is captured by the plate-bound anti-IFN-γ antibody around the secreting cell. The footprint of the captured IFN-γ will eventually be visualized as an ELISPOT, with its size and density reflecting the amount of cytokine produced by the cell during the assay's entire duration. Spot size and density are thus critical parameters that one using ELISPOT image analysis must take into consideration. The kinetics of the cytokine production also is reflected by the spot morphology, that is, their density and general shape. For example, a rapid secretion rate will produce large, fuzzy spots, whereas the slow-but-steady release of cytokines will result in smaller, denser spots. Evaluating both the size and the morphology of these spots is therefore crucial to performing accurate ELISPOT analysis.

Spot size and morphology frequently allow researchers to distinguish cytokine production by different cell types within mixed cell populations. For example, when interleukin (IL)-10 production by human peripheral blood mononuclear cells (PBMC) is measured in ELISPOT assays, most of the "antigen-induced" spots are not T-cell derived (as would be expected) but are produced by macrophages in response to lipopolysaccharide contamination of the antigen. Such macrophage-derived IL-10 spots are considerably smaller than the IL-10 spots generated by antigen-specific T-cells *(1)*. Although the lipopolysaccharide-induced macrophage-derived spots provide no information on specific immunity, the antigen-induced T-cell-derived IL-10 spots do because they indicate the presence of T regulatory cells. To measure the latter, the former need to be excluded from the counting results by setting appropriate size thresholds. **Figure 1** illustrates this point on the example of a tumor necrosis factor (TNF)-α assay; the small spots were macrophage-derived, whereas the large spots were generated by T-cells (A. Y. Karulin, and P. V. Lehmann, unpublished results). Similarly, small and faint IL-6 spots are produced by macrophages, whereas antigen-specific T-cells produce larger, "juicier" spots (unpublished data). ELISPOT image analysis must therefore be capable of distinguishing different spot sizes and morphologies to provide information relevant for T-cell diagnostics.

The antigen dose affects the cytokine secretion rate of T-cells. Stimulation of a T-cell clone with a high dose of the nominal antigenic peptide induces stronger cytokine production in the individual T-cells (i.e., it triggers larger and/or denser spots) than does the stimulation of the same clone with low dose peptide *(2)*. Therefore, when stimulated with a single antigen dose, as is frequently the case in ELIPSOT assays, high-avidity T-cells within the PBMCs produce larger spots than low-avidity clones. Confirming this notion, increased T-cell co-stimulation was shown to result in increased per cell productivity *(3)*. In diseases such as HIV, the cytokine productivity per cell can be reduced,

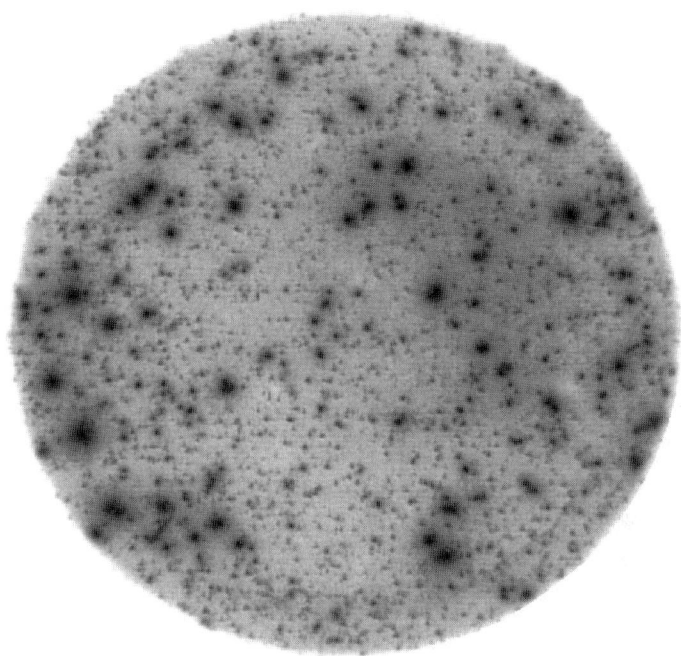

Fig. 1. ELISPOT morphologies as exemplified in a TNF-α assay. Two types of spots are seen, each after log-normal size/density distributions. The large diffuse spots were generated by antigen-induced T-cells, as shown by cell separation experiments; these spots were absent in the medium control wells. The fuzzy morphology of these spots results from a high per cell TNF-α productivity rate. In contrast, cell separation experiments showed that the small spots were generated by macrophages, and these spots were present in medium control wells. The pristine morphology of these spots is caused by the slow but continuous release of the TNF-α. Image analysis can be used to recognize these different morphologies, and thus, to clearly distinguish between T-cell and macrophage-derived spots. (As an aside, the figure also illustrates elevated background intensity in areas of increased secretory activity—the result of an ELISA effect as the cytokine is captured from the supernatant.)

resulting in smaller spots *(4)*. One advantage of ELISPOT assays is their ability to determine whether decreased net cytokine production in disease states is caused by a decreased number of cytokine-secreting T-cells or from reduced per cell productivity by unchanged frequencies of T-cells. To compensate for physiological and pathological variations in per cell productivity, ELISPOT image analysis tools must therefore be versatile, with the ability to permit fine-tuning of the image processing parameters.

Central to ELISPOT image analysis is the standardization of assay conditions, which can directly impact the spot morphology. Consider the affinity of the capture antibodies: a capture antibody with low affinity will produce fainter, more diffuse spots than a capture antibody of high affinity. Accordingly, the morphology of ELISPOTs can vary considerably when different antibodies (or even different concentrations of the same antibody) are used for coating.

The assay duration also can influence the spot morphology. The spots grow in size and density when the assay duration is prolonged, and the cells secrete continuously, as is the case for T-cell-derived IFN-γ *(2)*. The outcome is different, however, when there is an early burst of production that comes to a halt before the assay is terminated. In such cases, the spot size will continue to grow even after the production of the cytokine has stopped (because of lateral cytokine diffusion caused by the reversibility of its interaction with the membrane antibodies) their intensity will fade, however, because of the dilution of the cytokine. The temperature during development, and the nature of the enzymatic reaction will also define the spot morphology. The red spots developed with HRP-AEC differ fundamentally from the blue NBT/BCIP spots, with the former being more pristine (despite ALPH detection being more sensitive than HRP) and having a faint background, whereas the latter is more dramatic and fuzzy with a more heavily stained background.

Once an ELISPOT assay has been standardized, however, the interassay variability of spot morphology becomes negligible. Although spots of different cytokine ELISPOT assays will continue to look different even after standardization, the same counting parameters can be used assay after assay *(4,5)*. Therefore, apart from allowing counting parameters to be fine-tuned, ELISPOT image analysis tools must also allow the user to employ the same parameters for different assays, so as to permit objective comparison of the results of different assays.

1.2. ELISPOT Counts

One key piece of information to be gained from ELISPOT assays is the frequency of antigen-specific T-cells within the entire sample cell pool, as measured by the number of cells that engage in cytokine production after antigen stimulation. This frequency reflects the clonal size of the antigen-specific T-cells and, therefore, the magnitude of T-cell immunity. Obviously then, one prerequisite for obtaining correct frequency information is that both the image acquisition and the assay must be optimized for single-cell resolution.

In all T-cell cytokine ELISPOT assays, a wide spectrum of spot sizes and densities can be seen. Thus, when analyzing ELISPOT results, cut-off values need to be set for the minimum spots sizes and densities to be counted. The maximum spot size must likewise be defined so that clusters of cells can be identified as such. The minimum and maximum "gates" set will critically affect the number of spots counted. For this reason, one of the main goals of ELISPOT image analysis has been to establish absolute criteria for gating, thereby eliminating the "ghost of subjectivity" that has haunted ELISPOT counts in the pre-image analysis age.

The simplest experimental model that can be used to establish ELISPOT gating criteria involves the use of a T-cell clone that produces IFN-γ. These T-cells were activated by the nominal peptide on a clonal population of antigen-presenting cells (APCs) that cannot express IFN-γ *(2)*. In such experiments, conducted over a wide range of plated T-cells, the number of T-cells per well closely matched the numbers of spots detected. Even though the T-cells and APC were clonal, the spot sizes varied over a wide range! Closer analysis of the spot size distribution showed that they followed a log normal distribution. When the peptide dose was lowered, the per cell productivity (the mean spot size/density) decreased, but the size distribution still followed a log normal pattern. Similarly, when the assay duration was changed, the mean spot sizes/density varied, but the log normal distribution remained. In all subsequent studies of human and murine cells *(6)*, for clonal and bulk populations, for all cytokines measured (IL-2, IL-3, IL-4, IL-5, IL-6, IL-10, and IFN-γ) and in granzyme B assays *(7)*, this log normal distribution of ELISPOTs was noted. Therefore, by measuring the size/density of a multitude of individual spots, the statistical qualities of the spot size distributions can be established, allowing the software to set absolute criteria for the minimum and maximum size gates.

Having established these distributional properties, clusters of cells can be recognized, and the numbers of cells constituting these clusters can be calculated. By rooting ELISPOT image analysis in these objective statistical principles, one can establish absolute criteria for ELISPOT counting, thus eliminating subjectivity and elevating ELISPOT to an exact science.

1.3. Hardware Requirements

One limiting factor in the accuracy of ELISPOT image analysis is the hardware used for image acquisition. There is a common misconception that the pixel resolution of the camera is the key factor in determining image quality. This is an overly simplistic view. A fine-grain film alone does not provide pristine photographs unless the optics, the illumination, and many other fine details also are optimized. ELISPOT readers need to be high-end optical instruments to permit accurate analysis of ELISPOT images at single-cell resolution. In addition, such readers must feature precise robotic motion control, so as to accurately position and capture the membrane surface. The identity of the wells is of regulatory concern and must be verified by slip-proof, encoder-controlled stages and by faithfully recording the well position on each plate during image acquisition. Finally, to ensure consistent performance at single-cell resolution, regular machine calibration is required.

2. Materials

1. ImmunoSpot® Series 3B Analyzer (CTL, Cleveland, OH).
2. ImmunoSpot® 4.0 Software (CTL).
3. SpotMap™ 4.0 Software (CTL).

3. Methods

3.1. Scanning

In the first step of ELISPOT analysis, an ImmunoSpot Analyzer scans and saves image files (basically, digital photographs) of individual ELISPOT wells on a plate. The machine progresses automatically from well to well, using optical feedback to automatically center on each well, thus compensating for irregularities in the plate geometry. (ELISPOT plates are manufactured using a high-temperature molding process, and are prone to deform as they cool down.) Digital encoders keep track of the precise position of each well, thus helping to confirm well identify and positioning. In addition, the software keeps track of the encoder information, the time stamp and the identity of the operator. Systems also can be set up with access limitations for an added measure of security.

The end point of the fully automated scanning process for an ELISPOT plate is a tamperproof set of 96 image files, each representing a digital photograph of one well from the original 96-well plate. Scanning can also be performed using 12- and 24-well formats. The saved files allow users to document and analyze ELISPOT assays long after the original plates have decayed, and to reproduce the analysis results. While "live" analysis of images (that is, without saving them to a disk file) is also possible, it is not recommended because this obscures the transparency and reproducibility of the results, and thus violates good scientific and laboratory practice.

Suggestions for scanning:

- We recommend that plate images that belong together (e.g., plates from the same experiment) be stored in the same folder. The software allows the user to handle all the plates in a folder as a single unit; that is, the user can instruct ImmunoSpot to process all plates within a given folder, instead of tediously loading each one individually. Grouping such plates together can expedite all phases of the work: counting, quality control, and data export.
- We recommend that the scanned images be kept on a read/write storage device, such as the hard drive of the computer on which the counting and quality control will be done.

3.2. Analysis

The saved image files can then be processed on the Analyzer itself or on remote workstations equipped with the ImmunoSpot software. The dissociation of scanning and analysis enables work to proceed more efficiently by permitting an indefinite number of users to analyze images independently, without tying up the core machine.

3.2.1. Automated Analysis

In the first step of analysis, counting parameters are defined, and these parameters are used for analyzing all wells within an assay. This permits the objective comparison of results from different wells or plates of an assay.

The main steps of automated counting are as follows:

1. Loading the plate images. Virtually any number of plates can be loaded at this stage, due to the flexible software design.
2. Defining the counting parameters. The software provides default parameters that have been carefully selected to provide reasonably accurate counts for most ELISPOT assays. If strict, scientific counting precision is not a requirement, one can directly proceed to the spot counting stage.

Accurate counting, however, requires instructing the software about the nature of the spots to be counted (*see* **Note 1**). As discussed above, the spot characteristics can vary considerably, depending on the assay conditions and the cytokines under examination (*see* **Note 2**). For this reason, the ImmunoSpot software has been designed to allow fine-tuning of the counting parameters using a simple two-stage process.

Step 1: Sampling the Spot Morphology Using the "SmartSpot" Feature

By clicking on a spot, the software will analyze and "learn" to recognize the cardinal features of this spot, after which it can proceed to examine all other spots for these features (**Fig. 2**). Although establishing the appropriate counting parameters for the respective spot type is fully automated (and therefore objective and reproducible), the parameters can be manually fine-tuned for morphology, sensitivity, and a multitude of other criteria (*see* **Note 3**).

Step 2 : Gating Using the "Autogate" Feature

After the spot morphology has been defined, the software can be instructed to count all spots that exhibit this morphology (**Fig. 3**). In the process, the spot size distribution parameters can be established. A minimum of 500 spots need to be counted in this way to accumulate enough information for an accurate statistical analysis of the spot size distribution (but *see* **Note 4**). Typically, it takes sampling of approx 10 wells (a process that takes less than a minute) to sample at least 500 spots (*see* **Note 5**). By hitting the "Autogate button," the user can automatically set the lower and upper gate values (that is, spot size thresholds) based on the log-normal distribu-

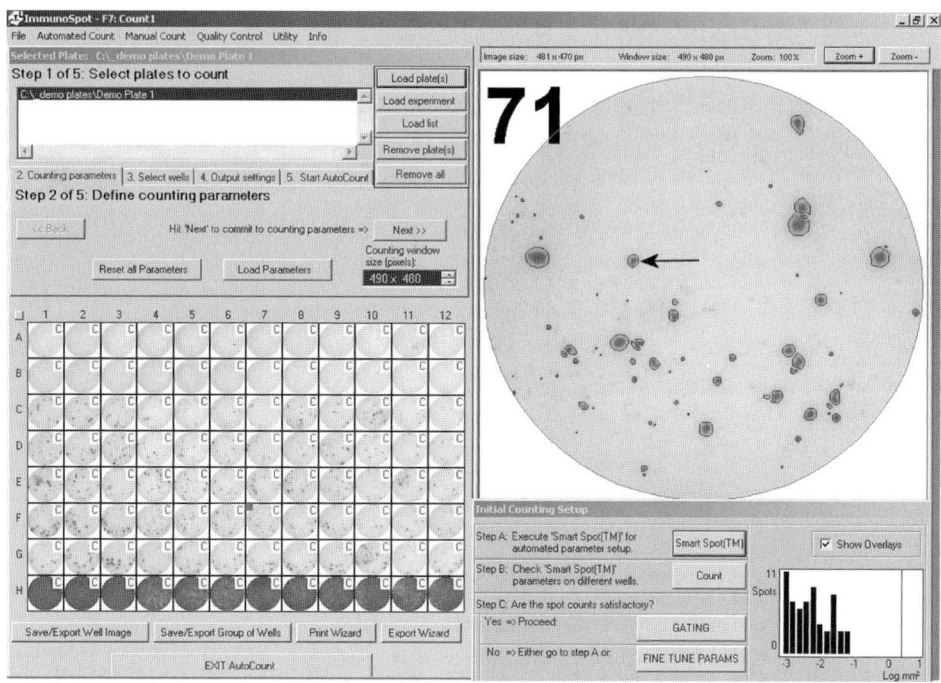

Fig. 2. Recognition of ELISPOT morphology. The first step of the analysis process is to "teach" the software the morphology of the spots of interest. This can be done by sampling characteristic spots (highlighted by arrow). The software then identifies all spots of the same morphology, irrespective of size, and marks each spot recognized as shown. In the subsequent step, the size distribution of the spots will be analyzed.

Fig. 3. Establishing the minimal/maximal spot size/density to be counted by auto-gating. The size/density distribution of the spots in the current well is captured in the left histogram (labeled "CURRENT"), which shows 45 spots as being recognized. More accurate information on the size/density distribution of ELISPOTS in the assay can be obtained by sampling multiple wells, as captured in the cumulative histogram on the right-hand side. (In this example, the cumulative histogram was generated by sampling 513 spots in 12 wells.) The "Autogate" feature uses the distributional properties of this cumulative data to compute the minimum and maximum spot limits or "gates" (indicated by the vertical lines in the histograms). When the actual well shown is recounted with these limits in place, one small spot is "gated out," resulting in the spot count of 44 shown at the top. Once the morphology and size/density criteria are established, the software applies these very same parameters to the automated counting of any number of wells.

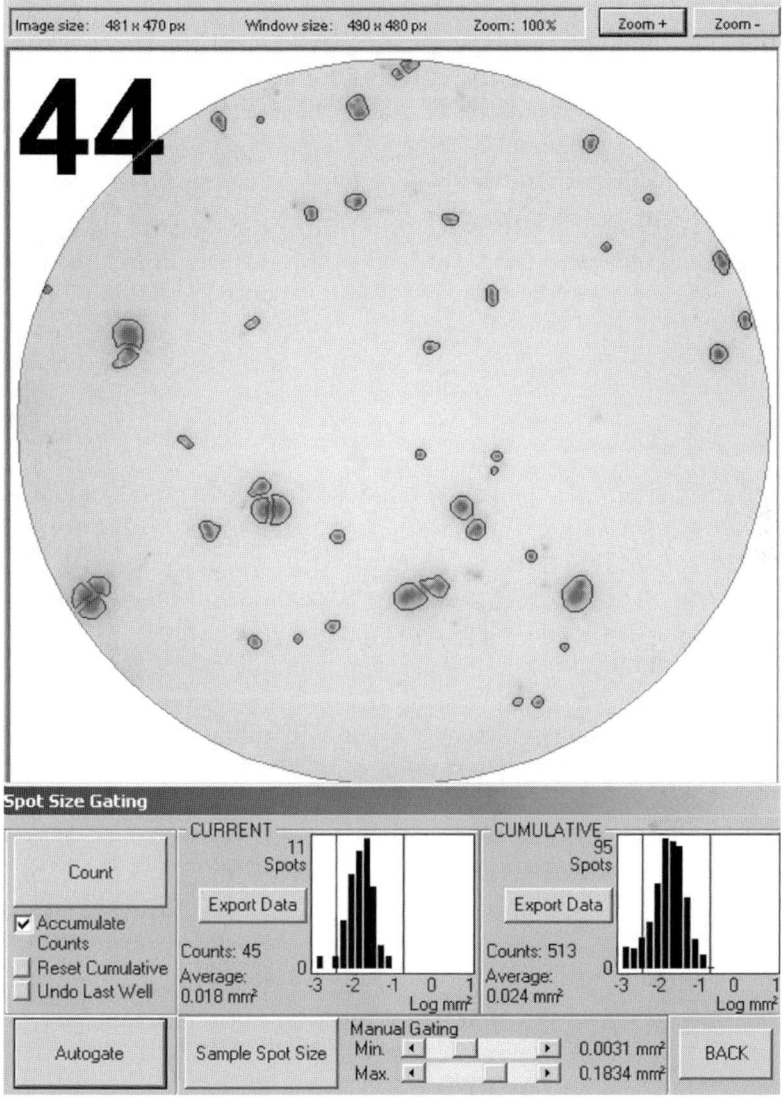

tion properties of the spot-size distribution. The Autogate feature thus allows objective, statistics-based criteria to be used in setting the minimum and maximum spot sizes allowed.

Spots smaller than specified by the minimum gate are ignored, that is, they are excluded from the final spot count. Spots larger than the maximum gate value are either counted as clusters or are excluded altogether. The latter option permits counting small spots generated by cell type "A," whereas gating out the larger spots generated by cell type "B."

The same counting parameters cannot be applied if results of different assays are to be counted in the same run. For this reason, the software allows the user to define an indefinite number of distinct counting parameter sets (templates) for any array of wells, arranged in any pattern on the loaded plates.

3. Automated counting. Once the parameters have been established and assigned to the wells, the software automatically counts spots on any number of plates or sections thereof (*see* **Note 6**). Overlays of the raw image files and of the counting results are saved for each well, as are the counting parameters. The results of the counting process thus become transparent, documented, and easily reproduced for subsequent verification in the quality control step.

3.2.2. Quality Control

Because ELISPOT assays can contain artifacts (e.g., contaminants, damaged or leaking membranes; *see* **Note 7**), the results of the automated counting need to be subjected to quality control. A menu in the ImmunoSpot software allows the user to view image overlays that indicate which spots have actually been counted, and to make corrections as needed (**Fig. 4**). In the example shown, a cluster of cells (resulting from the clumping of cells by free DNA in a freeze-thawed PBMC sample) has been excised, and the software has calculated how many spots would occupy that cluster, assuming that these spots have the same average size and density distribution as those within the rest of the well.

To ensure Good Laboratory Practice compliance, all changes made are recorded and annotated. This allows the principal investigator or regulatory agency to determine at a glance whether the counting results are accurate and appropriate. As part of this documentation, the software also generates a set of plate and well image files that can be helpful in preparing presentations, writing publications or discussing the results. Direct PowerPoint export options also make it convenient for the user to arrange groups of wells for presentation or documentation purposes.

3.3. ELISPOT Data Management

ELISPOT assays can require high degrees of throughput. It is not uncommon for a single well-trained team to test hundreds of samples each day for reactivity to hundreds of antigens (*see* **Note 8**). However, even a small assay can contain a veritable flood of information. Just three 96-well plates, for example, require storing 864 image files—raw images, counting overlays, post-quality

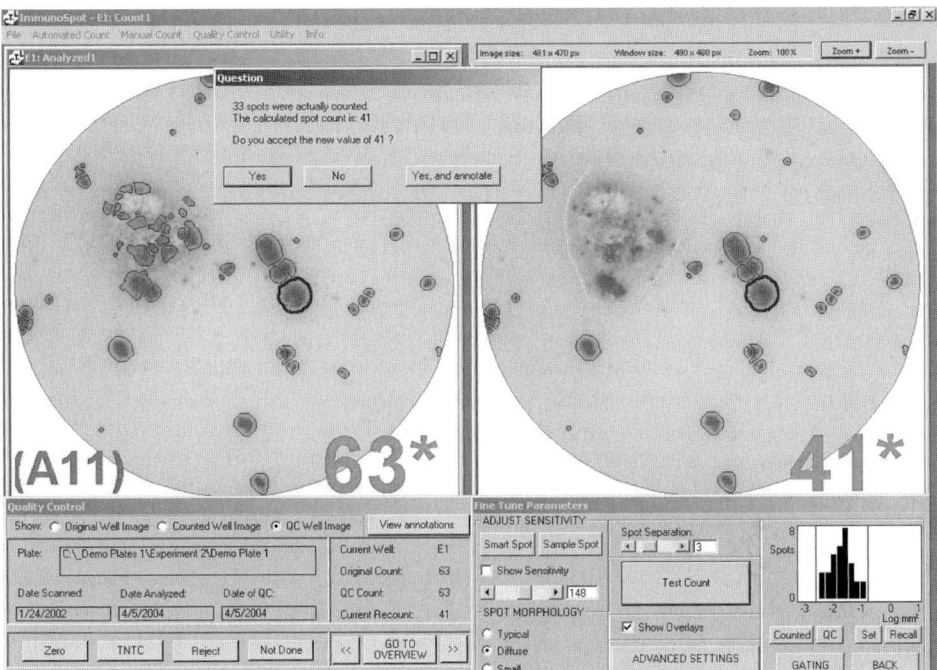

Fig. 4. Quality control step in ELISPOT analysis. The image on the left-hand side shows the counting results obtained in automated analysis mode on a well that contains an artifact (in this case, a cell clump caused the spot cluster in the upper left-hand quadrant). The cluster was treated as a group of individual spots, resulting in a spot count of 63. In control mode, this artifact-containing region can be outlined (on the right) and excluded from the analysis. The software then normalizes the spot count by correcting for the size of the unselected region. In the example shown, 33 spots were actually detected, and this count was increased by eight to compensate for the unselected area, resulting in an adjusted spot count of 41. (The asterisk beside the spot count of 41 indicates that this is a recalculated value, rather than a direct measurement.) In keeping with good laboratory practice, the software saves and annotates all such subjective adjustments made to the objective automated count. This same example also contains a spot near the center of the well that exceeds the upper gate threshold. This spot was automatically outlined in bold during automated analysis (the smaller bold outline in the left and right image), indicating that it was treated as a cluster. In such cases, the software automatically calculates the number of spots required to generate such a cluster based on the average spot size and density distribution , and re-computes the spot count accordingly. (The asterisk beside the spot count of 63 likewise indicates that this value was recalculated, as does the automatically generated A11 annotation code.)

control images, along with records of the counting parameters, the numbers of spots counted, and the spot size/density statistics for each well. This information needs to be linked to the assay information, that is, to the source of the cell material tested (e.g., PBMC of donor 'Z'), to the number of cells per well (so that frequencies can be normalized "per million"), to the antigens tested and their concentration, and to the cytokines measured. Thus, even a small three-plate assay, there can be more than 4320 sets of data that need to be linked together.

The ImmunoSpot software's SpotMap module was specifically designed to manage these data. For each well, and for each plate, the software allows the user to document the assay conditions: which cells were plated in which numbers, which antigens were used to challenge the cells and in which concentrations, and what cytokine was measured (**Fig. 5**). Custom routines help in laying out multi-plate experiments. The software even calculates the amounts of reagent needed for each assay. Once the counting results are available, they can be quickly linked to the other assay parameters. At a click of a button, even the most complex ELISPOT assay can be evaluated, the statistics calculated, and the requested information compiled in an Excel spreadsheet, allowing the results to be represented in virtually any desired format.

4. Notes

1. Occasionally, T-cells move around during the assay, causing the ELISPOTs to develop "tails." This is especially true when T-cells which have been preactivated in vivo or in vitro, as this makes the cells particularly mobile. Such spots can be counted accurately by decreasing the "spot separation tolerance" value.
2. Occasionally, white dots can develop in the middle of the spots. These result from the substrate peeling off, for example, when the flow rate of the plate washer is too high, or the plates are banged too hard while washing. The "Fill Holes" feature can be used mask the white dots away, allowing the spots to be counted accurately.
3. Occasionally, the background coloration can be darker in some parts of the well as the result of leakage in the membrane. This problem can be avoided by decreasing the concentration of Tween used. Additionally, the user can typically compensate by adjusting the "Background Balance" parameter.
4. The background coloration is increased over the entire membrane in a well if the number of cytokine-producing cells (i.e., the number of spots) is high. This is caused by an ELISA effect; that is, cytokine that is not captured around the secreting cell escapes in the supernatant and binds as a "carpet" over the entire well surface. "Autolight" adjustment compensates for the increased background coloration. Some protein antigens can also cause high uniform background by non-specifically binding the detection antibody.
5. Sometimes, the number of spots in the medium background is high for all samples because of the stimulatory effects of serum. Even brief exposure to nontested serum, for example, during washes or during freezing, can drive up the background

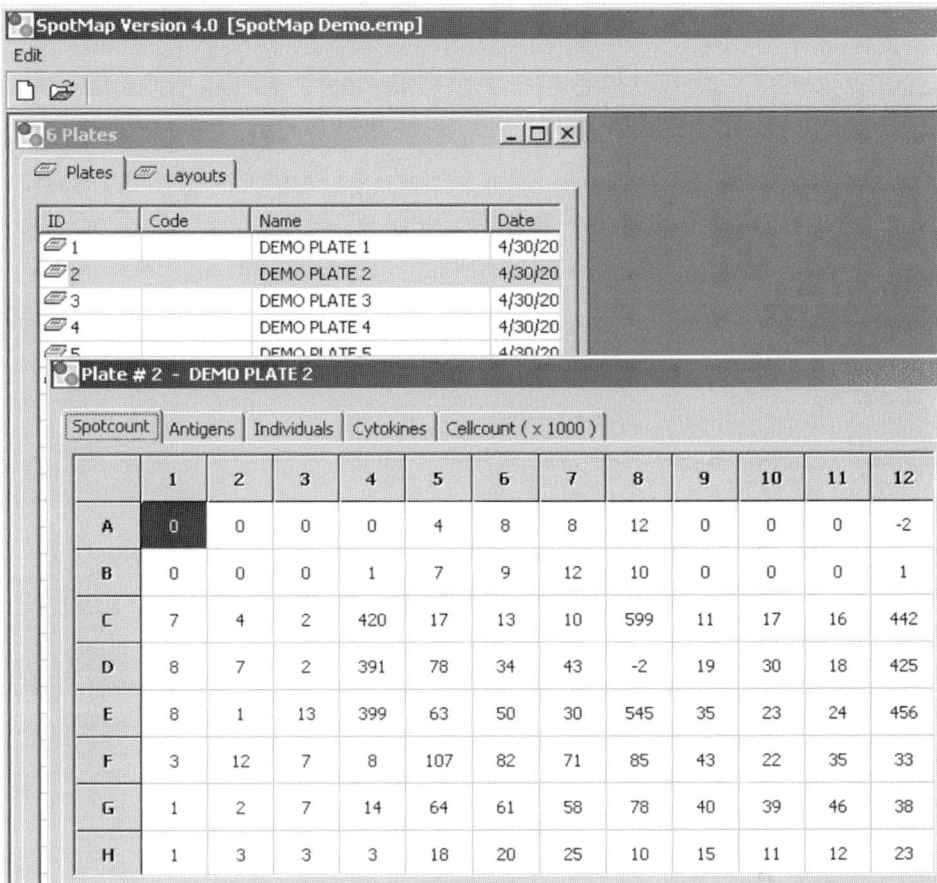

Fig. 5. ELISPOT data management. The spot counts in 96-well format are linked to the plate layout. For each well, the antigen, the test subject, the cytokine and the number of cells plated are specified. All this data is linked and processed for exporting into a database.

intensity. Occasionally, the number of medium background spots is high for a single individual out of several tested. This is a common finding for individuals undergoing a clinical or subclinical infection or other massive immune stimulation in vivo. Some assays, such as IL-6, IL-10, and TNF, tend to give high background coloration in general because of the activation of macrophages on the membrane of the ELISPOT plate. Such background spots are frequently smaller than the antigen-induced spots produced by T-cells, and can be gated out.

6. Occasionally, the counting parameters established can produce valid spot spots for most test subjects, but not for others. For example, spots that are either smaller or larger than usual can be seen with particularly low (or high) avidity T-cell responses,

or if co-stimulation is decreased (increased). This is one reason why, as part of any ELISPOT analysis, the researcher should have the option of viewing both the raw images and the counting results in quality control mode. This allows the researcher to judge whether the counting parameters established do indeed apply to the all subjects under examination. If recounting of any given subject becomes necessary, the altered parameters are automatically annotated by ImmunoSpot®, thus drawing attention to the atypical spot morphology or other image characteristics.

7. On occasion, the well images contain artifacts caused by membrane damage, for example, when the membrane is accidentally scratched with the pipet. The affected area can be excised in QC mode, and using the *normalization* algorithm, the spot count is recomputed. This renormalization is performed by computing the number of spots required to fill the excised area, using the average spot size and distribution density in the rest of the well. The same technique also can be used to correct for cell clustering. For example, if the testing was performed in triplicate and a cluster is found in one of the wells, this cluster can be excised and the spot count can be normalized. In both cases, these corrections are automatically recorded by the software in the form of annotations added to the well records.

8. Never blindly trust ELISPOT counts, whether from your own laboratory or from others! Overlays of both the raw images and the counting results are a simple and transparent way of understanding the assay results and judging the counting accuracy. *Well surveys* containing this information can be printed or exported into graphics files or PowerPoint presentations, allowing assessment to be performed at a glance. The direct side-by-side display of medium control and antigen wells can speak volumes about the quality of the assay and the spot counting.

Acknowledgments

I would like to thank all those who worked in my Case Western Reserve University-based laboratory on establishing the scientific foundations of cytokine ELISPOT assays. At the postdoctoral level, these are (in alphabetical order): Drs. Don Anthony, Beate Berner, Thomas Forsthuber, Peter Heeger, Alexey Karulin, Damian Kovalovsky, Patrick Ott, Clara Pelfrey, Frauke Rininsland, Stephan Schwander, Oleg Targoni, and Magdalena Tary-Lehmann. Several of my graduate students at Case have made major contributions in our ELISPOT efforts as well: Wolf Bartholomae, Jan Baus, Kamruz Darabi, Marcus Dittrich, Julia Eisenberg, Kristina Feldmann, Judith Gottwein, Robert Guerkov, Thomas Helms, Bernhard Herzog, Maike Hesse, Harald Hofstetter, Thomas Kleen, Christian Kreher, Haydar Kuekrek, Anke Lonsdorf, Kai Loevenbrueck, Stephan Quast, Tarvo Rajasalu, Tobias Schlingman, Britta Stern, and Hualin Yip. At Cellular Technology Limited, I am indebted to the R&D and programming efforts of Johannes Albrecht, Tameem Ansari, Andras Bakos, Ben Matthes, Mark Novak, Jerry Perchinske, Carey Shive, Endre Tary, Norma Sigmund, John Truden, Dean Velasco, and Wenji Zhang.

References

1. Guerkov, R. E., Targoni, O. S., Kreher, C. S., Boehm, B. O., Herrera, M. T., Tary-Lehmann, M., et al. (2003) Detection of low-frequency antigen-specific IL-10-producing CD4+ T-cells via ELISPOT in PBMC: cognate vs. nonspecific production of the cytokine. *J. Immunol. Methods* **279**, 111–121.

2. Hesse, M. D., Karulin, A. Y., Boehm, B. O., Lehmann, P. V., and Tary-Lehmann, M. (2001) A T-cell clone's avidity is a function of its activation state. *J. Immunol.* **167**, 1353–1361.

3. Ott, P. A. Berner, B. R., Herzog, B. A., Guerkov, R., Yonkers, N. L., Boehm, B. O., et al. (2004) CD28 costimulation enhances the sensitivity of the ELISPOT assay for detection of antigen-specific memory effector CD4+ and CD8+ cell populations in human diseases. *J. Immunol. Methods* **285**, 223–235.

4. Helms, T., Boehm, B. O., Assad, R. J., Trezza, R. T., Lehmann, P. V., and Tary-Lehmann, M. (2000) Direct visualization of cytokine-producing, recall antigen-specific CD4 memory T-cells in healthy individuals and HIV patients. *J. Immunol.* **164**, 3723–3732.

5. Kreher, C. R., Dittrich, M. T., Guerkov, R., Boehm, B. O., and Tary-Lehmann M, M. (2003). CD4+ and CD8+ cells in cryopreserved human PBMC maintain full functionality in cytokine ELISPOT assays. *J. Immunol. Methods.* **278**, 79–93.

6. Karulin, A. Y., Hesse, M. D., Tary-Lehmann, M., and . Lehmann, P. V. (2000) Single-cytokine-producing CD4 memory cells prevail *in vivo*, in type 1/type 2 immunity. *J. Immunol.* **164**, 1862–1872.

7. Rininsland, F. H., Helms, T., Asaad, R. J., Boehm, B. O., and Tary-Lehmann, M. (2000) Granzyme B ELISPOT assay for *ex vivo* measurement of T-cell immunity. *J. Immunol. Methods* **240**, 143–155.

8

High Resolution as a Key Feature to Perform Accurate ELISPOT Measurements Using Zeiss KS ELISPOT Readers

Wolf Malkusch

Summary

The enzyme-linked immunospot (ELISPOT) assay was originally developed for the detection of individual antibody secreting B-cells. Since then, the method has been improved, and ELISPOT is used for the determination of the production of tumor necrosis factor (TNF)-α, interferon (IFN)-γ, or various interleukins (IL)-4, IL-5. ELISPOT measurements are performed in 96-well plates with nitrocellulose membranes either visually or by means of image analysis. Image analysis offers various procedures to overcome variable background intensity problems and separate true from false spots. ELISPOT readers offer a complete solution for precise and automatic evaluation of ELISPOT assays. Number, size, and intensity of each single spot can be determined, printed, or saved for further statistical evaluation. Cytokine spots are always round, but because of floating edges with the background, they have a nonsmooth borderline. Resolution is a key feature for a precise detection of ELISPOT. In standard applications shape and edge steepness are essential parameters in addition to size and color for an accurate spot recognition. These parameters need a minimum spot diameter of 6 pixels. Collecting one single image per well with a standard color camera with 750×560 pixels will result in a resolution much too low to get all of the spots in a specimen. IFN-γ spots may have only 25 μm diameters, and TNF-α spots just 15 μm. A 750×560 pixel image of a 6-mm well has a pixel size of 12 μm, resulting in only 1 or 2 pixel for a spot. Using a precise microscope optic in combination with a high resolution (1300×1030 pixel) integrating digital color camera, and at least 2×2 images per well will result in a pixel size of 2.5 μm and, as a minimum, 6 pixel diameter per spot. New approaches try to detect two cytokines per cell at the same time (i.e., IFN-γ and IL-5). Standard staining procedures produce brownish spots (horseradish peroxidase) and blue spots (alkaline phosphatase). Problems may occur with color overlaps from cells producing both cytokines, resulting in violet spots. The latest experiments therefore try to use fluorescence labels as a marker. Fluorescein isothiocyanate results in green spots and Rhodamine in red spots. Cells producing both cytokines appear yellow. These colors can be separated much easier than the violet, red, and blue, especially using a high resolution.

Key Words: ELISPOT reader; ELISPOT method; spot teacher; camera resolution; pixel size; double labels; fluorescence markers.

From: *Methods in Molecular Biology, vol. 302: Handbook of ELISPOT: Methods and Protocols*
Edited by: A. E. Kalyuzhny © Humana Press Inc., Totowa, NJ

1. Introduction

Despite the pharmaceutical and medical success achieved so far, it is possible to treat only one third of the 30,000 known diseases. In addition, many diseases, such as rheumatism, Alzheimer's disease, AIDS, and cancer, remain incurable *(1)*. Research throughout the world is focused on cancer. The use of genetic engineering provides scientists with an increasingly better understanding of what malfunctions participate in the generation of malignant tumors on a molecular level. With regard to possible therapies, immunological methods make us particularly optimistic. In this field, the cytokines are of special interest. These proteins are substances created by a variety of cell types contributing to the activation of cells.

Tumor cells have a special characteristic. Like most cells, they have a structure that is specific only for their cell type. This is called a tumor-associated antigen. This structure makes it different from all the other body cells and therefore permits identification *(2)*. Where identification is possible, we are hopeful a direct remedy will soon be possible.

Cytostasis agents used in standard chemotherapy also affect healthy, proliferating body cells. This results in the known side effects of chemotherapy. The new therapy approach endeavors to use the characteristic structure of tumor cells to specifically attack the tumor itself *(3)*. Background: The tumor cells produce proteins—so-called antibodies. These antibodies all feature the same specific structure and also are expressed by the tumor. These tumor antibodies can be isolated from the patient's blood with relative ease and are used, so to speak, as the tumor-specific structure for immunization.

A certain cell type in the human immune system, the antigen-presenting cell, commonly known as a "dendritic cell," is able to present an antigen to the so-called T-cells. The T-cells then proliferate and, in the form of cytotoxic "killer cells," specifically attack the structure "shown" to them before by the antigen-presenting cell *(4)*. To measure the success of therapy, the specific defense cells, which were activated by the therapy, must be detected and recorded in a suitable manner. The T-cells that attack the cells of interest circulate in the patient's blood. The enzyme-linked immunospot (ELISPOT) assay is used to count the number of these specific "killer cells" before and after therapy, providing a measure for the immunologic reaction of the therapy.

In the field of immunology research, more and more molecules secreted in response to immunization are checked whether or not they can be used to develop direct therapy methods. The enzyme-linked immunospot (ELISPOT) technique, meanwhile, became the method to evaluate the effect of these molecules to immune cells on a single cell level. Fields of application are the determinations of therapy successes with regard to immunological responses in cancer diseases, AIDS, Alzheimer's disease, asthma, and so on. Immunological response does not always mean therapeutical success, but it is a first step in this direction.

Meanwhile, more than 150 cytokines are isolated and detected *(5)*, and their number is increasing nearly on a daily basis. Because cytokines play an important role during inflammation and diseases, they are the best tool to measure the activation of immune cells. Whereas the previously used enzyme-linked immunosorbent assay (ELISA) method only allows one to measure the concentration of cytokines in the supernatant of a cell culture, the ELISPOT method allows one to detect of cytokine production on a single cell level.

2. The ELISPOT Method

The ELISPOT assay was developed for the detection of individual antibody-secreting B-cells *(6)*. Since then, the method has been improved in a way that cells can be detected producing only approx 100 molecules of a protein (i.e., cytokine) per second *(7)*. These high protein concentrations in the cell surrounding will be detected with specific antibodies. When the ELISPOT method is used, the created spots only show an imprint of those cells originating them. The advantage of this method is the fact that the spots are long lasting, and they can be evaluated visually as well as by means of image analysis. Using image analysis, one can automatically determine the number, size, and intensity of spots with an increased objectivity.

For the ELISPOT determination of the production of tumor necrosis factor (TNF)-α or interferon (IFN)-γ or interleukin (IL)-4, IL-5 and more by a single cell, the bottoms of nitrocellulose microtiter plates are covered with antibodies directed against the cytokine of interest. All antibodies not bound to the nitrocellulose are then washed away. The covered wells are now filled with lymphocytes to be tested and negative controls. The T-cells on the membranes secrete the cytokines, which will bind to the antibodies. After a certain incubation time, the cells will be washed out, and the bound cytokines will be marked with a second antibody. This complex will finally be marked with a staining substance. Now single spots can be counted under the microscope *(8,9)*.

In this way, the spots created by the ELISPOT method are only a reprint of those cells that originally created them. Some advantages are that specimens are not dangerous and that they can be kept for a long time. Furthermore, they can be evaluated, either visually or by means of image analysis, and the evaluation may be repeated for control purposes whenever required *(7)*.

The ELISPOT method consists of five steps: (1) adding a cytokine specific antibody to the nitrocellulose membranes of a microtiter plate; (2) exclusion of nonspecific absorption of other proteins; (3) adding cytokine secreting cells in various concentrations; (4) addition of a second anti-cytokine antibody; and (5) detection of the antibody–cytokine complex.

At present two staining techniques are used. The alkaline phosphatase (AP) marker produces blue spots whereas the horseradish peroxidase marker produces brownish ones. The method also is used for the detection of secretions of specific subgroups of lymphocytes or T-cells from peripheral blood, as well as of monocytes and granulocytes *(7)*.

The immunochemically stained cytokine spots are scanned with a 3-chip CCD color camera on an incident light microscope with motorized stage and auto-focus control. The images are digitized with 24-bit color resolution and evaluated by the KS ELISPOT system. The region for evaluation is determined using the mouse on the system monitor. The definition of the start and end point is sufficient to begin the routine measurement *(10)*.

All results of the analysis are displayed immediately on the screen. Number, size, and intensity of each single spot can be printed or exported to a file for future statistic evaluation or graphic presentation. All well images can be saved for re-evaluation or documentation purposes.

2.1. Common Problems With ELISPOT Specimens

The ELISPOT assay faces two background problems. First, a variable background intensity of the nitrocellulose may be observed. To overcome this problem, a specific algorithm had been developed, taking into account the varying background conditions in the region when detecting the spots *(10)*. It is no longer necessary to adjust different threshold values for various positions of a well.

With the aid of a unique spot learning procedure, all system parameters, necessary for a correct spot recognition, are adjusted by simple cursor clicks to desired spots. A further advantage of this new method is the improvement of measurement reproducibility.

The second problem is the occurrence of small and very dark spots that were not generated by secreted cytokines. In the visual evaluation, these spots are differentiated from "true" spots by their sharper edges. True spots always have a dark center with fading color intensity towards the edges. False spots are usually small with a homogeneous intensity. Functions for the differentiation of true spots also have been implemented in the detection algorithm of the KS ELISPOT software *(10)*. The shape can be tested to decide whether or not a spot exists. Cytokine spots are always round but because of "floating edges" with the background they have a "non-smooth" borderline.

Depending on the magnification used, each experiment can be defined in this way under the tool setup. The spot definition itself is performed by automatic color segmentation. This algorithm first determines all probable spot positions after the acquisition of a complete well image. Each of these positions will then be checked via edge steepness and form factors for its spot probability.

All these conditions demand certain minimum and maximum magnifications for accurate use of the system. The recognition of differences in edge intensities needs a certain minimum spot diameter regarding the pixel resolution. Acquiring a complete well as one single image often lacks of necessary resolution. This method therefore should only be applied on specimens with mean or large spot diameters. Whenever the acquisition of only one image per well will produce exact results because of only larger spots, the advantage of these systems is the extremely high evaluation speed. An example for this group of instruments is the KS ELISPOT compact, using a stereo microscope Stemi 2000 as input source.

Whenever spots decrease to less than certain minimum diameters, the KS ELISPOT compact may be used in a scanning mode, similar to the KS ELISPOT, that is, using the Axioplan 2 light microscope as input source. This will result in a well image composed from several single images with a higher resolution, which is then evaluated with the correct pixel resolution without any edge problems using a circular measurement frame.

A check for all selected settings can be performed for a complete well. Using the teaching algorithms, all parameters responsible for the determination of a spot (spot diameter, color, saturation, contrast, shape, slope) can be trained by simply clicking to spots. In a similar way false spots can be eliminated from the parameter set. The finally accepted settings can be saved for the routine usage in different experiments under separate names.

2.2. Important Requirements of an Automatic ELISPOT Evaluation System

1. Easy handling: The system should be developed for use in routine laboratories and easily adaptable to new preparation methods. In the KS ELISPOT, this is achieved by the unique spot learning procedure.
2. Complete package: All system parts, such as the microscope, scanning stage, control unit, computer, and software, should be available from one source to guarantee a maximum performance.
3. Minimum adjustments: Especially for the routine usage, only a few adjustments should be necessary to start the evaluation.
4. Overview or full-resolution image: During evaluation, the permanent display of the overview well image under evaluation is essential for control purposes. The access to the full resolution image, including the display of results, also is very important.
5. Storage of measurement results: Spot results (number, single areas, and single intensities) have to be stored in a simple file format like text files for external processing and graphic presentation.

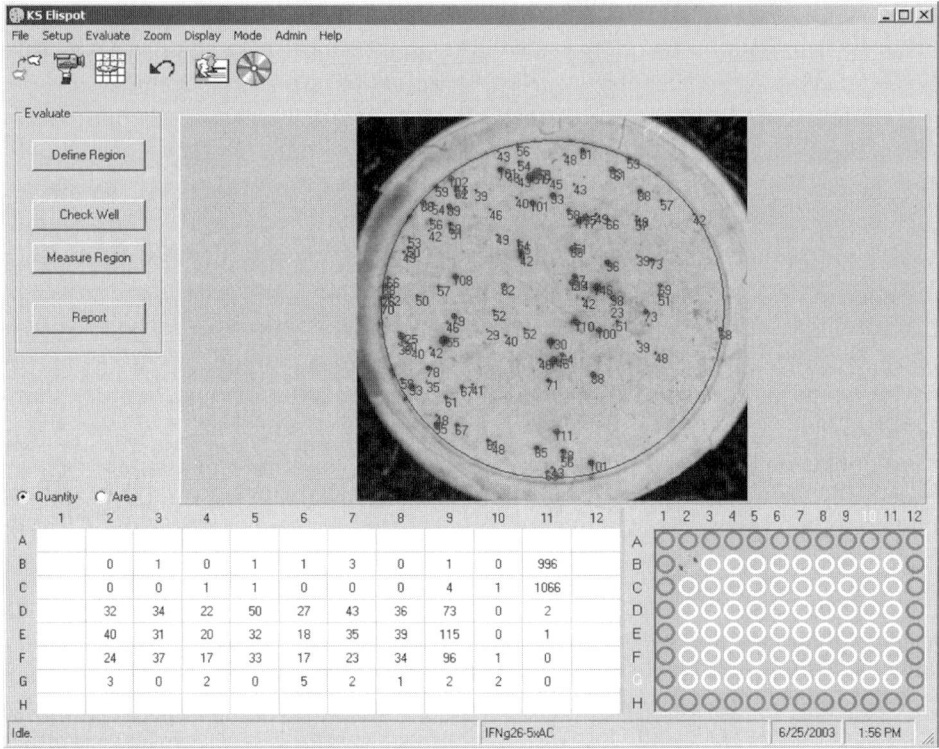

Fig. 1. Example of an easy-to-use interface: only four buttons for routine measurements, a window for the display of well images during the measurement, a plate field for the definition of the region of interest, and a result table for the immediate display of count results.

6. Data output: All measurement results should possibly be stored in an internal format to be saved and externally processed. As an alternative reports like a Word protocol may be produced showing experiment description, measurement data, and parameter settings, as well as a plate image of all measured wells and a plate image with all detected spots in each measured well.
7. Storage of images: Images of all measured wells should be saved for re-evaluation and documentation purposes, if needed.

An example of the tool buttons to be used for routine evaluations are listed below (*see also* **Figs. 1** and **2**):

1. **Define Region:** Defines the plate region for evaluation, resulting in the number of wells to be measured. At the same time the well positions and the stage co-ordinates are synchronized.

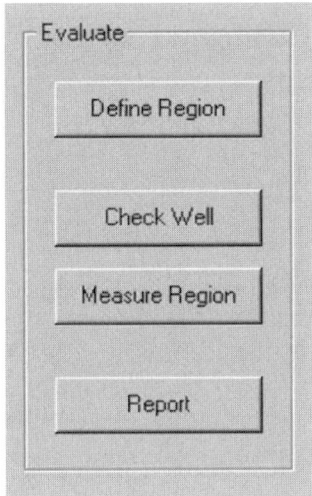

Fig. 2. Only four buttons are used for routine applications.

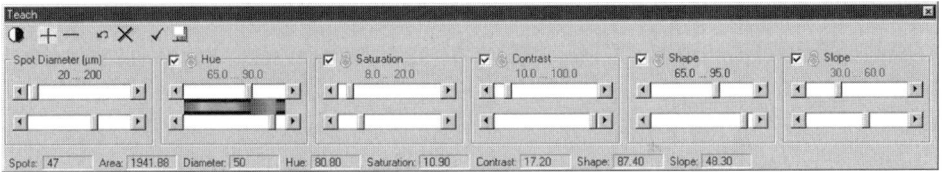

Fig. 3. Example for a spot teacher. Spots in the well image will be selected by simple mouse clicks to adapt the sliders to their values.

2. **Check Well**: Performs a test evaluation of one selected well to check the system settings and to perform the learn procedure.
3. **Measure Region:** Evaluates all selected wells of a microtiter plate.
4. **Report:** Displays results page with the option to print and save the data.

2.3. Learning Method for Spot Recognition

At least when setting up an ELISPOT reader for the first time, it will be necessary to train the software in the appearance of the spots inside the specimen. A tool like a spot teacher is extremely helpful because it uses a specifically developed learning method to recognize the spots. This learning algorithm extends all its parameters for spots that were selected from clicks on not-yet-recognized spots using the cursor and the left mouse key (*see* **Fig. 3**). In this way all spot parameters are adapted automatically. Spot definitions created in

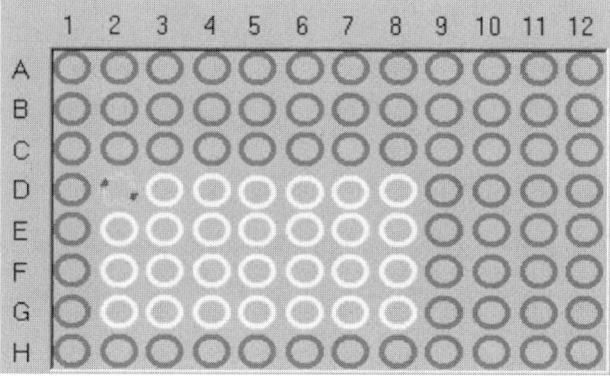

Fig. 4. The region of interest per plate can be defined by simply marking the wells inside that region with the mouse.

this way can be stored for future use. For a better control all recognized spots have to be displayed graphically together with their diameters.

2.4. Routine Use

An easy way to define the region for evaluation may be performed in a plate table on the screen using the mouse (*see* **Fig. 4**). To start a measurement the definition of only the start and end well should be necessary. It should be possible to eliminate not required positions from the measurement by mouse clicks. The possibility to define various regions per plate with different configurations for the evaluation can be very helpful.

For longer-lasting evaluations, the immediate presentation per well of the results of an analysis (spots per well) is absolutely necessary for the control. During the evaluation period, the indication of the likely time for the complete evaluation of the entire plate is advantageous when permanently updated, together with the estimated time for the completion of the evaluation—as a value and as a progress bar.

A general background correction is useful with specimens having extremely high background staining and at the same time high and dense spot numbers (this is often the case with positive controls). This helps to recognize the spots to an acceptable amount (*see* **Fig. 5**). Variable local spot background settings offer alternatives for a better spot recognition. Variable internal image resolutions allow the best compromise between evaluation accuracy and system speed.

2.5. Performance

After the definition of the overall region to be measured, one single configuration file should define all necessary settings for the system (*see* **Fig. 6**). The

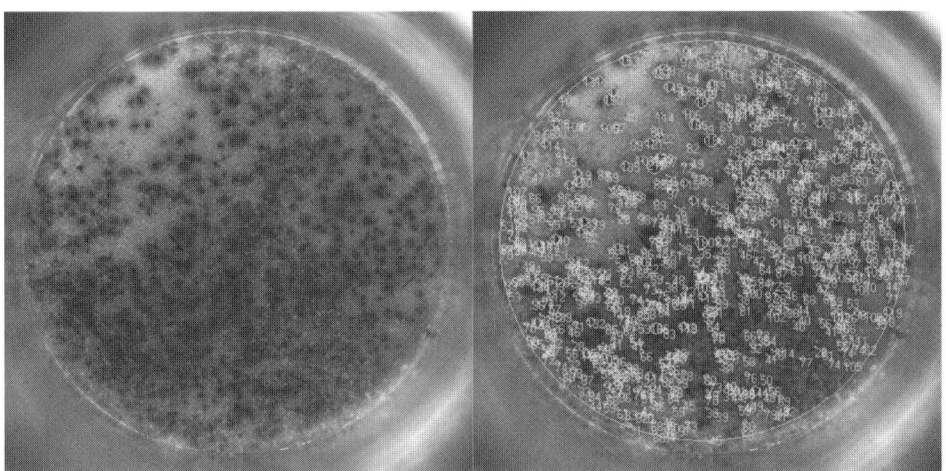

Fig. 5. Typical well image from a positive control. Spots are very dark on an intense background. Partially, the spots and the background are confluent. Using special correction functions the spots still can be recognized in an acceptable amount.

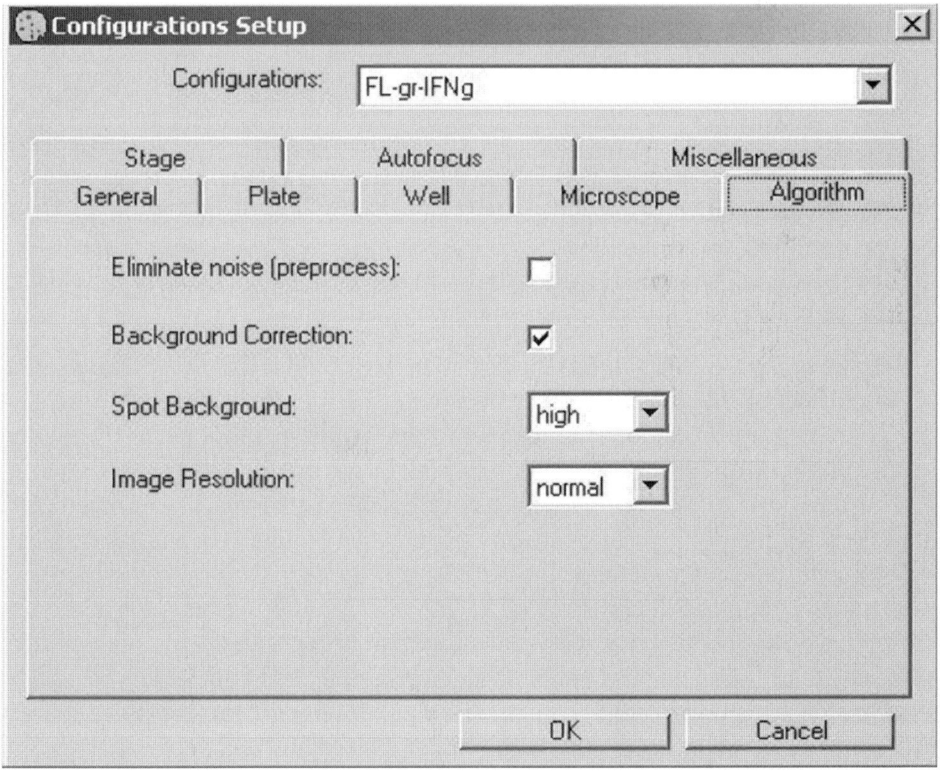

Configurations Setup ×

Configurations: FL-gr-IFNg ▼

| Stage | Autofocus | Miscellaneous |
| General | Plate | Well | Microscope | Algorithm |

Eliminate noise (preprocess): ☐

Background Correction: ☑

Spot Background: high ▼

Image Resolution: normal ▼

OK   Cancel

Fig. 6. All parameters for the evaluation of an ELISPOT experiment will be defined in a configuration window on several tab-sheets. Each configuration can be saved for a future usage.

possibility of multiple definitions within one plate to evaluate various regions with differently defined setups is a nice feature that helps facilitate evaluations.

Before the measurement of a whole plate or even of a series of plates, it is advisable to first test the system settings of the used configurations on some selected positive and negative wells for their correct function. In case modifications become necessary, a teach function that allows one to adapt all relevant spot parameters by simple mouse-clicks to the wrongly recognized spots in the well is extremely helpful.

Usually by means of image analysis, there are six parameters that will describe best a real spot (*see* **Fig. 3**) and separate it from other particles in the specimen, like dirt, debris, scratches, and others. They are listed below:

1. The **Spot Diameter** may be preset to a minimum and maximum value. The smallest expected spot size is about 25 μm for human IFN-γ cytokines. TNF-α and usually mouse IFN-γ cytokines produce much smaller spots, as small as 15 μm in diameter.
2. **Hue** is the definition of color values with lower and upper limit. Values range continuously from red (90–15) via yellow (15–40) and green (40–65) to blue (65–90) on a scale from 0 to 100.
3. **Saturation** defines the color saturation with lower and upper limit.
4. **Contrast**: defines the overall contrast range for spots. This should be a position independent relative contrast, taking into account also the local surroundings of the spot.
5. The **Shape** is the definition of the form factor. High values represent an ideal circle and lower values describe shapes more and more differing from circles.
6. The **Slope** defines the edge steepness of spots. Ideal spots usually have values in the mid range, and higher values mean steeper edges.

For the control of the settings, access to the raw data is essential. The defined limit values should also be printed later with the measurement results so that one is able to repeat the measurements at a later time or allow the re-evaluation of them on a different machine.

It is advisable to already have all measurement results available immediately after the scanning of the plate. Sometimes it may be helpful to have all well images recorded for repeated evaluations in the phase of finding the best settings for the evaluations of an experiment series of plates. This also is of advantage when a new evaluation of the identical wells in a plate is necessary, as with double stains, where IFN and IL are labeled at the same time, and red, blue, and violet spots have to be measured in each well. The re-evaluation for the different colors on stored images will be much faster, when no new scanning of the specimen is necessary.

This will be even of more importance when fluorescence labels for the different cytokines are used because fluorescence markers are extremely sensitive to exposure time, especially with ultraviolet light. In this case, a first run to record the images will allow multiple evaluation runs for adapting the settings and measurement variables and finally perform all evaluation steps in the different channels.

2.6. Specimens

Suitable specimens are original 96-well microtiter plates, filtration plates with nitrocellulose membranes, or nitrocellulose membranes transferred to sticky foils. Some users prefer to remove the membranes from the plates and stick them on a plastic foil before the measurement to avoid reflections from the well side-walls during image acquisition. In this case, the specimen should be mounted very flush with the plate of the stage (best applied with spray adhesive and a roller squeeze).

Special motor stage plates with markers for the driving ranges are necessary to use with peeled-off nitrocellulose membranes; for the use with complete nitrocellulose plates, suitable specimen holders needs to be available as well.

3. A Short Introduction to the KS ELISPOT Software

KS ELISPOT compact is a system with a complete solution for precise and automatic evaluation of ELISPOT assays. It has been developed especially for use in routine laboratories with regard to a high-speed throughput of specimens. Using the KS ELISPOT system (**Fig. 7**) with a light microscope with incident illumination, motor stage, and automatic focusing unit, the aim is high resolution for highest measurement precision and accuracy. The images of the cytokine spots are taken in both cases with a true color camera.

After the digitization, the KS ELISPOT software proceeds with automatic data processing. For routine usage, easy handling is essential. Special importance was attached to the design of the user interface to reduce the user elements to a minimum. Finally, only four buttons are used (*see* **Fig. 2**), which encourages minimal training times to learn the system. Special types of specimen holders guarantee the usage of either whole plates or removed nitrocellulose membranes on a sticky foil. Finally, all data may be transferred directly into a spreadsheet program for further evaluation or graphical display.

First, the region for evaluation in the microtiter plate field on the screen of the system must be marked with the cursor. Next the stage position has to be synchronized with the program coordinates. This will be performed with the **Define Region** button. The system is now ready for the evaluation of a plate.

Using the **Check Well** function, the system settings can be tested for their functioning before the evaluation of a complete plate . Using the Check Well function is always a prerequisite to use the **Teach** algorithm to adapt the spot recognition or to define a new configuration setting for multiple evaluation patterns.

The test well can be selected and addressed by a single right mouse click in the plate overview. The stage will move to the center position of the selected

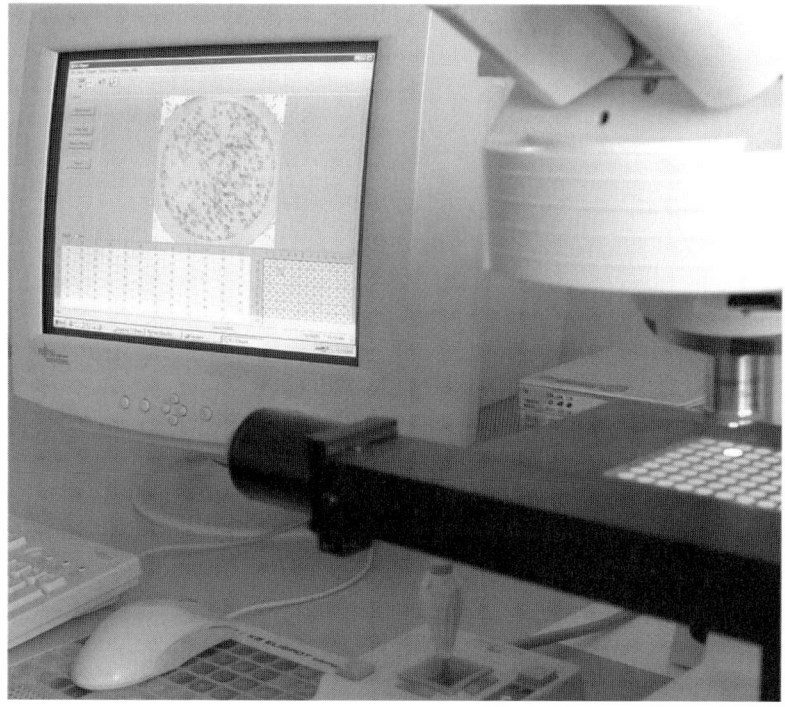

Fig. 7. KS ELISPOT reader.

well and perform a meander scan. From all images, one complete well image will be generated, displayed in a reduced resolution, and evaluated in full resolution. The evaluation result will be displayed in the overlay plane of the image. On the resulting image a check of the limit settings with the **Teach** algorithm is possible.

The **Measure Region** function will start the evaluation of the complete plate. The stage will move to the center of the first well position and initiate the autofocus. Then the well position will be scanned in a meander mode. From all fields, a complete well image will be generated and displayed in a reduced mode in the image field of the user interface.

After a short time, the result will be displayed in the overlay (spot indication and spot diameter). The stage will now move to the next well position and the sequence is repeated until the last well is evaluated. All rejected positions will be skipped. With each stage movement, an estimation is performed for the duration of the complete evaluation procedure and shown in a separate window. After the evaluation of the last well position, the stage will automatically return to the start position.

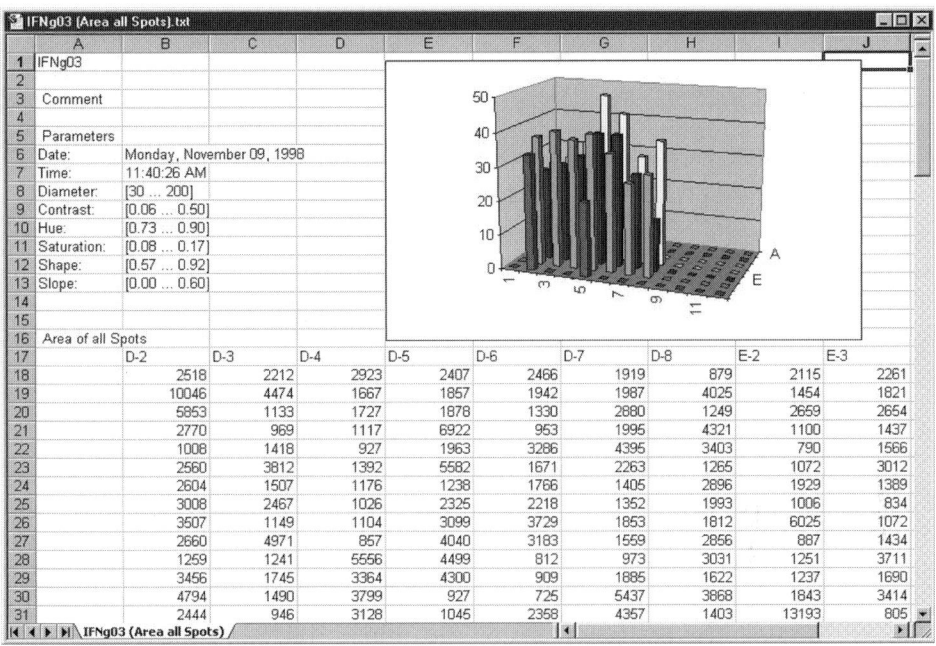

Fig. 8. Example for exported raw data into an Excel spreadsheet with a graphic presentation of the spot size distribution.

A mouse click on the **Report** button will open a separate result window. These results may be printed or saved in different data files. In case single well images are needed in higher resolution, all well images may be kept temporarily as ".JPG" files with display resolution. As an alternative the images may be stored with full resolution.

3.1. Results

Number, size, and intensity of each single spot can be printed via the results page, or stored in data files for further statistical evaluations or graphic displays (*see* **Fig. 8**). Images of single wells can be stored for documentation purposes. Apart from title, date, and spot size limits, all single spot values are stored in well number order. For each well the number of spots, the areas and the intensities are available. In addition to the measurement data all adjustment parameters will be stored.

As an alternative to the internal data format, a result protocol may be automatically generated in Microsoft Word. This protocol exists of a title page with the general experiment description, the results of spot counting and area measurements, as well as the parameter settings for the spot definition. Furthermore, a color overview image of all evaluated wells in a plate is created and additionally an overview image showing all recognized spots in each measured well.

4. Why Is Resolution so Important?

Usually a standard camera in combination with a framegrabber will result in a single image size of 750×560 pixel. KS ELISPOT is collecting several images per well that are combined to one common well image for the evaluation.

A digital integrating camera like the AxioCam MRc has a resolution of 1300×1030 pixels and offers excellent options for the correction of colors and background directly in the camera itself. Using this camera resolution, fewer images per well are necessary for identical overall resolution. More than three standard camera images are necessary for the same resolution as one single Axiocam MRc image.

All reader adjustments end in a certain possible minimum as well as maximum magnification for the use of the system. To recognize a difference in edge intensities, a minimum diameter of the spots is necessary with regard to the pixel resolution. Using just one camera image per well will result in insufficient resolution. Evaluating a series of single images with higher magnification will cause problems with those spots that cut the image edges as well as with those spots located close to the well edge. Therefore, KS ELISPOT first generates one high-resolution image of a complete well from multiple camera images. This image will then be evaluated with correct pixel resolution, without any edge problems, using a circular measurement frame.

The decisive advantage of the Zeiss KS ELISPOT system in comparison with other evaluation methods is the direct increase of the value of scientific data. Results are clearly more reproducible than from manual evaluations. This was shown in a recent investigation at the Sloan Kettering Memorial Hospital in New York (S. Janetzki et al., submitted), comparing the results from different scientist among each other, counting manually and using an ELISPOT reader. Finally, using a reader, in addition to the counting, area and intensity values of each single spot are available.

Compared with other systems, another advantage of the KS ELISPOT system is its resolution. The images are taken with a Zeiss microscope and, as a minimum, 12 images are collected for each well. Collecting only one single image per well will result in a resolution that is too low to include all spots in a specimen (*see* **Fig. 9**). In experiments with IFN-γ, the spot size be as small as 25 μ diameter, and in TNF-α, applications, 15 μ. A 750 ± 560 pixel image of a well can only resolve a pixel size of approx 12 μ. This is absolutely not sufficient to measure most of the smaller spots, especially in the case of TNF-α. This is only guaranteed using a precise microscope optic with a good resolu-

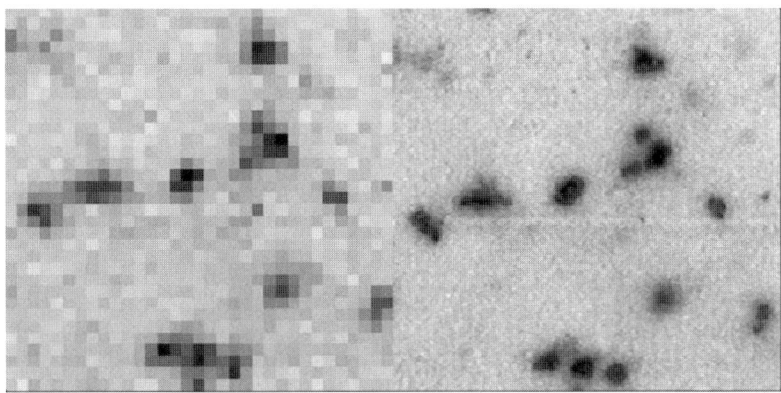

Fig. 9. Resolution comparison in an example of a TNF-α ELISPOT specimen using a standard camera. Left: resolution of one image per well. Right: resolution of the identical area with 12 images per well.

tion, resulting in pixel sizes between 2.0 and 2.5 μ, giving at least a 6 pixel diameter even of the smallest spots *(2)*.

In a recently performed comparison of different readers at the University Clinics in Würzburg (S. Schaed et al., unpublished data), using 1, 4, or 12 images per well, it was shown that one image per well resolution cannot produce accurate results for spots smaller than 50 μ diameter. For experiments with IFN-γ at least four images with a standard camera are necessary to get accurate and reproducible results from spots with 30 μ diameter. With 12 images per well, all possible spot sizes are always recognized correctly, even below a diameter of 20 μ. Using a high-resolution digital camera like the Axiocam MRc, the number of images per well that are needed to obtain accurate results for the smaller spots can be reduced to one from four images per well.

5. Future Developments in the Field of ELISPOT Evaluations

5.1. Measurement of Double Cytokine Production by Standard Staining Methods

To detect more than one cytokine produced by a cell at the same time, different approaches have been tried. Currently two different staining methods are used, as mentioned earlier. The horseradish peroxidase method will produce brownish spots, and the alkaline phosphatase method blue ones. If a cell is producing both cytokines, the result will be violet spots.

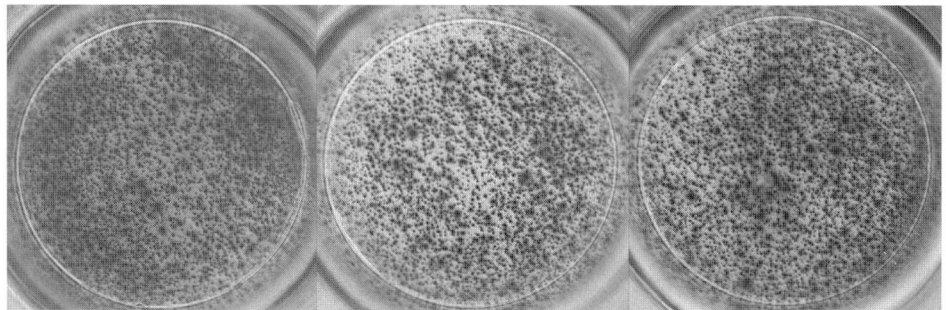

Fig. 10. ELISPOT specimen with similar number of spots per well. Some wells are labeled with a red marker (left), others with a blue marker (middle), and some with both markers resulting in violet spots (right). Result of counts per well with teaching on each color and cross check: Setting 1: color 88–97 (red), setting 2: color 57–62 (blue), setting 3: color 82–95 (violet). *See* **Color Plate 3** following page 50.

Well	Setting	Count	Setting	Count
A2 (red)	1	597	2	0
A4 (blue)	2	672	1	0
A6 (violet)	3	660	1	462
A6 (violet)	3	660	2	0

In the test setups parameters could be defined to clearly separate the blue from the red spots, and the blue from the violet spots. But there was always an overlap of more than 60% between the red and the violet spots. This overlap is too big to accept any result as a clear separation between cells only producing one cytokine from those producing both cytokines.

These tests have been performed on special specimen prepared by MabTech to test the usability of standard staining procedures for the detection of multiple labeled cells (*see* **Fig. 10**). Without better new staining procedures these methods will not result in good and valid data, as long as no clear separation between the colors can be guaranteed.

5.2. Measurement of Double Cytokine Production by Fluorescence Methods

To measure fluorescent specimen an appropriate filter is necessary in the microscope, and the camera must be capable to integrate signals. Using a Sony DXC 950 or a Hitachi HV-C20A the shutter time must be set to an adequate integration time. For the trigger signal, a special cable may be necessary

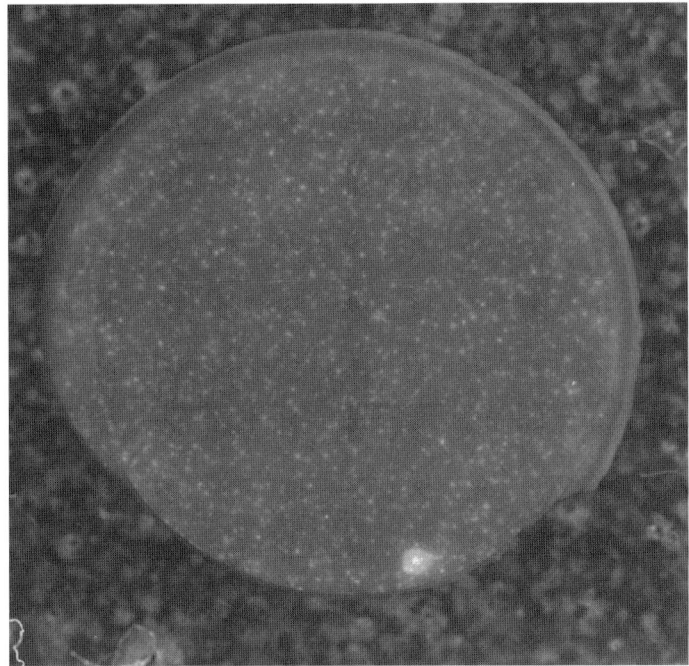

Fig. 11. ELISPOT specimen labeled with fluorescence markers for IFN-γ (fluorescein isothiocyanate, green spots) and IL-5 (rhodamine, red spots). Spots from cells expressing both cytokines appear yellow. Count results: red spots, 229; green spots, 291; yellow spots, 64. *See* **Color Plate 4** following page 50.

between camera and frame-grabber. The integration time for these cameras is only adjustable at the cameras themselves and cannot be controlled by the evaluation software. This solution therefore is not really user-friendly.

The AxioCam MRc as an alternative is an integrating digital camera that can be adjusted to an optimum fluorescence exposure time directly from the evaluation software. Various settings for different experiments may be saved in their own configuration files, and easily recalled when necessary. The handling of this system is much easier and can easily be used in a routine environment.

Fluorescent signals usually result in bright spots on dark background (*see* **Fig. 11**). This is different to the usually expected dark spots on a bright background from standard specimen. A special functionality is therefore necessary to enable the ELISPOT software to perform the measurement. Fluorescence images in addition are not as bright as standard images and require a special illumination for the different wavelengths to produce the requested signals.

But once the system is configured in the correct way, the separation between the signals is much easier and more accurate as with the mixed conventional staining method.

IFN-γ will be labeled with a green fluorescence dye (fluorescein isothio-cyanate), and IL-5 with a red dye (rhodamine). Those spots from cells producing both cytokines will appear yellow. These colors can be separated much easier than the violet between red and blue resulting with the conventional stainings. The method is currently under development at the European Hospital Georges Pompidou in Paris *(11)*, but the first results presented at the first French ELISPOT workshop in September 2003 are more than promising.

References

1. Malkusch, W. (1999) Impfstoffe gegen Tumorzellen gesucht. *Bioforum* **22**, 274–276.
2. Malkusch, W. (2000) A control for vaccines against tumour cells. *Eur. Hospital* **9**, 16.
3. Meidenbauer, N., Harris, D., Spitler, L., and Whiteside, T. (2000) Generation of PSA-reactive effector cells after vaccination with a PSA-based vaccine in patients with prostate cancer. *Prostate* **43**, 88–100.
4. Schaed, S., Klimek, V., Panageas, K., Musselli, C., Butterworth, L., Hwu, W. J., et al. (2002) T-cell responses against Tyrosinase 368–376 (370D) peptide in HLA*A0201+ melanoma patients: randomized trial comparing incomplete Freund's adjuvant, granulocyte macrophage colony-stimulating factor, and QS-21 as immunological adjuvants. *Clin. Cancer Res.* **8**, 967–972.
5. Roberts, S. S. (1988) Cytokine research proliferates despite growing pains. *Science* **5394** V282, 1727–1736.
6. Czerkinsky, C., Andersson, G., Ekre, H-P., Nilsson, L-A., Klareskog, L., and Ouchterlony, Ö. (1988) Reverse ELISPOT assay for clonal analysis of cytokine production. I. Enumeration of gamma-interferon-secreting cells. *J. Immunol. Methods* **110**, 29–36
7. Klinman, D. M., and Nutman, T.B. (1994) ELISPOT assay to detect cytokine-secreting murine and human cells, in: *Current Protocols in Immunology*, Suppl. 10, Unit **6.19**, 1–8.
8. Herr, W., Schneider, J., Lohse, A. W., Meyer zum Büschenfelde, K. H., and Wölfel, T. (1996) Detection and quantification of blood-derived CD8+ T-lymphocytes secreting tumor necrosis factor α in response to HLA-A2.1 binding melanoma and viral peptide antigens. *J. Immunol. Methods* **191**, 131–142.
9. Janetzki, S., Palla, D., Rosenhauer, V., Lochs, H., Lewis, J. J., and Srivastava, P. K. (2000) Immunization of cancer patients with autologous cancer-derived heat shock protein gp96 preparations: a pilot study. *Int. J. Cancer* **88**, 232–238.
10. Malkusch, W. (1999) KS ELISPOT: A new approach to control the efficiency of therapies. *Biomedical Products* **7**, 90–92.
11. Gazagne, A., Claret, E., Wijdenes, J., Yssel, H., Bousquet, F., Levy, E., et al. (2004) A Fluorospot assay to detect single T-lymphocytes simultaneously producing multiple cytokines. *J. Immunol. Methods* **283**, 91–98.

II

ELISPOT ASSAY IN ANIMAL MODELS

9

Improving the Sensitivity of the ELISPOT Analyses of Antigen-Specific Cellular Immune Responses in Rhesus Macaques

K. Jagannadha Sastry, Pramod N. Nehete, and Bharti Nehete

Summary

The close similarities in hematopoietic and immune systems of humans and rhesus macaques (*Macaca mulatta*) make them the desired nonhuman primate animal model for developing vaccines against infectious diseases relevant to humans. The best example is the simian immunodeficiency virus infection in macaques as a model for AIDS, resulting from infection of humans by HIV-1. Vaccine efficacy against viruses depends on priming cell-mediated immunity by the use of sensitive assays that can accurately detect even small levels of antigen-specific T-cell responses that may otherwise be missed easily. With this in mind, we developed the dendritic cell enzyme-linked immunospot (DC-ELISPOT) protocol by incorporating antigen-pulsed DCs to stimulate lymphocytes, as opposed to the conventional ELISPOT assay, which cultures mixtures of various low-level populations of indigent antigen-presenting cells and responding lymphocytes with antigens. In rhesus macaques immunized with an HIV envelope peptide cocktail vaccine, the DC-ELISPOT protocol enabled more accurate enumeration of the cellular immune responses, as antigen-specific inteferon-γ-producing cells, with up to 18-fold increase in detection sensitivity compared with conventional ELISPOT and elimination of false negative results. The increased sensitivity of DC-ELISOT protocol is further validated in tests determining recall antigen-specific responses in human volunteers after tuberculin skin testing.

Key Words: Dendritic cells; cytokine; rhesus macaques; nonhuman primates; cell-mediated immunity; HIV-1; SIV; AIDS; peptides.

1. Introduction

The primate system, consisting of rhesus macaques (*Macaca mulatta*), is a well-established and accepted model for developing vaccines against several infectious diseases that are relevant to humans because of the similarities in the organizations of the hematopoietic and immune systems. This is particularly true for studies related to acquired immunodeficiency syndrome (AIDS),

From: *Methods in Molecular Biology, vol. 302: Handbook of ELISPOT: Methods and Protocols*
Edited by: A. E. Kalyuzhny © Humana Press Inc., Totowa, NJ

which results from infection by the human immunodeficiency virus type 1 (HIV-1), because infection by the simian counter part of this virus, SIV, in rhesus macaques results in AIDS-like disease and thus is the best-studied model for determining immunogenicity and efficacy of HIV vaccine candidates. Priming cell-mediated immunity targeting T-cell responses against viral antigens is essential for the development of successful vaccine candidates. An accurate and sensitive methodology for functional analysis of antigen-specific T-cells is important for determining the strength and breadth of cell-mediated immunity induced by candidate vaccines. A key functional aspect of T-cells is the secretion of cytokines upon encounter and proper activation by foreign antigens in the context of self-major histocompatibility complex (MHC) molecules on antigen-presenting cells (APCs). The enzyme-linked immunospot (ELISPOT) assay is a sensitive technique for the detection of antigen-specific cytokine secretion at a single cell level (*see* **Note 1**; **refs.** *1–6*). The conventional ELISPOT assay, as currently practiced for enumerating antigen-specific T-cells in peripheral blood or tissue biopsies, relies on resident macrophages/monocytes, among others, as APCs. However, it is well established that dendritic cells (DCs) are the most potent APCs and are usually referred to as professional APCs. This is because DCs display a high density of MHC class I and II molecules, co-stimulatory signals through the B7 family members, and adhesion molecules that facilitate contact with lymphocytes *(7).* Thus, DCs can present antigens to both CD4+ and CD8+ T-lymphocytes, and they may be the only APCs capable of stimulating naïve T-cells *(8,9).* Monocytes in the mononuclear cells can be stimulated in vitro to differentiate into functional DCs, thereby offering a good source of these professional APCs for functional immune assays. We took advantage of this extraordinary strength of DCs to enhance the detection sensitivity of the functional antigen-specific T-cells for the ELISPOT analysis and developed a modified protocol referred here as DC-ELISPOT assay (**Fig. 1** and **ref.** *10*). We demonstrated the superiority of DC-ELISPOT methodology over the conventional ELISPOT assay by analyzing cellular immune responses primed by a peptide-cocktail HIV vaccine in rhesus macaques, where the DC-ELISPOT assay enabled elimination of false-negative results from the ELISPOT assay (**Fig. 2**). Thus, the DC-ELISPOT assay will be preferable over the ELISPOT to realize the true strength and breadth of vaccine-induced antigen-specific T-cell responses in primates *(10,11).* The DC-ELISPOT protocol also was used for analyzing influenza-specific interferon (IFN)-γ-producing cells in human volunteers receiving the flu shots and was found to yield more sensitive readout of the immune response compared with the conventional ELISPOT assay (*see* **Note 2**; **ref.** *10*).

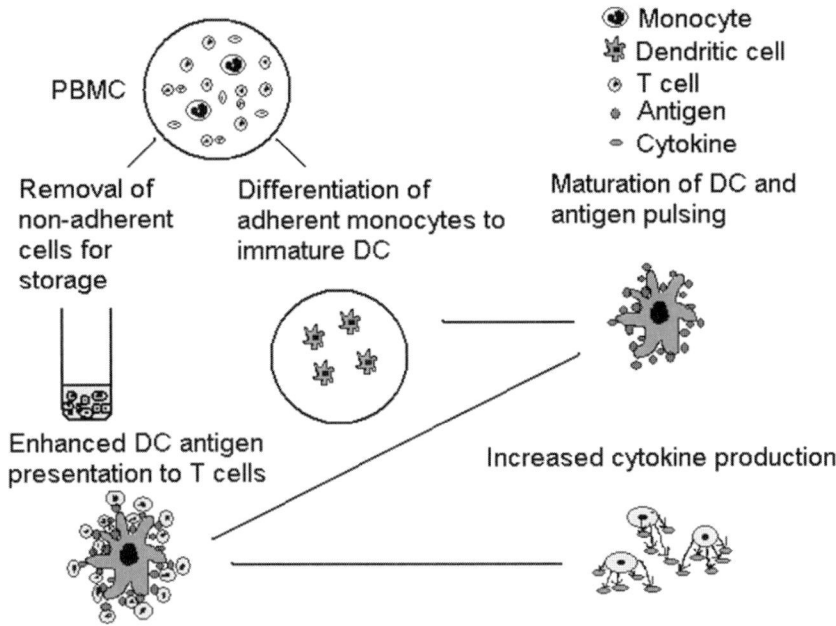

Fig. 1. Schematic representation of DC-ELISPOT assay protocol.

1.1. Principle of the ELISPOT Assay

The ELISPOT technique requires the quantitative sandwich enzyme-linked immunosorbent assay (ELISA) methodology for the quantitation of individual cells secreting specific cytokines. The ELISPOT assay uses two high affinity cytokine specific antibodies directed against different epitopes on the same cytokine molecule. Monoclonal or polyclonal antibody that is specific for the cytokine of interest is precoated onto an ELISPOT plate, that is, a 96-well microtiter plate with polyvinylidene difluoride-backed membrane bottom. Mixture of cells containing both APCs and responding lymphocytes are stimulated by incubating with the antigen of interest in a humidified CO_2 incubator at 37°C for 36–48 h. During this incubation period, the immobilized antibody in the immediate vicinity of the antigen-responding cells binds to the secreted cytokine. After washing away the cells, biotinylated polyclonal antibody that is specific to the second epitope of the cytokine will be added followed by treatment with avidin-HRP conjugate or alkaline phosphatase conjugated to streptavidin and chromogen AEC substrate or 5-5-bromo-4-chloro-3-indolyl phosphate/Nitroblue tetrazolium substrate to develop red or blue color spots, respectively, representing

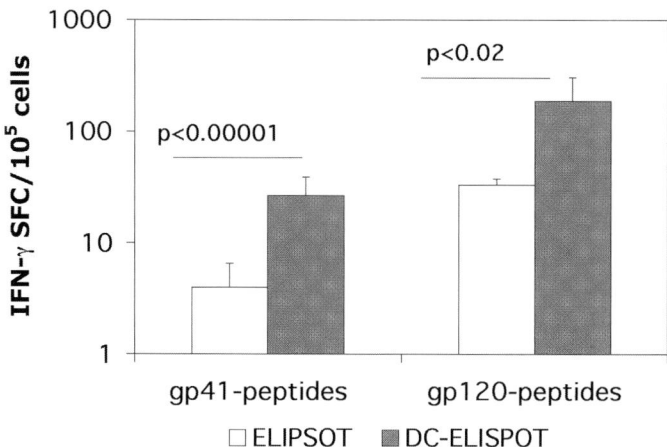

Fig. 2. DCs reveal enhanced antigen-specific immune responses by lymphocytes in rhesus macaques immunized with an HIV envelope peptide cocktail vaccine (mixture of peptides from gp41 and gp120). The results shown are the number of IFN-γ SFCs in freshly isolated PBMCs stimulated in vitro with gp41-peptides or gp120-peptides (corresponding to the vaccine cocktail) in the conventional ELISPOT assay or cocultures of peripheral blood lymphocytes and autologous monocyte-derived DC pulsed with the peptides in the DC-ELISPOT assay. Each bar represents an average number of IFN-γ SFCs in six different vaccinated rhesus macaques. The number of IFN-γ SFCs obtained using the unrelated negative control peptide for stimulation were subtracted from that of vaccine-peptide stimulation. The *p* value shown in each panel represents the significance of difference between the ELISPOT and DC-ELISPOT protocols for the average number of IFN-γ SFCs observed in response to activation by the vaccine peptides (two-tailed student *t*-test).

an individual cytokine-secreting cell. The spots can be counted with an automated ELISPOT reader system (KS Elispot, from Carl Zeiss, Inc. Thornwood, NY; *see* **Note 3**) or manually using a stereomicroscope. The results are generally represented as cytokine spot-forming cells (SFCs) per the total number of input cells in each of the wells of the ELISPOT plate.

1.2. Conventional ELISPOT Assay

The conventional ELISPOT assay uses peripheral blood mononuclear cells (PBMCs) or cell suspensions prepared from tumors or tissue biopsies as a source for both APCs as well as cytokine secreting cells responding to antigen-specific stimulation. Thus, in this procedure, a heterogeneous mixture of cells containing both APC and responding lymphocytes is stimulated by incubating with the antigen of interest in the ELISPOT plates *(12–15)* (*see* **Tip 1**).

1.3. DC-ELISPOT Assay

The DC-ELISPOT assay differs from the conventional ELISPOT assay in that it contains an additional step where DCs are prepared from the cell preparations (PBMCs or tissue biopsy) by cytokine-mediated differentiation and maturation of plastic-adherent monocyte population and incubated with antigens of interest for enabling antigen uptake and presentation (**Fig. 2**). Graded amounts of antigen-loaded mature DCs are co-cultured with responding lymphocytes in the ELISPOT plates. The DC-ELISPOT methodology also allows for specifically testing the phenotype of the responding lymphocyte population by preselecting the CD4$^+$ and CD8$^+$ T-cells (*see* **Note 4** and **Tip 2**).

2. Materials

1. Histopaque 1077 (cat. no. 1077-1, Sigma Chemical Co., St. Louis, MO).
2. Hanks balanced salt solution (HBSS; cat. no. H-1641, Sigma).
3. IL-4 (cat. no. 204-IL-025, R & D Systems).
4. Granulocyte macrophage-colony stimulating factor; (GM-CSF Leukine, Immunex).
5. TNF-α (cat. no. 210-TA, R & D Systems).
6. MultiScreen plates (MAIP S 45-10; Millipore) (*see* **Note 5**).
7. 1 mg Mouse anti-Human IFN-γ (cat. no. NON0007, BioSource International).
8. Rabbit anti-Human IFN-γ Biotin Conjugate (cat no. NON0007, BioSource International).
9. CD4$^+$ cell isolation kit (prod. no. 113.03 & 113.04, Dynal).
10. CD8$^+$ cell isolation kit (prod. no. 113.05 & 113.06, Dynal).
11. Avidin peroxidase.
12. 3-amino-9-ethyl-carbazole (AEC).
13. Tween-20.
14. *N,N*-Dimethyl-Formamide (DMF).
15. Peroxidase solution B (cat. no. 5065-00, KPL).
16. Multiscreen plate clear sealing tape (cat. no. MATA HCL00, Millipore).
17. Eli-Puncher Kit (ZellNet Consulting, Inc.; www.zellet.com/services).
18. RPMI complete culture medium.

The materials for ELISPOT assay also are available from other vendors (*see* **Note 6**).

3. Methods

3.1. Isolation of PBMCs From Macaque Whole Blood

1. Dilute heparinized blood with HBSS at 1:2 or 1:3 ratios in 50-mL centrifuge tubes (*see* **Note 7**).
2. Slowly layer the Ficoll (Histopaque-1077) underneath the diluted blood by placing the tip of the pipet containing Ficoll at the bottom of the tube and releasing the solution slowly. Alternately, layer the diluted blood onto the Ficoll carefully.

Centrifuge the tubes at $400g$ for 30 min at 18° to 22°C using swinging bucket rotor with no brakes (*see* **Note 8**).

3. The tubes should have three distinct layers of solutions. Aspirate the upper layer containing plasma and platelets either by using vacuum or pipet. Transfer the white/opaque interface of mononuclear cells to another 50-mL centrifuge tube. Just below the layer of mononuclear cells (PBMCs) is the Ficoll layer that contains granulocytes. Collection of Ficoll layer along with the PBMC will increase the granulocytes contamination.

4. Wash the PBMCs three times by adding excess amount of HBSS and centrifuging at $400g$ for 10 min.

5. Re-suspend the PBMC in 10 mL of RPMI-1640 containing 10% fetal bovine serum (FBS; complete medium), and determine the viable cell count by the standard trypan blue dye–exclusion method.

3.2. Preparing Monocyte-Derived DCs From PBMCs

1. Add the PBMCs at 10×10^6 cells/3 mL of complete RPMI in each well of a 6-well plate.

2. Incubate the cells at 37°C for 2 h in the CO_2 incubator. The monocytes will attach to the bottom plastic surface during this period.

3. Swirl the cells after 1 h during the incubation.

4. Gently remove the nonadherent cell population and collect in another tube. Spin, wash, and freeze in aliquots for future use as responder cells (these cells can be used either directly or after further processing to obtain pure populations of CD4$^+$ and CD8$^+$ cells; *see* **Note 9**).

5. Wash the adherent cell monolayer of monocytes three times to remove the any residual nonadherent cells.

6. Add medium containing GM-CSF and IL-4 (each at 1000 U/mL) to the adherent monolayer of cells (monocytes) and incubate for 6 d at 37°C in the CO_2 incubator (*see* **Note 10**). Add GM-CSF and IL-4 (each at 1000 U/mL) every alternate day. The adherent monocytes will differentiate into immature DC with distinct morphology.

7. On the sixth day, gently scrape the loosely attached immature DC, wash, and re-suspend at 2×10^4 cells/ml in complete media containing GM-CSF and IL-4 (each at 1000 U/mL). Add TNF-α (1000 units/mL) for inducing maturation and incubate overnight with antigens and control reagents at required concentrations in 20 µL of volume (*see* **Note 11**; **refs. *10,16,17***).

3.3. Separation of CD4$^+$ and CD8$^+$ T-Cells From PBMCs

1. Thaw previously frozen non-adherent cells (responder cells collected in **step 4** in **Subheading 4.2.**). Wash once and re-suspend in cold phosphate-buffered saline (PBS) containing 2% FBS. Use 5×10^6 cells/ml for the isolation of CD4$^+$ cells, and 10×10^6 cells/mL for CD8$^+$ cell. Keep the cells on ice.

2. From the CD4$^+$ cell Isolation Kit (Dynal, cat. no. 113.04), take out 72 µL of beads for 5×10^6 cells/mL or from CD8$^+$ cell Isolation Kit (Dynal; cat. no. 113.06), take

72 µL of beads for 10×10^6 cells/mL. Wash the beads three times with 1–2 mL of PBS containing 2% FBS using a Magnetic separation device (Dynal MPC) to remove sodium azide. Resuspend in a 1-mL volume in PBS containing 2% FBS.

3. Add the re-suspended beads (72 µL/mL) to each tube.
4. Incubate at 2–8°C for 20 min on rotator with low speed.
5. The CD4⁺ and CD8⁺ cells will attach to the respective beads. Collect the cells on the beads with magnetic separation and wash three times with PBS containing 2% FBS.
6. Add immediately 100 µL of RPMI 1640 containing 1% FBS per 10×10^6 cells.
7. Add 10 µL of DETACH a BEAD solution per 100 µL of cell suspension.
8. Incubate for 45 to 60 min at room temperature with gentle mixing.
9. The positively selected cells will detach from the beads, and can be collected by removing the magnetic beads with magnetic separation and washing the detached cells three times with PBS containing 2% FBS.
10. Remove the DETACH a BEAD solution by centrifuging the cells at 300–500g.
11. Re-suspend the cells to the concentration of 1×10^6 cells/mL and add 1×10^5 cells/ well in a 100-µL volume for co-culturing with the antigen-loaded DCs in the ELISPOT plates.

3.4. DC-ELISPOT Assay

3.4.1. Day 1: Preparation of the ELISPOT Plates and APCs

3.4.1.1. COATING THE ELISPOT PLATE WITH THE PRIMARY ANTIBODY TO THE CYTOKINE

1. Wash the plates once with 100 µL/well of 70% alcohol and then three times with sterile water. Invert the plate and gently tap on paper towels to remove all the water. Do not let the plate dry until the end of the experiment.
2. Make 0.1 M sodium bicarbonate, pH 9.5, and filter sterilize. Dilute the anti-IFN-γ antibody to 1 µg/mL concentration in this buffer and coat individual wells of the ELISPOT plate with 100 µL of the antibody (*see* **Note 12**).
3. Incubate the plate overnight at 4°C. (Incubation at 4°C is recommended but can be incubated minimum 4 h at room temperature or 2 h at 37°C.)

3.4.1.2. PREPARATION OF ANTIGEN-LOADED DCs

1. The day 6 cultures of monocyte-derived immature DCs (as in **step 7** in **Subheading 3.2.**) are washed and resuspended at 2×10^4 cells/mL in complete media containing GM-CSF and IL-4 (each at 1000 U/mL). Aliquots of 100 µL cells in individual wells of 96-well U-bottom plates are incubated with TNF-α (for maturation) and antigens and control reagents, at required concentrations (in 20 µL of volume), in complete medium for 18–24 h at 37°C in CO_2 incubator.

Example of a typical 96-well plate set up:

Wells:	1, 2, and 3	4, 5, and 6	7, 8, and 9	10, 11, and 12
Row A	Cells alone	Con A	negative control	test antigen

3.4.2. Day 2: Antigen-Specific Stimulation Of Lymphocytes

3.4.2.1. PREPARATION OF ELISPOT PLATES FOR COCULTURES

1. Remove anti-IFN-γ antibody from ELISPOT plates by gently decanting the plates without using vacuum.
2. Wash off unbound antibody with 200 μL of PBS/well; incubate for 5 min; decant the washing buffer, repeat twice.
3. Block the wells with 200 μL of complete medium for a minimum of 2 h at room temperature or at least 1 h at 37°C (*see* **Note 13**).

3.4.2.2. PREPARATION OF RESPONDER T-CELLS FOR COCULTURES

1. Responder cells (nonadherent T-cells collected from PBMCs as in **step 5** in **Subheading 3.2.**) can be used directly or after purifying subpopulation of CD4+ and CD8+ T-cells (as described in **Subheading 3.3.** using magnetic beads; *see* **Note 14**).
2. Wash cells in complete medium, count, and re-suspend at a final concentration of 1 to 2×10^6 cells/mL in complete medium (*see* **Note 15**).

3.4.2.3. COCULTURING STIMULATORS AND RESPONDERS

1. Add 100 μL of responder cells/well (to get 1×10^5 to 2×10^5 cells/well) in plate that already has the stimulated DC (as in **Subheading 4.4.1.2.**). Now these cocultures from this plate are ready to be transferred to the ELISPOT plate.
2. Decant blocking medium from the ELISPOT plate (continuation of **step 3** in **Subheading 4.4.2.1.**).
3. Gently transfer the cocultures from the 96-well U-bottom plates to ELISPOT plate.
4. Incubate for 36 to 48 h at 37°C in 5% CO_2 (*see* **Note 16** and **Tip 3**).

3.4.3. Day 3: Staining of Enumeration of Antigen-Specific Cytokine-Producing Cells

3.4.3.1. PREPARATION OF ELISPOT PLATES FOR STAINING AND SPOT DEVELOPMENT

1. Remove the cells from the wells and wash the plate six times with PBS containing 0.05% Tween-20.
2. Add 100 μL/well of biotinylated anti-IFN-γ antibody in PBS (50 to 125 ng/well).
3. Incubate for 3 h at room temperature and wash the plate six times with PBS containing 0.05% Tween-20.

3.4.3.2. DEVELOPMENT OF ELISPOT PLATES

1. Add 100 μL/well of 2 μg/mL of Avidin-HRP in PBS (make fresh solution).
2. Incubate for 1.5 to 2 h at room temperature.
3. While the plates are in the incubator, prepare the AEC substrate by dissolving 3 mg of AEC in 200 μL of DMF buffer: 9.8 mL of 0.1 *M* sodium acetate (prepare 0.1 *M* sodium acetate, pH 5.0, using acetic acid to adjust pH), 0.2 mL of AEC in DMF, 40 μL of 3% H_2O_2. Store the solution in a dark place until use.

3.4.3.3. COLOR DEVELOPMENT

1. Wash the plate six times with PBS Tween-20.
2. Add 100 µL of AEC substrate and store in the dark for 3 h to overnight.
3. Wash the plate for four to five times using wash bottle or under steady stream of RO water or by a multichannel device.
4. Remove the water by taping the plate to dry and let it air dry in inverted position until it is completely dry (*see* **Tips 4** and **5**).

3.4.4. Membrane Removal

The unique design of the MultiScreen plate allows for performing the ELISPOT assay in the 96-well filter plate because each individual filter disc can be removed for subsequent analysis. The removal of the membranes from the bottom of the MultiScreen plate is important because it avoids shading problems during the automated reading process caused by reflection of light from the walls of the individual wells in the plate and assures a straight well surface. Generally, in the plate the well surfaces are concave that can be problematic for focusing during high-resolution evaluation. On a more practical side, the membranes can be stored easily compared to the plates (*see* **Tip 6**).

3.4.4.1. MEMBRANE REMOVE PROTOCOL

1. Attach the sealing tape to the bottom of the plate and ensure that the tape sticks to the edge of the wells by applying pressure to the tape with a metal devise such as scissors, forceps, etc.
2. Hold the plate with your left hand.
3. Insert the Molecular Devices harvesting filter inside the first well and, while pushing out the membrane, use another finger to stop the filter movement by holding it behind the well you are pushing out.
4. Total time for removing the membranes is approx 2–3 min.
5. Store tape in a plastic sheet protector along with protocol and evaluation results.

3.5. Data Analysis and Reporting

As detailed in **Subheading 3.** (and **Note 6**), there are automatic computer-based enumeration programs available. Once the plates are analyzed either by these computer based methods or manually, the data are reported as SFCs per total input responder cells (*see* **Tip 6**). Alternately, several reports in the literature (analyzing human, mouse or primate responses) converted the data to report as SFCs per 10^6 input cells in which case typically the average number of spots counted for each treatment are extrapolated by multiplying with the appropriate factor (e.g., if the total input cells are 1×10^5, then the multiplication factor is 10 to report the number of SFCs/10^6 cells). There are also variations in the literature reports with regard to the cut-off values for determining positive responses.

In general a minimum of 5 SFCs/10^5 input cells is considered background and the positive value will be at least twice to this background value (i.e., 10 SFCs/10^5 input cells). If the background values exceed 5 SFCs/10^5 input cells, the positive value will be twice to that of back ground or at least 5 SFCs/10^5 input cells above the background value.

4. Notes

1. Although the most common use of the ELISPOT assay has been for the detection of antigen-specific INF-γ-producing cells, this methodology is suitable for any cytokine for which two specific antibodies (for capture and detection) are available. Makitalo et al. *(20)* described the successful application of ELISPOT methodology for the detection and enumeration of lymphocytes responding to antigen-specific stimulation for the production of several cytokines that included IL-2, IL-4, IL-5, IL-6, IL-12, IL-13, and GM-CSF using blood samples from rhesus as well as cynomolgus macaques.

2. Another strategy for enhancing antigen-specific IFN-γ production by lymphocytes from rhesus macaques includes the use of IL-15 in the conventional ELISPOT methodology as described by Calarota et al *(19)*. These authors showed that IL-15 at concentrations ranging between 0.5 and 2.5 ng/mL was able to substantially enhance antigen-specific IFN- γ production by rhesus macaque lymphocytes without unduly increasing the background levels.

3. Alternate source for the automated analysis system is ImmunoSpot Series 1 Analyzer, from Cellular Technology, Ltd. Cleveland, OH.

4. Alternately, T-cells can be enriched from PBMC samples by depletion of HLA-DR- and CD11b-expressing cells by MACS selection (Miltenyi Biotech) as reported by Mehlhop et al. *(17)*.

5. The 96-well plates for the ELISPOT assay can also be obtained from Cellular Technology, Cleveland, OH (cat. no. M200) as described by Calarota et al. *(19),* but according to Kumar et al. *(6)* plates other than the MAIP S 4510 from Millipore (Bedford, MA) will have problems associated with high background or a complete absence of signal.

6. The ELISPOT reagents (capture antibody and detection antibody) and color-development reagents as individual reagents or kits along with precoated plates are also available from R&D System (www.RnDSystem.com), DiaPharma Group (www.diapharma.com), and Mabtech, Cincinnati, OH. The Human IFN-g ELISPOT Kit from BD Biosciences shows cross-reactivity with nonhuman primates (www.bdbiosciences.com).

7. Blood collected in sodium heparin is preferred to ethylenediamine tetraacetic acid according to Pahar et al. *(18)* because fewer cytokine producing cells were observed by these authors from macaque blood samples anticoagulated with ethylenediamine tetraacetic acid .

8. Alternately, Makitalo et al. *(20)* reported that CPT vacutainer tubes containing sodium heparin (from Becton Dickinson) were as efficient as Ficoll for the isolation of PBMCs from macaque blood samples.

9. The preferred composition of the medium for storing the cells in liquid nitrogen is RPMI medium with 20% FBS and 10% dimethyl sulfoxide *(6)*. Alternately, we stored the cells in FBS with 10% dimethyl sulfoxide in liquid nitrogen and recovered high percentage of viable cells for assay at later time points.

10. The concentration of IL-4 can be between 100 and 1000 U/mL for the induction of DC differentiation *(17)*.

11. A number of reagents can be used for inducing maturation of DCs. These include monocyte-conditioned medium at 50% level, CD40-ligand (CD40-L) at a 1:300 dilution, Poly I:C at 50 μg/mL, PGE2 (10^{-6} *M*) along with TNF-α (5 ng/mL), or a cocktail of PGE2 (10^{-6} *M*)/TNF-α (5 ng/mL)/IL-1β (10 ng/ml)/IL-6 (20 ng/mL). Of the various stimuli used for the maturation of macaque DC, the cocktail of PGE2/TNF-α/IL-1β/IL-6 is effective more consistently *(17)*. Immunophenotype of functional DC maturation induced by these stimuli should show stable up-regulation of CD25, CD40, CD80, CD83, CD86, HLA-DR, DC-LAMP (CD208), and DEC-205 (CD205).

12. Based on literature reports, the concentration of IFN-γ antibody for coating the plates varies between 1 and 15 μg/mL *(6,10,19,20)*. However, we have consistently observed 1 μg/mL to be adequate for optimal signal from responding lymphocytes *(10)*.

13. The medium for blocking the wells in the ELISPOT plate can have fetal bovine serum at concentrations between 2 and 10% *(6,19,20)*, but we have always used 10% serum in RPMI medium at room temperature with consistent results *(10)*.

14. The efficiency of cryopreserved cells was found to be equal to that of fresh cells for antigen-specific cytokine production and analysis by the ELISPOT assay *(6,18)*.

15. In general the number of responder in each well as reported by several groups ranged between 0.3–4 × 10^5, but Kumar et al. *(6)* reported that 2 × 10^5 is an optimum cell number because they observed ≤ 1 × 10^5 cells yielded decreased number of positive spots whereas >2 × 10^5 cells gave higher background.

16. The incubation period for the detection of cytokine producing cells in the ELISPOT assay is usually between 24 and 48 h at 37°C. Kumar et al. *(6)* compared these time points and reported that 24 h as optimum because at longer incubation times the intensity of spots was compromised by a tendency for a precipitate to form over the more intense spots making the counting difficult. In our experience, we observed that enumeration of spot forming cells after 36–48 h was without any technical problems *(10)*.

Tips

1. Several important precautions are taken during setting up of DC-ELISPOT plates. For example, care should be taken not to touch and puncture the membrane on the bottom of the ELISPOT plate wells while pipetting cells and reagents.

2. Determine the optimal responding cell concentration by testing different cell numbers (e.g., 10^3 to 10^6 cells per well) in the first experiment. Overloading will not leave sufficient space in the wells of the ELISPOT plate and results will be compromised.

3. To avoid streaks and ambiguous spots, do not disturb the ELISPOT plate during incubation and do not stack the plates in the incubator. Place each ELISPOT plate individually on the shelf to allow an even distribution of heat to each plate and to avoid edge effects.

4. High background can be avoided by optimization of concentration of detection antibody and avidin-HRP and washing steps with PBS-Tween. Monitor the color development carefully. Do not overdevelop, as this will lead to high background. After the completion of experiment, dry the plates at room temperature only. Dry the plate longer if necessary. The speed at which the plate completely dries depends on the relative humidity in the environment. Drying at higher temperature will cause cracking of membrane filters.

5. Store color-developed, dried plates in a sealed plastic bag protected from light to avoid color reduction that can be caused by air and light.

6. When scanning a plate in the ImmunoSpot Analyzer, make sure the filter is completely inserted and aligned into the base.

Acknowledgments

This work was supported in part by funds from National Institute of Allergy and Infectious Diseases AI 42694 and AI 46969. All culture media were produced by the Central Media Lab, and all the synthetic peptides were prepared in the Synthetic Antigen Core Facility, which were both supported by funds from National Institute of Health Grant, CA 16672. Special thanks are also due to Ms. Andrea Cass for assistance with collecting the literature reports cited.

References

1. Helms, T., Boehm, B., Asadd, R., Trezza, R., Lehmann, P., and Tary-Lehmann, M. (2000) Direct visualization of cytokine-producing recall antigen-specific CD4 memory T-cells in healthy individuals and HIV patients. *J. Immunol.* **164**,3723–3732.

2. McCutcheon, M., Wehner, N., Wensky, A., Kushner, M., Doan, S., Hsiao, L., et al. (1997) A sensitive ELISPOT assay to detect low-frequency human T-lymphocytes. *J. Immunol. Methods* **210**,149–166.

3. Fujihashi, K., McGhee, J., Beagley, K., McPherson, D., McPherson, S., Huang, C.-M., et al. (1993) Cytokine-specific ELISPOT assay: single cell analysis of IL-2, IL-4, and IL-6 producing cells. *J. Immunol. Methods* **160**,181–189.

4. Ronnblom, L., Cederblad, B., Sanderberg, K., and Alm, G., (1988) Determination of herpes simplex virus-induced alpha interferon-secreting human blood lymphocytes by a filter immuno-plaque assay. *Scan. J. Immunol.* **2**, 165–170.

5. Czerkinsky, C., Anderson, G., Ekre, H., Nilsson, L., Klareskog, L., and Ouchterlony, O. (1988) Reverse ELISPOT assay for clonal analysis of cytokine production. *J. Immunol. Methods.* **110**, 29–36.

6. Kumar, A., Weiss, W., Tine, J. A., Hoffman, S. L., and Rogers, W. O. (2001) ELISPOT assay for detection of peptide specific Interferon-? secreting cells in rhesus macaques. *J. Immunol. Methods.* **247**, 49–60.

7. Steinmann, R. M. (1991) The dendritic cell system and its role in immunogenicity. *Annual Rev. Immunol.* **9**, 271–296.
8. Carvalho, L. H., Hafalla, J. C. R., and Zavala, F. (2001) ELISPOT assay to measure antigen-specific murine CD8+ T-cell responses. *J. Immunol. Methods.* **252**, 207–218.
9. Romero, P., Cerottini, J. C., and Waanders, G. A. (1998) Novel methods to monitor antigen-specific cytotoxic T-cell responses in cancer immunotherapy. *Mol. Med. Today.* **4**, 305–312.
10. Nehete, P. N., Gambhira, R., Nehete, B. P., and Sastry, K. J. (2003) Dendritic cells enhance detection of antigen-specific cellular immune responses by lymphocytes from rhesus macaques immunized with an HIV envelope peptide cocktail vaccine. *J. Med. Primatol.* **32**, 67–73.
11. Nehete, P. N., Chitta, S., Hossain, M. M., Hill, L, Bernaky, B. J., Baze, W., et al. (2002) Protection against chronic infection and AIDS by HIV envelope peptide-cocktail vaccine in a pathogenic SHIV-rhesus model. *Vaccine* **20**, 813–825.
12. Schmittel, A., Keilholz, U., Bauer, S., Kuhne, U., Stevanovic, S., Thiel, E., et al. (2001) Application of the IFN-? ELISPOT assay to quantify T-cell responses against proteins. *J. Immunol. Methods.* **247**, 17–24.
13. Schmittel, A., Keilholz, U., and Scheibenbogen, C. (1997) Evaluation of the interferon-? ELISPOT-assay for quantification of peptide specific T-lymphocytes from peripheral blood. *J. Immunol. Methods.* **210**, 167–174.
14. van der Maide, P.H., Grosnestain, R.J., de Lable, M.C.D.C., Henney, J., Pala, P., and Siaoul, M. (1995) Enumeration of lymphokine-secreting cells as a quantitative measure for cellular immune responses in rhesus macaques. *J Med Primatol.* **24**, 271–281.
15. Carter, L.L., and Swain, S.L. (1997) Single cell analyses of cytokine production. *Curr. Opin. Immunol.* **9**, 177–182.
16. O'Doherty, U., Ignatius, R., Bhardwaj, N., and Pope, M. (1997) Generation of monocyte-derived dendritic cells from precursors in rhesus macaque blood. *J. Immunol. Methods.* **207**, 185–194.
17. Mehlhop, E., Villamide, L. A., Frank, I., Gettie, A., Santisteban, C., Messmer, D., et al. (2002) Enhanced in vitro stimulation of rhesus macaque dendritic cells for activation of SIV-specific T-cell responses. *J. Immunol. Methods* **260**, 219–234.
18. Pahar, B., Li, J., Rourke, T., Miller, C. J., and McChesney, M. B. (2003) Detection of antigen-specific T-cell interferon gamma expression by ELISPOT and cytokine flowcytometry assays in rhesus macaques. *J. Immunol. Methods* **282**, 103–115.
19. Calarota, S. A., Otero, M., Hermanstayne, K., Lewis, M., Rosati, M., Felber, B. K., et al. (2003) Use of interleukin 15 to enhance interferon-gamma production by antigen-specific stimulated lymphocytes from rhesus macaques. *J. Immunol. Methods* **279**, 55–67.
20. Makitalo, B., Anderson, M. A., Arestrom, I., Karlen, K., Villinger, F., Ansari, A., et al. (2002) ELISpot and ELISA analysis of spontaneous, mitogen-induced and antigen-specific cytokine production in cynomolgus and rhesus macaques. *J. Immunol. Methods* **270**, 85–97.

10

Feline Cytokine ELISPOT

Issues in Assay Development

Sushila K. Nordone, Rosemary Stevens, Alora S. LaVoy, and Gregg A. Dean

Summary

The enzyme-linked immunospot (ELISPOT) assay is a sensitive and relatively simple assay for detecting secreted cellular products such as cytokines and has become an invaluable immunological tool. The ELISPOT has been used extensively in human and murine research but has only recently been used to assess the feline immune system. For researchers studying feline disease or using the cat as a model of human disease, the quantification of cytokine-producing cells by ELISPOT is an invaluable technique for investigations of disease immunopathogenesis and vaccine efficacy. For example, use of the interferon (IFN)-γ ELISPOT to measure the frequency of antigen-specific T-cells during feline immunodeficiency virus (FIV) infection or after immunization with candidate FIV vaccines is of particular interest. This application of the ELISPOT may serve to expand the utility of FIV as a model for human immunodeficiency virus. Broader applications of the ELISPOT should further our understanding of feline diseases and be useful in the rational development of more efficacious vaccines and therapeutic modalities for the enhancement of feline health. This chapter discusses important parameters of ELISPOT design that will enable researchers to develop and analyze the feline-specific assays within their own laboratory.

Key Words: ELISPOT; feline; IFN-γ; IL-2; IL-4; cytokine; T-cell; FIV; infection; immunization; quantitation; frequency; spot-forming cell; peptide.

1. Introduction

The enzyme-linked immunospot (ELISPOT) assay is a rapid, simple, and highly sensitive assay used to quantify the frequency of cells secreting a protein of interest *(1–5)*. Although the assay has been in use for more than a decade, it is a relatively new assay within the field of feline research. To date, the only feline protein quantified by this method is the cytokine interferon (IFN)-γ; how-

From: *Methods in Molecular Biology, vol. 302: Handbook of ELISPOT: Methods and Protocols*
Edited by: A. E. Kalyuzhny © Humana Press Inc., Totowa, NJ

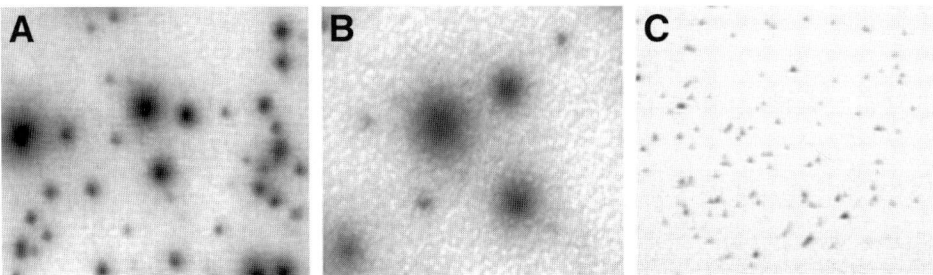

Fig. 1. True IFN-γ SFCs vs artifact. True IFN-γ SFCs from feline PBMCs stimulated with 50 ng/mL PMA and 300 ng/mL ionomycin are shown at 400X magnification (**A**). Close up of center spots shown in **A** (**B**). Artifact from a failed IFN-γ ELISPOT assay is shown at 250X magnification (**C**). Note dark center and diffuse outer ring of spots in (**A**) and (**B**) vs small spots of uniform intensity and elongated protein aggregates observed in (**C**).

ever, the potential uses for this assay are unlimited. Because of the lack of established protocols for individual feline proteins, this chapter is designed to be a guide for investigators seeking to set up the assay within their own laboratories. Drawing from information gained from other species and our own experiences, we cover the critical parameters in ELISPOT development with an emphasis on the reasoning behind the methodology. A detailed list of reagents and buffers will be given; however, because the assay is adapted to measure a variety of secreted proteins, these reagents may need to be tailored to meet individual needs.

The assay in its simplest form includes five steps: (1) coating a polyvinylidene diflouride-backed microtiter plate with a purified capture antibody; (2) blocking of plates to prevent nonspecific adsorption of additional proteins; (3) incubation of cells; (4) addition of a biotinylated detection antibody which binds to a different epitope than the capture antibody; and (5) colorimetric development of protein-antibody complexes (**Fig. 1**). Resulting spots are permanently attached to the membrane and represent the footprints of secreting cells. Quantitation of spot-forming cells (SFCs) can then be determined using an inverted microscope or electronically with one of several commercially available spot counters.

1.1. Initial Considerations in Assay Development

1.1.1. Antibody Selection

Reagents for use in feline research remain extremely limited and for the most part continue to be the result of in-house development. Therefore, the most critical piece of information for feline researchers is that not all antibodies are suitable for use in ELISPOT. Antibodies identified for use in enzyme-linked immunosorbent assay (ELISA) or Western blotting may not work well in the ELISPOT format. Further challenging the development of this assay is that two antibodies recognizing different epitopes on the same protein must be identified. Therefore, antibodies must be screened in multiple combinations at various concentrations to identify the optimal antibody pair. Cross-reactive antibodies against protein from other species (such as anti-canine IFN-γ) should not be ruled out in cases of high genetic and protein homology. For example, we have had success coating with 5 μg/mL polyclonal canine IFN-γ (R&D Systems, cat. no. AF781) and detecting with 1 μg/mL biotinylated polyclonal feline IFN-γ (R&D Systems, cat. no. BAF764), or using feline IFN-γ ELISPOT development module antibodies (R&D Systems, cat. no. SEL764) at the recommended dilutions.

The range of antibody concentration derived from other species as well as for feline IFN-γ is 5 to 0.1 μg/mL for coating and detection, and is most affected by whether the antibodies are monoclonal or polyclonal. It is easiest to start at the high end of antibody concentration, identify those pairs that are most promising, and then titer antibody concentrations. Antibodies used in this assay should be of high affinity, purified (by protein G or protein A affinity chromatography) and free of particulate matter or other contaminants.

If the ultimate goal of the assay is to determine the frequency of cells secreting a cytokine during an ex vivo antigen-specific recall response, the screening of antibody pairs must include both mitogen activation as a positive control and the antigen to which the animals have been previously exposed. The reason for this lies in the frequency of cells responding to mitogen vs antigen. In some situations, antibody combinations at a given concentration may work extremely well in detecting mitogen-induced protein but are unable to adequately detect antigen-specific secretion. It is also important to include wells that receive no stimulation or contain no cells to reveal artifactual spots should they occur.

1.1.2. Spot Characterization

All spots are not equal, and it is of paramount importance during the assay development phase that the investigator is observing real spots before proceeding with research. True spots have a dense center with a light outer ring caused by the diffusion of protein from the secreting cell (**Fig. 1A, B**). Artifactual spots are usually uniform in intensity and size and can vary from being very dark to light or wispy in appearance (**Fig. 1C**). A trained observer should be able to distinguish between real and artifactual spots if there is a question. Artifactual spots can be caused by the aggregation of antibodies used in coating and detection or incomplete removal of cells from the plate before the addition of detection antibody. If artifactual spots are observed, the following steps can be taken to troubleshoot the cause: (1) the detection antibody and streptavidin-HRP can be filtered using a low protein binding filter; (2) the addition of a "no cells" negative control may help to identify the source of the precipitate; or (3) before the addition of biotinylated antibody, make additional washes with wash buffer followed by distilled water to ensure lysis of any cells adhering to the membrane. Finally, an understanding of anticipated results can help in gauging whether or not the results are accurate.

Background spots are a further distinction in that they are true spots reflecting protein secretion but are not caused by any specific stimulus (seen as spots in unstimulated wells). These can be problematic when measurement of antigen-specific responders is desired. In cytokine ELISPOT, timing of cell harvest can be critical in minimizing nonspecific spot formation. Timing varies with the cytokine tested. Tightly regulated responses such as IFN-γ should have minimal background in unstimulated wells (<10 spots) within 2–3 wk after immunization but may remain elevated during natural infection with a pathogen. For example, we have observed peripheral blood mononuclear cells (PBMCs) from chronically infected FIV+ cats to have an average of 50 spontaneous (i.e., antigen nonspecific) IFN-γ spot-forming cells (SFCs)/10^6 cells in contrast to PBMCs of specific pathogen-free cats, which produce no detectable spontaneous IFN-γ. T helper cell cytokines such as interleukin (IL)-2 and IL-4 in the murine system have been know to persist much longer after immunization and have a higher residual "background" in the absence of ex vivo antigen restimulation. In the case of helper cytokines, "bystander" spots can also be observed. These spots are small, dense spots surrounding larger spots with the dense center and light outer ring. Incubation time can affect the occurrence of "bystander" or nonspecific spots, with longer incubations leading to an increased incidence of background. In most cases, empirical determination of optimal incubation time can limit the occurrence of these spots.

1.1.3. Antigen Restimulation

Purity of peptides and proteins used in antigen-specific restimulation can be a decisive factor in quality of data. Peptide synthesis is not 100% efficient, leading to the generation of a range of truncated or deletion sequences. Furthermore, organic compounds used as scavengers during synthesis can induce nonspecific reactivity if not removed from the final product. Therefore, synthetic peptides should be purified by high-performance liquid chromatography to ensure removal of most chemical scavengers and nonfunctional sequences. Resulting purified peptides should be analyzed by mass spectrophotometry and have a final purity of >85%. Companies such as SynPep

Corporation (Dublin, CA) that are dedicated to peptide production are often the best source when purchasing peptides.

Depending on solubility, peptides can be reconstituted in either sterile, distilled water or dimethyl sulfoxide (DMSO) to a stock concentration of 10 mg/mL. If using DMSO, it is advisable to maintain the final concentration of DMSO in culture to <0.25% to avoid inhibition of cellular activity. Once solubilized, peptides can be further diluted to 2 mg/mL in tissue culture medium to use as a working stock solution. If using pools of peptides, peptides should be combined in equimolar amounts.

Ex vivo antigen restimulation requires optimization of antigen concentration before conducting experiments. The range of final peptide concentration in the assay ranges from 10 to 0.1 μg/mL, with the most commonly used being 1–2 μg/mL. Peptide concentration affects responses in a dose-dependent manner, with excessive peptide leading to inhibition of responses. Therefore, optimal concentration must be determined empirically for each new peptide or peptide pool. When full-length proteins or peptide pools of 15 mers are used for restimulation, it may be necessary to preculture cells with peptide for a 12- to 24-h period before adding to the ELISPOT plate to allow for efficient processing and presentation.

2. Materials

1. Coating buffer: Sterile endotoxin-tested phosphate-buffered saline (PBS), pH 7.2–7.4. Coating antibody concentration should be determined empirically as described in the **Subheading 1.1.1.**
2. Blocking buffer: For convenience, we use tissue culture medium (LBT) containing 10% fetal bovine serum (FBS). LBT medium: RPMI 1640 supplemented with 10% FBS, 15 mM HEPES, 1 mM sodium pyruvate, 4 mM L-glutamine, 10 IU/mL penicillin, 55 μM 2-ME and 10 μg/mL streptomycin. An alternate blocking buffer is endotoxin-tested PBS, pH 7.2–7.4, containing 1–5% bovine serum albumin or FBS.
3. Diluent for detection antibody and streptavidin-HRP: Endotoxin-tested PBS, pH 7.4, with 10% FBS. Store at 4°C; stable for 1 wk.
4. Streptavidin–HRP (1 mg/mL): dilute 1:500–2000 in dilution buffer. Stock solutions of streptavidin–HRP may vary extensively from lot to lot and, therefore, the final concentration should be determined empirically before using new batches. An appropriate concentration of streptavidin–HRP will allow for the development of dark centered, robust spots while not causing a red cast on the well membrane. Minimizing the red tint on the membrane is critical when using automated counting in that smaller spots cannot be distinguished from a red background.
5. Wash buffer: PBS, pH 7.2–7.4, with 0.05% Tween-20. Add 500 μL of Tween-20 to 1 L of PBS and stir for 15 min. Store at 4°C; stable for 1 mo.
6. Detection substrate stock solutions: 0.4% 3-amino-9-ethylcarbazole (AEC; cat. no. A6926, Sigma, St. Louis, MO). Dissolve 0.4 g of AEC in 100 mL of *N,N*-dimethylformamide. Store at room temperature, stable for 1 yr. 0.1 M sodium

acetate buffer: dissolve 8.20 g of sodium acetate in 1 L of water. Adjust to pH 5.2 using HCl and filter sterilize. Store at room temperature; stable for 1 yr. Discard buffer if sodium acetate precipitates out of solution over time. Thirty percent hydrogen peroxide (H_2O_2): Protect from light; store at 4°C.

3. Methods

All procedures should be performed using aseptic technique in a laminar flow hood.

3.1. Plate Preparation

There are two types of polyvinylidene diflouride (PVDF) microtiter plates most commonly used for ELISPOT; both produced by Millipore (Billerica, MA). MSIPS4W (10 or 50) are nonsterile white plates that allow greater protein diffusion, yielding qualitatively better looking spots. ELISPOT-certified ELIIP10SSP plates are sterile white plates that restrict protein diffusion but do not affect quantitative outcome. Because ELISPOT requires short culture times and includes the use of antibiotics in culture medium, nonsterile plates can be used without complication.

3.1.1. Coating Plates With Capture Antibodies

Pre-wet PVDF ELISPOT plates with 50 µL of 70% methanol/well. Flick out the methanol and wash wells three times with 200 µL/well sterile endotoxin-tested PBS. Dilute the capture antibody to the appropriate concentration (*see* **Subheading 1.1.1.**) in sterile endotoxin-tested PBS, making 5 mL/plate. Add 50 µL/well and gently tap plate on each side to ensure complete coverage of membrane. Incubate overnight at 4°C or 3 h at 37°C. Flick out coating antibody and wash wells two to three times with 200 µL of PBS. Add 200 µL/well blocking buffer and incubate at 37°C in 5% CO_2 for 30–60 min.

3.2. Cell Preparation

Single cell suspensions used in ELISPOT, whether isolated from tissue or peripheral blood, must be free of all cellular debris and fibrous material to allow sufficient contact of cells with the microtiter plate membrane (*see* **Note 1**). When isolating PBMCs, it is critical to use an anticoagulant that will not inhibit or alter cellular function (*see* **Note 2**). If frozen cells are to be thawed for use in ELISPOT, a preculture period is recommended (*see* **Note 3**). It is advisable to use fresh cells whenever possible, particularly when using cells derived from mucosal tissue. Optimal cell density will have to be empirically determined based upon tissue type and peptide stimulation. Minimally, 2×10^5 cells per well should be used to allow appropriate cell contact for optimal immune response. We suggest using 5×10^5 cells per well to ensure maximal cell-to-cell contact and increase the likelihood that antigen-specific responses will be detected.

3.3. Plate Setup

To maximize plate usage and minimize cost, a representative row of a suggested plate setup is shown below.

	1	2	3	4	5	6	7	8	9	10	11	12
	Ag in Culture			**No Ag**		**PMA**	**Ag in Culture**			**No Ag**		**PMA**
A	Cat 1	Cat 1	Cat 1	Cat 1	Cat 1	Cat 1	Cat 2	Cat 2	Cat 2	Cat 2	Cat 2	Cat 2

3.3.1. Filling Plate

1. Flick out blocking buffer. If using blocking buffer other than tissue culture medium, rinse wells twice with medium to equilibrate membrane.
2. Using an eight-channel pipetor, pipet 50 µL of 2X antigen into columns 1–3, 7–9.
3. Fill wells 4–5 and 10–11 with 50 µL of tissue culture medium.
4. Fill positive control columns 6 and 12 with 50 µL of mitogen diluted to 2X concentration (for phorbol myristate acetate (PMA)/ionomycin, this is 100 ng of PMA and 600 ng of ionomycin).
5. Resuspended cells to 1×10^7 cells/mL. Sterile 12-channel disposable reservoir liners (Corning cat. no. 4877) are an easy and convenient way to dispense cells for pipetting. Pour cells into individual row of liner and, using a multichannel pipetor with six tips, dispense 50 µL of cells to add 5×10^5 cells/well for each cat, making sure cells remain in suspension while pipetting.
6. Once all cells are added, gently but firmly tap the sides of the covered plate with your hands/knuckles to get an even distribution of cells in the wells. Plates should be tapped four times on each side. Do NOT rotate plate in a circle or rock back and forth because this will lead to an uneven distribution of cells. Avoid splashing by holding cover and plate firmly while tapping.

3.4. Plate Incubation

Incubate plates at 37°C in 5% CO_2. Incubation times need to be determined empirically for individual peptides and proteins used in re-stimulation. The objective is to maximize spot size while avoiding ex vivo expansion. The standard range of incubation is 12–36 h. Limiting incubation time to 36 h precludes cellular expansion and therefore should not alter the frequency of responders. Increasing incubation time within the range of 12–36 h allows for greater protein secretion, and thus larger spot size. As previously mentioned, in the case of helper cytokines such as IL-2, longer incubation times may lead to the production of "bystander" spots unrelated to antigen specificity. During incubation, the plate should not be moved or bumped. Jarring the plate may interfere with spot formation.

3.5. Detection Antibody Incubation

Work can now proceed outside of a laminar flow hood. At the end of the incubation, remove cells from plate by gently flicking; **do not use excessive force by tapping plates while removing cells or wash buffer**. Wash plates three times with PBS, pH 7.2–7.4, followed by three washes with PBS–Tween wash buffer. Although plate washers or squirt bottles can be used, force during washing may lead to detachment of protein–antibody complexes (visualized as a "doughnut" at the time of development). Using a multichannel pipet to add wash buffer and gentle flicking will adequately remove cells and wash wells. Blot outer rim of well to remove excess wash buffer. Dilute biotinylated detection antibody in PBS–10% FBS, making 5 mL/plate. The appropriate concentration of detection antibody should be determined in advance as outlined in **Subheading 1.1.1.** Add 50 μL of diluted antibody to each well. Incubate overnight at 4°C or at room temperature for 4 h.

3.6. Detection of SFCs and Their Quantification

Wash plates 6X with PBS–Tween wash buffer. Gently blot outer rim to remove excess wash buffer. Dilute streptavidin–HRP 1:500–2000 (*see* **Subheading 2.**) in PBS–10% FBS. Add 50 μL to each well and incubate for 1 h at room temperature. Wash plates 3X with PBS–Tween wash buffer then 3X with PBS. Blot excess liquid from outer rim of plate. Make final substrate solution as follows (10 mL/plate; make immediately before use):

Final Amount	AEC Stock	Sodium acetate buffer	30% H_2O_2
10 mL	0.67 mL	10 mL	10 μL
20 mL	1.34 mL	20 mL	20 μL
30 mL	2.01 mL	30 mL	30 μL
40 mL	2.68 mL	40 mL	40 μL
50 mL	3.35 mL	50 mL	50 μL

Combine indicated amounts of AEC and sodium acetate stock solutions, vortex, and filter through a 0.45-μm filter. Add indicated amount of 30% H_2O_2 and vortex. Immediately add 100 μL of substrate solution to each well. Incubate for 10–60 min at room temperature, constantly monitoring for spot development. Stop reaction once spots appear well developed (dark red) and prior to excessive reddening of the membrane. To stop development, remove plastic backing from microtiter plate and place under a gentle stream of tap water. Wash thoroughly both inside wells and on the back of the membrane, and air dry for 24 h before attempting to quantify. An extended drying period may be necessary if the membrane has a red appearance. Drying should be done in the dark to prevent bleaching of spots.

Streptavidin–alkaline phosphatase paired with 5-5-bromo-4-chloro-3-indolyl phosphate/Nitroblue tetrazolium as a substrate can be used as an alternate development method. This combination leads to the development of blue rather than red spots as with AEC. Blue spots tend to be uniformly dark in appearance on ELISPOT-certified plates and can be harder to troubleshoot if they appear in "no cell" or "medium alone" control wells. In contrast, diffusion rings of blue spots are evident on MSIPS4W plates.

Spots can be counted using 10X to 30X magnification on an inverted microscope or with an automated counter (ImmunoSpot Analyzer, Cellular Technologies Ltd., Cleveland, OH) once membranes are dry. To decrease bias in cell quantitation, it is advisable to have a "blinded" counter if performing a manual count, or to select automated counting of plates (rather than manual counting of individual wells) when using a mechanized counting program. Even the best run assays may have light background spots in wells that do not warrant inclusion in SFC counts, but could inadvertently be counted in known positive wells by a biased operator. Automated counting programs allow the operator to set the level of sensitivity such that these spots are not counted, thus preventing biasing of final data. Raw data should be normalized to SFC per million cells. For example, 200 SFC counted in a well containing 5.0×10^5 cells would be multiplied by 2 to adjust the number to 400 SFCs/10^6 cells. Plates should be stored at room temperature and protected from light after development to prevent bleaching of spots.

4. Notes

1. For a debris-free preparation of PBMCs, remove plasma and dilute cells 1:4 in sterile PBS and layer each 20 mL of diluted blood over 12 mL of Histopaque 1077 (Sigma, St. Louis, MO) in a 50-mL conical tube. The greater dilution of blood combined with the large surface area of Histopaque in a 50-mL conical tube will ensure a cleaner cell preparation by eliminating fibrinogen and most of the large platelets found in feline blood. Platelets tend to clump and will prevent an accurate cell count. Centrifuge at 350g for 30 min at room temperature **with no brake**. Collect the buffy coat found at the interface of Histopaque and PBS and transfer into a new 50-mL conical tube. Fill with sterile PBS and centrifuge at 200g for 10 min. The 200g spin will further eliminate platelets and other contaminants. Because the cell pellet is not firmly packed, aspirate off liquid. Wash a second time in 50 mL of sterile PBS at 350g, this time decanting liquid. Resuspend pellet in medium and count. For debris-free preparation of splenocytes, tissue should be homogenized to release cells. Following lysis of red blood cells, cells can be diluted

1:4 in sterile PBS add layered over Histopaque and processed as described for PBMC to ensure removal of fibrous material and cellular debris.

2. Anticoagulants used during blood collection can affect the ability of cells to function in downstream assays *(6,7)*. Ethylenediamine tetraacetic acid , which inhibits coagulation thorough calcium chelation, may impair cytokine induction during antigen-specific re-stimulation. To restore functionality after ethylenediamine tetraacetic acid collection, incubate cells overnight or wash cells extensively. Conversely, heparin may lead to increases in non-specific cytokine production. Acid citrate dextrose or sodium citrate may be a good choice for anticoagulant in that there are no reported deleterious effects on cellular function.

3. High viability of cells upon thawing is critical to the success of ELISPOT. 10% DMSO/90% FBS works well as freezing medium. Freezing medium should be at room temperature when added to cells to maximize cell recovery and functionality *(8)*. Place tubes on ice immediately and transfer to chilled cryofreezing container. Quick-thaw cryotubes one at a time in a 37°C water bath. Slowly pipet the cells drop-wise into complete medium and centrifuge at 350g for 10 min. Decant medium from pellet, resuspend to 2.5×10^6 cells/mL, and culture overnight in polypropylene tubes (Falcon, cat. no. 2059 or 2063) at 37°C in 5% CO_2. Macrophages and other adherent cells do not adhere to polypropylene plastic, thereby enhancing the recovery of cells *(9)*. The following day, count cells and resuspend to 1×10^7 cells/mL in both medium containing restimulation peptide and an additional aliquot in medium alone. Culture in polypropylene tubes overnight. The overnight culture in the presence of antigen is designed to facilitate antigen processing and presentation with cells in close proximity *(9)*. The following day cells can be directly added to wells containing 50 µL/well medium (rather than 2X antigen) without recounting or addition of supplemental antigen. PMA/ionomycin is added when unstimulated cells are added to the microtiter plate.

References

1. Czerkinsky, C., Andersson, G., Ekre, H. P., Nilsson, L. A., Klareskog, L., and Ouchterlony, O. (1988) Reverse ELISPOT assay for clonal analysis of cytokine production. I. Enumeration of gamma-interferon-secreting cells. *J. Immunol. Methods* **110,** 29–36.
2. Czerkinsky, C., Andersson, G., Ferrua, B., Nordstrom, I., Quiding, M., Eriksson, K., et al. (1991) Detection of human cytokine-secreting cells in distinct anatomical compartments. *Immunol. Rev.* **119,** 5–22.
3. Czerkinsky, C. Nilsson, C. L. A., Nygren, H., Ouchterlony, O., and Tarkowski, A. (1983) A solid-phase enzyme-linked immunospot (ELISPOT) assay for enumeration of specific antibody-secreting cells. *J. Immunol. Methods* **65,** 109–121.
4. Moller, S. A. and Borrebaeck, C. A. (1985) A filter immuno-plaque assay for the detection of antibody-secreting cells in vitro. *J. Immunol. Methods* **79,** 195–204.
5. Sedgwick, J. D., and Holt, P. G. (1983) A solid-phase immunoenzymatic technique for the enumeration of specific antibody-secreting cells. *J. Immunol. Methods* **57,** 301–309.

6. De Jongh, R., Vranken, J., Vundelinckx, G., Bosmans, E., Maes, M., and Heylen, R. (1997) The effects of anticoagulation and processing on assays of IL-6, sIL-6R, sIL-2R and soluble transferrin receptor. *Cytokine* **9,** 696–701.

7. Riches, P., Gooding, R., Millar, B. C., and Rowbottom, A. W. (1992) Influence of collection and separation of blood samples on plasma IL-1, IL-6 and TNF-alpha concentrations. *J. Immunol. Methods* **153,** 125–131.

8. Kreher, C. R., Dittrich, M. T., Guerkov, R., Boehm, B. O., and Tary-Lehmann, M. (2003) CD4+ and CD8+ cells in cryopreserved human PBMC maintain full functionality in cytokine ELISPOT assays. *J. Immunol. Methods* **278,** 79–93.

9. Schmittel, A., Keilholz, U., Bauer, S., Kuhne, U., Stevanovic, S., Thiel, E., and Scheibenbogen, C (2001) Application of the IFN-gamma ELISPOT assay to quantify T-cell responses against proteins. *J. Immunol. Methods* **247,** 17–24.

11

Measuring T-Cell Function in Animal Models of Tuberculosis by ELISPOT

Vanja Lazarevic, Santosh Pawar, and JoAnne Flynn

Summary

Enzyme-linked immunospot (ELISPOT) was originally developed from an enzyme-linked immunosorbent assay (ELISA) to detect and measure the frequency of individual cells that produce cytokines in response to antigenic stimulation. ELISPOT assay is more sensitive than the enzyme-linked immunosorbent assay or intracellular cytokine staining. Increased sensitivity of ELISPOT is particularly advantageous when studying T-cell-mediated responses in *Mycobacterium tuberculosis* infection because antigen specific T-cells can occur at a low frequency in vivo. This method has been successfully used to analyze *M. tuberculosis* immune responses in humans in addition to murine and nonhuman primate models of tuberculosis.

Key Words: *Mycobacterium tuberculosis*; IFN-γ; ELISPOT; bronchoalveolar lavage; lung; T-cells.

1. Introduction

T-cells play a critical role in the control of *Mycobacterium tuberculosis* infection. Most of the T-cell-mediated immunity is mediated through the production of interferon (IFN)-γ by CD4 and CD8 T-cells, and the robustness of the immune response to *M. tuberculosis* infection often is measured by the frequency of IFN-γ-producing T-cells. There is a growing need for the development of more sensitive assays that accurately assess the function of activated T-cells in subjects infected with *M. tuberculosis*. The enzyme-linked immunospot (ELISPOT) assay has been successfully used to estimate the frequency of *M. tuberculosis*-specific IFN-γ-producing T-cells (*1–7*). In an ELISPOT assay, isolated T-cells are stimulated with *M. tuberculosis*-infected or antigen-pulsed dendritic cells (DCs). The local production of IFN-γ by stimulated cells is then

From: *Methods in Molecular Biology, vol. 302: Handbook of ELISPOT: Methods and Protocols*
Edited by: A. E. Kalyuzhny © Humana Press Inc., Totowa, NJ

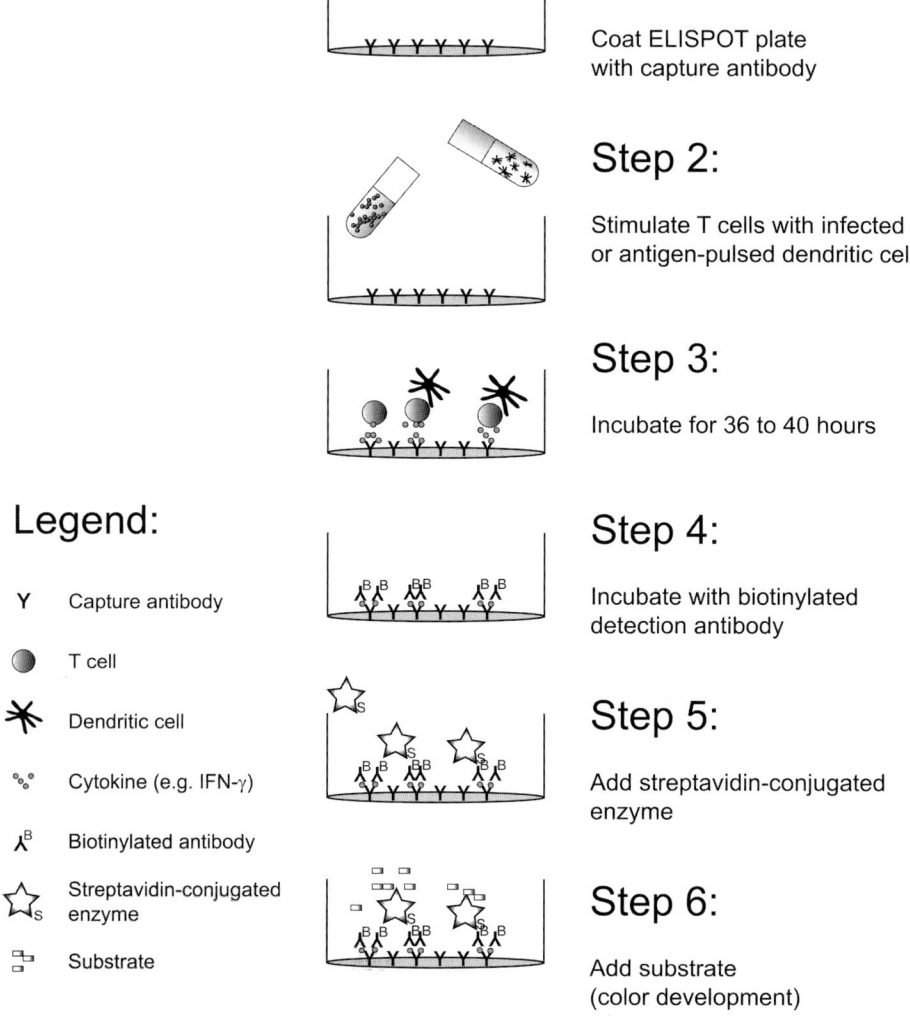

Step 1:

Coat ELISPOT plate
with capture antibody

Step 2:

Stimulate T cells with infected
or antigen-pulsed dendritic cells

Step 3:

Incubate for 36 to 40 hours

Legend:

Y   Capture antibody

●   T cell

✳   Dendritic cell

°₀°   Cytokine (e.g. IFN-γ)

入ᴮ   Biotinylated antibody

☆ₛ   Streptavidin-conjugated
      enzyme

▭   Substrate

Step 4:

Incubate with biotinylated
detection antibody

Step 5:

Add streptavidin-conjugated
enzyme

Step 6:

Add substrate
(color development)

captured between the coating and biotinylated detection antibodies. The final step is addition of a substrate that, after enzymatic cleavage by streptavidin-conjugated alkaline phosphatase or horseradish peroxidase, is converted into an insoluble product. The result is the formation of colored spots that can be enumerated by inverted microscope or by using an ELISPOT reader. The outline of the protocol is depicted in **Fig. 1**. In this chapter, we have included the protocol procedure for ELISPOT assays that are used in murine and primate models of tuberculosis.

Fig. 1. *M. tuberculosis*-specific ELISPOT procedure. Step 1, High protein binding immobilon-P plates are pretreated with 70% ethanol. The plates are coated with anti-IFN-γ antibodies overnight at 4°C. Steps 2 and 3, T-cells isolated from *M. tuberculosis*-infected mice or monkeys are stimulated with uninfected, *M. tuberculosis*-infected or antigen-pulsed dendritic cells for 40 h. Step 4, After a series of washes, plates are incubated with biotinylated anti-IFN-γ antibody for 2 h at 37°C. Step 5, Following secondary antibody incubation, plates are washed and a streptavidin-conjugate enzyme, such as horseradish peroxidase or alkaline phosphatase, is added for 1 h at room temperature. Step 6, After series of washes, a substrate is added, which is converted into insoluble precipitate that is visualized as red or blue spots. Once the spots are visible, the reaction is terminated by addition of deionized water, and 2% paraformaldehyde to inactivate *M. tuberculosis*. The plates are dried overnight in the dark, and spots are enumerated using an inverted microscope or ELISPOT reader.

2. Materials

2.1. Murine IFN-γ ELISPOT Assay

1. 70% Ethanol.
2. 1X tissue culture phosphate-buffered saline (PBS).
3. Millipore MultiScreen 96-well MAIPS4510 plates (Millipore Corp.).
4. Capture, rat anti-mouse IFN-γ antibody (clone R4-6A2; BD Pharmingen).
5. Blocking buffer (DMEM, 15% certified fetal bovine serum [FBS]).
6. T-cell medium (TCM): DMEM, 10% certified FBS, 1 mM sodium pyruvate, 2 mM L-glutamine, 25 mM HEPES, and 50 μM 2-ME.
7. NH$_4$Cl-Tris solution.
8. Murine interleukin (IL)-2; Boehringer Mannheim).
9. Concanavalin A (Sigma-Aldrich).
10. Uninfected and *M. tuberculosis*-infected bone marrow-derived DCs.
11. ELISPOT wash buffer: 0.1% Tween 20 in 1X PBS.
12. Detection, biotinylated anti-mouse IFN-γ antibody (clone XMG 1.2; BD Pharmingen).
13. Incubation buffer: 0.5% bovine serum albumin, 0.1% Tween-20 in 1X PBS.
14. Vectastain ABC Kit (Vector Laboratories Inc).
15. 3-Amino-9-ethylcarbazole (AEC) peroxidase substrate kit (Vector Laboratories Inc).
16. 2% Paraformaldehyde (PFA).
17. Inverted microscope, or an ELISPOT reader (e.g., C.T.L. Cellular Technology Ltd).

2.2. Monkey IFN-γ ELISPOT Assay

1. Millipore MultiScreen 96-well MAIPS4510 plates (Millipore Corp).
2. 70% Ethanol.

3. 1X Tissue culture PBS.
4. Certified FBS.
5. IMDM (BioWhittaker).
6. RPMI-1640 (BioWhittaker).
7. DCM (DC medium containing IMDM + 10% FBS).
8. Human recombinant granulocyte macrophage-colony stimulating factor (GM-CSF; Sigma).
9. Human recombinant IL-4 (Sigma).
10. Human Serum AB (huABS; Gemini Bioproducts).
11. 1 *M* HEPES buffer solution (Gibco-BRL).
12. L-glutamine (BioWhittaker).
13. Phorbol 12,13 dibutyrate (PDBu; Sigma).
14. Ionomycin dissolved in ethanol or dimethyl sulfoxide (Sigma).
15. Anti-monkey CD3 antibody (clone FN-18, Biosource International).
16. Anti-Monkey/Human Interferon-γ ELISpot kit (MABtech).
17. ELISPOT wash buffer (0.05% Tween-20 in 1X PBS).
18. AEC peroxidase substrate kit (Vector Laboratories Inc.).
19. 1% PFA.

3. Methods
3.1. Murine IFN-γ ELISPOT Assay

There are three parts to the ELISPOT method that are described in **Fig. 1**. In the first step the ELISPOT 96-well plate is coated with capture anti-IFN-γ antibody and the stimulator cells are prepared. In the second part the isolated cells from the lungs and lymph nodes of *M. tuberculosis*-infected mice are cultured with uninfected, *M. tuberculosis*-infected or antigen-pulsed dendritic cells for 40 h. Finally, the IFN-γ-producing T-cells are visualized after stepwise incubation of plates with biotinylated anti-IFN-γ antibody, streptavidin-conjugated enzyme, and substrate. All the steps are carried out under sterile, and biosafety level 3 restricted conditions.

3.1.1. Coating of ELISPOT Plates and Preparation of Stimulatory Cells

3.1.1.1. COATING OF ELISPOT PLATES

1. Pre-treat 96-well plates with 200 μL/well of 70% ethanol for 10 min.
2. Rinse the wells with 200 μL/well of tissue culture 1X PBS three times (5 min each wash).
3. Coat 96-well plates with 100 μL/well of 10 μg/mL solution of capture, anti-mouse IFN-γ antibody (clone R4-6A2) in 1X PBS. Incubate overnight at 4°C.

3.1.1.2. PREPARATION OF STIMULATOR CELLS

1. Bone marrow-derived DCs should be cultured for 5 d in the presence of 1000 U/mL of mGM-CSF and 1000 U of mIL-4/mL before their use as stimulator cells.

2. Suspend cells at 1×10^6/mL. Leave 5×10^6 of dendritic cells uninfected as a negative control. Infect 2×10^7 DCs with *M. tuberculosis* (from a liquid growing culture) at multiplicity of infection 3 overnight. Approximately 50–60% of DCs will die after *M. tuberculosis* infection.

3.1.2. Incubation of T-Cells From Infected Mice With Stimulator Cells

1. The next day, wash ELISPOT plates three times with 1X PBS. Block the plates with the blocking buffer for 2 h at room temperature.
2. In the meantime, euthanize *M. tuberculosis*-infected mice and obtain lung and lymph nodes.
3. Crush the lungs and lymph nodes in 5 mL of DMEM using 70-μm nylon cell strainers and 5-mL syringe plunger to obtain single cell suspension.
4. Spin cells at 470*g* for 5 min.
5. Decant the supernatant and suspend the lung cells in 4 mL and lymph nodes in 1 mL of NH_4–Tris solution for 2 min at room temperature to lyse red blood cells.
6. Wash cells with 1X PBS once and spin at 470*g* for 5 min.
7. Suspend cells in 1 mL of TCM and count.
8. Plate lung cells at 80,000 cells/well and lymph node cells at 150,000 cells/well in 100 μL of volume.
9. Incubate T-cells in duplicate for 40 h with the following conditions: negative control: TCM (50 μL/well); positive control: Concanavalin A (10 μg/mL final concentration; 50 μL/well); negative control: uninfected DCs (1:2 ratio DC:T; 50 μL/well); *M. tuberculosis*-infected DCs (1:2 ratio DC:T; 50 μL/well) or DCs that were pulsed with *M. tuberculosis*-specific antigen overnight at 10 μg/mL concentration (1:2 ratio DC:T; 50 μL/well).
10. Add IL-2 at the final concentration of 20 U/mL in 50 μL/well.
11. Incubate cells at 37°C, 5% CO_2.

3.1.3. ELISPOT Development

1. Empty ELISPOT plates into a basin containing undiluted, mycobactericidal detergent inside the biosafety level 3 containment hood.
2. Wash wells six times with ELISPOT wash buffer (5 min each wash).
3. Add 100 μL/well of 5 μg/mL of solution of the detection, biotinylated anti-IFN-γ antibody. Incubate at 37°C for 2 h or at 4°C overnight.
4. After the incubation, empty the ELISPOT plates, and wash with ELISPOT wash buffer six times, 5 min each wash.
5. Prepare Avidin Peroxidase Complex (10 mL of ELISPOT wash buffer plus one drop reagent A plus one drop reagent B) **at least 30 min** before use. Add 100 μL/well of avidin peroxidase complex and incubate the plate at room temperature for 1 h.
6. Wash plates six times with ELISPOT wash buffer (5 min each wash).
7. Wash plates three times with 1X PBS, 5 min each wash.

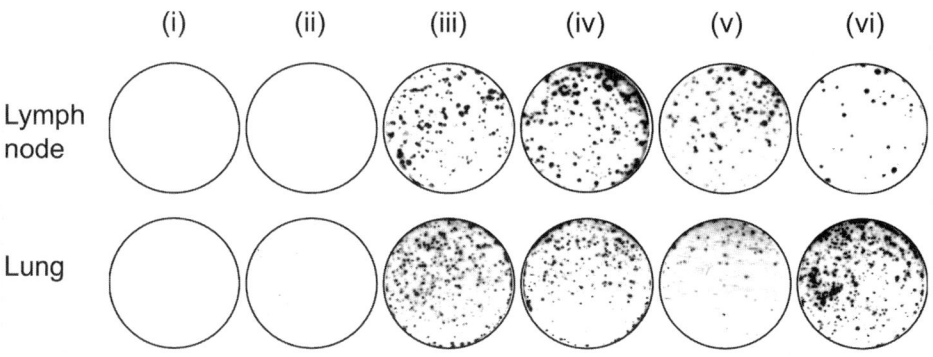

Fig. 2. *M. tuberculosis*-specific ELISPOT assay in a murine model of tuberculosis. Murine lymph node (150,000/well) and lung (80,000/well) cells were isolated from *M. tuberculosis* infected mice at 2 and 4 weeks after infection, respectively. Cells were cultured in: i) medium alone; ii) uninfected DCs; iii) *M. tuberculosis*-infected DCs; vi) *M. tuberculosis*-infected DC + blocking anti-MHC Class I antibody (CD4+IFN-γ+); v) *M. tuberculosis*-infected DC + blocking anti-MHC Class II antibody (CD8+IFN-γ+); or vi) DCs pulsed with *M. tuberculosis* ESAT-6 protein.

8. Prepare Vectastain AEC substrate (10 mL of deionized water plus four drops buffer stock plus six drops of AEC plus four drops of H_2O_2) immediately before use. Mix well.
9. Add 100 μL/well of AEC substrate. Incubate 4–10 min until spots develop.
10. Stop the reaction by adding 200 μL/well of deionized water.
11. Empty plates, and add 100 μL/well of 2% paraformaldehyde to inactivate *M. tuberculosis*. Incubate for 2 min at room temperature.
12. Empty plates and air-dry in the dark.
13. At this point, it is safe to take ELISPOT plates out of biosafety level 3 laboratory and enumerate the spots using an inverted microscope or an ELISPOT reader. The representative wells of murine lung and lymph nodes cells stimulated with *M. tuberculosis*-infected dendritic cells are shown in **Fig. 2**.

3.2. Monkey IFN-γ ELISPOT Assay on Peripheral Blood Mononuclear Cells

Stimulation of T-cells isolated from infected monkeys can be achieved in two ways: (1) If peripheral blood mononuclear cells (PMBCs) are used in ELISPOT assay, no additional antigen presenting cells are required as it is assumed that PBMCs already contain sufficient numbers of these cells. (2) However, if bronchoalveolar lavage cells (BALs) or lung cells are used in ELISPOT assay then additional antigen presenting cells are required.

The monkey-specific ELISPOT method comprises of the following steps: (1) coating of ELISPOT plate with "capture" antibody; (2) blocking step; (3) setting up ELISPOT; and (4) ELISPOT development.

3.2.1. ELISPOT Assay Procedure

3.2.1.1. COATING AN ELISPOT PLATE

1. Pretreat ELISPOT plates by adding 100 µL/well of 70% ethanol for 5 min at room temperature. Wash wells twice with 200 µL/well of sterile distilled water and twice with 200 µL/well of 1X PBS. Each wash should be for 3–5 min at room temperature.
2. Prepare anti-human/monkey IFN-γ "capture" antibody (clone GZ4) at 15 µg/mL concentration in 1X BS. Add 100 µL/well of the capture antibody to the ELISPOT plate. Incubate plates overnight at 4°C.

3.2.1.2. BLOCKING STEP

1. The next day, empty ELISPOT plates and drain any remaining liquid. Wash wells four times with 200 µL/well of 1X PBS; 3–5 min each wash.
2. Block ELISPOT plates with 200 µL/well of RPMI (containing 1% glutamine and 1% HEPES) + 10% huABS and incubate for 1–2 h in 37°C, 5% CO_2.

3.2.1.3. PREPARATION OF PBMCS AND SETTING UP THE ELISPOT ASSAY

1. Isolate PBMCs from monkey blood samples by standard Percoll density gradient centrifugation. ELISPOT assays can also be performed using frozen PBMCs with additional stabilization step (*see* **Note 1**).
2. Collect the buffy-coat layer (PBMCs) and transfer into a 50-mL tube.
3. Wash twice with 50 mL of 1X PBS.
4. Count PBMCs and plate for ELISPOT at 200,000 cells/150 µL/well.
5. Prepare *M. tuberculosis*-specific antigens in RPMI containing 10% huABS at the following concentration: (1) *M. tuberculosis* whole proteins, 2 µg/mL (final concentration); (2) *M. tuberculosis* peptides, 10 µg/mL (final concentration); (3) positive controls: PDBu + ionomycin (final concentration 50 nM and 10 µM, respectively) and anti-monkey CD3 antibody (1 µg/mL final concentration); (4) negative control: medium alone
6. Add 50 µL/well of antigen stimulators to the PBMCs in duplicate.
7. Incubate plates incubated for 40 h at 37°C, 5% CO_2.

3.2.1.4. ELISPOT DEVELOPMENT

1. Antibody solutions in the ELISPOT developing steps are made in PBS containing 0.5% certified FBS.
2. Empty plates and wash them with 200 µL/well of distilled water for 10 min at room temperature to lyse adherent cells.
3. Wash wells four times with 200 µL/well of ELISPOT wash buffer (3–5 min each wash).
4. Add 100 µL/well of 2.5 µg/mL biotinylated anti-human/monkey IFN-γ secondary antibody (clone 7-B6-1) and incubate for 2 h at 37°C, 5% CO_2.

(i)   (ii)   (iii)   (iv)   (v)   (vi)   (vii)

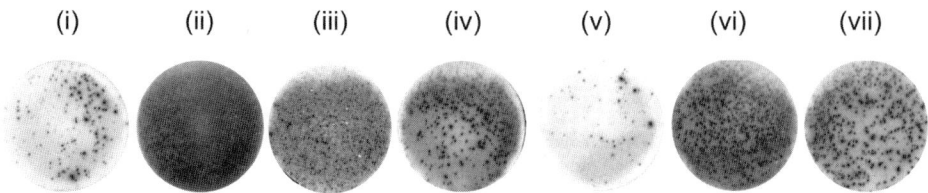

Fig. 3. *M. tuberculosis*-specific ELISPOT assay in a primate model of tuberculosis. These are representative wells of spots after incubation of 200,000/well of monkey BAL cells with: i) medium control; ii) PDBu + ionomycin control; iii) anti-monkey CD3 antibody; iv) *M. tuberculosis*-infected DC; v) uninfected DC; vi) DCs pulsed with Mtb culture-filtrate protein (CFP); vii) DCs pulsed with *M. tuberculosis* ESAT-6 protein.

5. Wash wells four times with ELISPOT wash buffer, add 100 µL/well of 1:100 dilution of streptavidin–horseradish peroxidase and incubate for 1 h at 37°C, 5% CO_2.
6. Wash wells four times with ELISPOT wash buffer followed by 1X PBS wash.
7. Add 100 µL/well of AEC substrate (sequential addition of four drops of buffer solution, six drops of AEC substrate, and four drops of H_2O_2 per 10 ml distilled water; mix well after each addition; *See* **Note 2**).
8. Incubate plates until brick red-colored spots develop (usually 3–7 min). The actual developing step is conducted in darkness. Terminate the reaction by washing plates two times with distilled water and twice with 1X PBS.
9. Empty plates and air-dry them overnight in the dark.
10. Spots can be counted either manually, using an inverted microscope, or by using an automated ELISPOT reader. Typical spot appearance in some of the stimulatory conditions and controls is exemplified in **Fig. 3**.

3.3 Monkey BAL/Lung Cell IFN-γ ELISPOT Assay

BAL/lung cell ELISPOT is technically similar to the PBMC ELISPOT except that DCs are added as supplementary antigen presenting cells, and fewer cells per well are used.

3.3.1. BAL/lung Cell IFN-γ ELISPOT Assay Procedure

3.3.1.1. PREPARATION OF DCs FROM FRESHLY ISOLATED PBMCs

1. DCs are generated in vitro from freshly isolated PBMCs 5–7 d before ELISPOT assay. It also is possible to generate dendritic cells from frozen PBMCs and use them for antigen presentation (*see* **Note 3**).
2. In a 6-well plate, suspend 1×10^7 PBMCs/well in 1–2 mL of DCM containing 1000 U/mL of human GM-CSF + 1000 U/mL of human IL-4.
3. After 3 d of culture, add 1 mL/well of fresh DCM containing 2000 U/mL of GM-CSF and 2000 U/mL of IL-4. The yield of DCs varies because the number of precursor cells present in PBMC will be different between animals.

4. The night before ELISPOT is set up, collect DCs and count. Typically $1-3 \times 10^5$ DCs are infected for 4 h with *M. tuberculosis* at a multiplicity of infection of 4. After a 4-h infection, the medium is changed to remove any extracellular mycobacteria, and cells are incubated overnight.
5. Infected and uninfected dendritic cells serve as controls in the experiment with *M. tuberculosis*-specific proteins and peptides.

3.3.1.2. PREPARATION OF BAL/LUNG CELLS FROM INFECTED MONKEYS

1. BAL cells are isolated from the airways of the lung by a bronchoscope. Very often this can lead to the contamination of samples due to the natural microflora present in the monkey upper respiratory tract. To avoid contamination problems, 1% penicillin is added to the culture medium (*see* **Note 4**).
2. Lung cells are obtained by necropsy by taking sections of the various lung lobes that are homogenized into a cell suspension using a semi-automatic tissue homogenizer (Medimachine, Pharmacia).
3. All BAL or lung cell procedures on *M. tuberculosis*-infected monkeys and ELISPOT procedures with these cells are carried out under sterile and biosafety level 3+ conditions.

3.3.1.3. SETTING UP BAL/LUNG CELL ELISPOT

1. Plate BALs (or lung cells) at 200,000/100 µL/well.
2. Then add 10,000 DCs/50 µL/well and antigenic stimulators/50 µL/well to make a total volume of 200 µL/well.
3. In wells containing infected and uninfected DCs or in PDBu + ionomycin controls, antigenic stimulators are replaced by 50 µL/well of RPMI-10%ABS.
4. Incubate plates for 40 h at 37°C, 5% CO_2.

3.3.1.4. DEVELOPING ELISPOT SPOTS

1. All developing steps are the same as in PBMCs ELISPOT with one exception. After reaction is stopped, add 200 µL/well of 1% PFA to the wells and incubate for 10 min at room temperature to inactive *M. tuberculosis*. This is followed by two washes with 1X PBS. If spots' resolution is not satisfactory, it could be due to poor coating of ELISPOT plates with the capture antibody or prolonged incubation of PBMCs and BAL cells with the stimulators (*see* **Table 1** and **Notes 5** and **6**).

4. Notes

1. ELISPOT assays also can be performed using liquid nitrogen-frozen PBMCs or BAL (or lung) cells with an additional step to stabilize cells from freeze-thaw cycle. Method described here is for PBMCs, but also can be used with BAL or lung cells. Cryovials with PBMCs (1×10^7 PBMCs/vial) are taken out of the liquid nitrogen and immediately placed under hot running tap water to quickly thaw the cells. Immediately suspend cells in 10 mL of RPMI medium to remove dimethyl sulfoxide from the cell freezing medium. Centrifuge cells at 450g for 5 min at 4°C,

Table 1
Troubleshooting Guide

Problem	Check if	What to do
No spots	• Wells were coated with the captureanti-IFN-γ antibody. • Stimulators were added in sufficient concentrations. • Biotinylated anti-IFN-γ antibody and streptavidin–HRP were added. • Substrate solution was mixed correctly.	• Add capture and biotinylated antibodies in a correct order and in sufficient concentrations. • Make substrate solution by adding reagents in a correct order.
b) Faint spots	• Biotinylated anti-IFN-γ antibodyand streptavidin–HRP were added. • Substrate solution was made correctly. • Incubation time after substrate addition was long enough for spot development.	• Add capture and biotinylated antibodies in a correct order and in sufficient concentrations. • Make substrate solution by adding reagents in a correct order. • Ensure that substrate incubation period was sufficient for spot development.
c) Overdeveloped wells with high background	• Blocking step was done for a required period of time. • Spot development after substrate addition was too long.	• Blocking should be 1–2 h long. • Longer incubation times with streptavidin–HRP will increase background. • Do not develop for extended periods and terminate reaction before background develops.

and suspend them in 10 mL of RPMI containing 1% glutamine, 1% HEPES, and 20% FBS. Plate cells in two wells in a six-well tissue culture plate. Incubate in 37°C, 5% CO_2 incubator overnight. The cells are ready to be used in the ELISPOT assay the next day. The recovery of viable cells is >90% with this method. Addition of human IL-2 (25 U/mL) does not significantly change the viability of PBMCs.

2. Horseradish peroxidase and AEC substrate/chromogen result in the production of red spots. Blue spots, which may be easier to visualize, can be obtained by using alkaline phosphatase and Vector Blue substrate (Vector Blue Alkaline Phosphatase Substrate Kit III; Vector Laboratories, Burlingame, CA).

3. It is also possible to generate dendritic cells from frozen PBMCs. After the quick thawing step, PBMCs are placed in RPMI + 20% FBS in a 6-well plate as

described above. Next day, harvest cells and centrifuge at 450g for 5 min at 4°C. Suspend approx 1×10^7/well of PBMCs in 1–2 mL of DC medium and culture for additional 5–7 d in the presence of human GM-CSF and IL-4. DCs generated from PBMCs work as good as purified dendritic cells in ELISPOT assay. Pure DCs do not survive the freeze-thaw cycle.

4. Occasional contamination problems can occur in BAL or lung cell ELISPOT assays. This problem can be solved by adding 1% penicillin into the culture medium. Streptomycin should be avoided because of its antimycobactericidal activity, which could interfere with *M. tuberculosis* infection of DCs.
5. Poor coating can lead to the formation of fused and messy spots, which can make accurate enumeration impossible. The resolution of spots can be improved by the pre-treatment of ELISPOT plates with 70% ethanol.
6. Another possible cause of the formation of the poorly defined spots in ELISPOT plate is a prolonged incubation of cells with the stimulators. The length of antigenic stimulation should not exceed 40 h.

References

1. Chapman, A. L., Munkanta, M., Wilkinson, K. A., Pathan, A. A., Ewer, K., Ayles, H., et al. (2002). Rapid detection of active and latent tuberculosis infection in HIV-positive individuals by enumeration of Mycobacterium tuberculosis-specific T-cells. *AIDS* **16**, 2285–2293.
2. Ewer, K., Deeks, J., Alvarez, L., Bryant, G., Waller, S., Andersen, P., et al. (2003). Comparison of T-cell-based assay with tuberculin skin test for diagnosis of Mycobacterium tuberculosis infection in a school tuberculosis outbreak. *Lancet* **361**, 1168–1173.
3. Flynn, J. L., Capuano, S. V., Croix, D., Pawar, S., Myers, A., Zinovik, A., et al. (2003). Non-human primates: a model for tuberculosis research. *Tuberculosis* (Edinb) **83**, 116–118.
4. Hill, P. C., Brookes, R. H., Fox, A., Fielding, K., Jeffries, D. J., Jackson-Sillah, D., et al. (2004). Large-scale evaluation of enzyme-linked immunospot assay and skin test for diagnosis of Mycobacterium tuberculosis infection against a gradient of exposure in The Gambia. *Clin Infect Dis* **38**, 966–973.
5. Lalvani, A., Pathan, A. A., Durkan, H., Wilkinson, K. A., Whelan, A., Deeks, J. J., et al. (2001). Enhanced contact tracing and spatial tracking of Mycobacterium tuberculosis infection by enumeration of antigen-specific T-cells. *Lancet* **357**, 2017–2021.
6. Pathan, A. A., Wilkinson, K. A., Wilkinson, R. J., Latif, M., McShane, H., Pasvol, G., et al. (2000). High frequencies of circulating IFN-gamma-secreting CD8 cytotoxic T-cells specific for a novel MHC class I-restricted Mycobacterium tuberculosis epitope in M. tuberculosis-infected subjects without disease. *Eur J Immunol* **30**, 2713–2721.
7. Tsukaguchi, K., Balaji, K. N., and Boom, W. H. (1995). CD4+ alpha beta T-cell and gamma delta T-cell responses to Mycobacterium tuberculosis. Similarities and differences in Ag recognition, cytotoxic effector function, and cytokine production. *J Immunol* **154**, 1786–1796.

12

Interferon-γ ELISPOT Assay for the Quantitative Measurement of Antigen-Specific Murine CD8⁺ T-Cells

Geoffrey A. Cole

Summary

Effective screening of new vaccines and immunotherapeutics requires assay methods that can provide quantitative measurement of cellular immune responses. The enzyme-linked immunospot (ELISPOT) is a sensitive technique for the detection of cytokine-producing cells at the single cell level. This assay is rapid and reproducible and permits the direct enumeration of low-frequency antigen-specific T-cells. This protocol describes in detail an interferon (IFN)-γ ELISPOT method for measuring antigen-specific murine CD8⁺ T-cells. Spleen cells from specific cytotoxic T-lymphocyte (CTL) peptide-primed mice are used source of CD8⁺ T-cells to demonstrate the utility of this technique. The assay procedure is facilitated by the use of a ready-to-use IFN-γ ELISPOT assay kit and it can be adapted to other model systems in which CD8⁺ T-cell responses are monitored.

Key Words: IFN-γ ELISPOT protocol; CD8⁺ T-cell frequency determination.

1. Introduction

Assay methods that provide simple, reliable, and quantitative measurement of specific CD8⁺ T-cell activity are essential for evaluating vaccine strategies and antigen-specific immunotherapy regimens. The enzyme-linked immunospot (ELISPOT) assay permits enumeration of specific T-cells by detection of antigen-stimulated cytokine secretion. Originally developed to enumerate antibody-producing cells *(1)*, the technique was later modified to quantify the frequency of antigen-specific T-cells *(2)*. The assay enables the direct, ex vivo detection of individual antigen-specific T-cells without any need for in vitro bulk sensitization and expansion steps such as those necessary in lymphoproliferative or ⁵¹Cr release assays. The assay provides greater sensitivity than ELISA *(3)* or intracellular cytokine staining *(4)* and, in contrast to MHC tetramer reagents *(5)*, it detects T-cells based on their functional response to antigen. The assay is rapid, simple to

From: *Methods in Molecular Biology, vol. 302: Handbook of ELISPOT: Methods and Protocols*
Edited by: A. E. Kalyuzhny © Humana Press Inc., Totowa, NJ

perform, requires no specialized instrumentation, and provides a quantitative measure of specific CD8⁺ T-cells. This technique has been applied to analyze the specificity, magnitude and kinetics of CD8⁺ T-cell responses induced by experimental vaccines, immunization protocols, and virus infections *(6–12)*.

This chapter describes a basic method for performing ELISPOT to detect antigen-specific interferon (IFN)-γ-secreting murine T-cells. It uses a commercially available ELISPOT kit to measure antigen-specific CD8⁺ T-cells in the spleen of mice immunized with specific CTL peptide epitopes.

2. Materials

2.1. Mice and Reagents

1. Mice: Female BALB/c mice were purchased from a commercial vendor (Jackson Laboratories, Bar Harbor, ME) and were 6–9 wk of age when immunized.
2. Synthetic peptides: RPQASGVYM (lymphocytic choriomeningitis virus nucleoprotein, amino acid residues 118-126), TPHPARIGL (*Escherichia coli* β-galactosidase, amino acid residues 876-884), and ISQAVHAAHAEINE (ovalbumin, amino acid residues 323-336) were synthesized by a commercial vendor (Multiple Peptide Systems, San Diego, CA) and were purified to ≥90% homogeneity. Peptides were dissolved in 100% dimethylsulfoxide at 20 mg/mL and were stored at –20°C.
3. Dimethylsulfoxide (DMSO; Mallinckrodt Baker, Inc., Phillipsburg, NJ).
4. Dulbecco's phosphate buffered saline (DPBS; Invitrogen Corp; Carlsbad, CA).
5. Freund's adjuvant, incomplete (IFA; Sigma, St. Louis, MO).
6. Culture medium: RPM1-1640 medium supplemented with 10% heat-inactivated fetal bovine serum (FBS), 10 mM HEPES, 2 mM L-glutamine, 50 µM 2-mercaptoethanol, 100 µg/mL penicillin G, and 100 U/mL streptomycin. All medium components except FBS and 2-mercaptoethanol were purchased from Invitrogen. Heat-inactivated FBS was purchased from Hyclone (Logan, UT). 2-mercaptoethanol was purchased from Mallinckrodt Baker, Inc. Filter sterilize the medium through a 0.22-µm disposable filter system (Corning, Inc., Corning, NY).
7. Wash medium: prepare RPMI-1640 with the same supplements as complete medium except FCS is omitted.
8. Trypan Blue stain, 0.4% (Invitrogen).
9. Concanavalin A (ConA; Sigma): reconstitute in complete medium and filter sterilize (0.2-µm Acrodisc syringe filter, Pall Corp., Ann Arbor, MI). Dispense 50-µL single-use aliquots and store at –20°C.
10. Mouse erythrocyte lysing kit (R&D Systems, Minneapolis, MN).
11. Mouse IFN-γ ELISPOT ready-to-use kit (R&D Systems).

2.2. Equipment and Supplies

1. Yale® 1 mL tuberculin syringes (Becton Dickinson, Franklin Lakes, NJ).
2. 18 gage Microemulsifying needle (Popper & Sons, New Hyde Park, NY).
3. Sterile 1.5-mL Microfuge tubes.
4. Needles, 23 and 27.5 gage with Luer-Lok® (Becton-Dickinson).

5. Sterile 15 and 50 mL centrifuge tubes (BD Falcon).
6. Sterile 20, 200, and 1000 μL micropipet tips.
7. Pipetors (e.g., P20, P200, and P1000; Rainin Instrument LLC, Woburn, MA).
8. Dissecting scissors and forceps. Store in 70% ethanol.
9. Frosted microscope slides (75 × 25 mm, Corning). Autoclaved.
10. Animal restrainer (e.g., Small Mouse Restrainer, Braintree Scientific, Braintree, MA).
11. Sterile 5, 10, and 25 mL pipets (BD Falcon).
12. Pipet-Aid (Drummond Scientific, Broomall, PA).
13. Laminar flow cabinet.
14. Hematocytometer (Leavy counting chamber, Hausser Scientific, Horsham, PA).
15. Reagent reservoirs (Corning).
16. Eppendorf Repeater® Pipetor and sterile Eppendorf Combitips® (Brinkmann, Westbury, NY).
17. 37°C humidified incubator with 5% CO_2 atmosphere.
18. Vacuum-operated 12-channel plate washer (Nunc-Immuno® Washer, Nalge Nunc, Rochester, NY).
19. Microscope.
20. Variable rotator shaker (Koala-Ty; Accurate Chemical and Scientific Corp; Westbury, NY).
21. Stereoscopic microscope with zoom and illumination source.
22. Hand-held push button counter (e.g., VWR hand tally counter).

3. Methods

The methods described below outline (1) peptide immunization of mice to induce antigen-specific IFN-γ-secreting splenic CD8+ T-cells; (2) isolation and preparation of splenocytes; (3) the ELISPOT assay procedure; and (4) evaluation of developed ELISPOT plates.

3.1. Immunization of Mice

The synthetic antigenic peptides and the immunization regimen outlined in **Subheading 3.1.** are not part of the ELISPOT procedure but serve to demonstrate the utility of this protocol. Two H-2L^d epitopes recognized by CD8+ T-cells were used as immunogens. RPQASGVYM is the immunodominant LCMV CTL peptide recognized by H-2^d mice *(13,14)*. TPHPARIGL peptide is a β-galactosidase epitope *(15)*. ISQAVHAAHAEINE is a cognate helper T-lymphocyte peptide that is employed to enhance CTL epitope immunogencity *(16,17)*.

1. Thaw peptides immediately before use. Peptide stocks may be freeze-thawed several times (as many as five). Mix equimolar amounts of the individual CTL peptides and T helper peptide in DPBS and combine with IFA at a 1:1 (vol/vol) ratio in a sterile Microfuge tube. Combine components to yield a final CTL peptide concentration of 1 mg/mL.

2. Vortex mixture until phases are mixed. Draw up the mixture into a dry 70% ethanol sterilized 1 mL glass syringe with an attached 23-gage needle. Remove the needle, expel as much air as possible, and attach the syringe to a dry 70% ethanol sterilized microemulsifying needle. Again expel air and attach the free end of the needle to a second sterile syringe.
3. Emulsify by repeatedly forcing mixture back and forth between syringes until it is white and homogeneous in appearance.
4. Push contents into one syringe and remove the microemulsifying needle. Attach a 27-gage needle and remove the air bubbles.
5. Restrain mice and inject 50 µL of emulsion subcutaneously at the base of the tail. Immunize two to three animals per treatment group.

3.2. Isolation and Preparation of Splenocytes

All spleen collection and processing procedures should be performed using aseptic technique. The antibody-coated plate provided with the ELISPOT kit, however, is not sterile. The presence of antibiotics in the culture medium and the short incubation period prevent microbial contamination from being a problem.

Harvest spleens from mice 1–3 wk after immunization. Depending on experimental design, spleens can either be processed separately or, as described here, prepared as two to three pooled organs per treatment group. Collect spleens from age-matched unprimed mice for use as an untreated control population and a source of antigen presenting "feeder" cells (*see* **Subheading 3.3.2.1.**). Keep cells on ice as much as possible throughout processing.

1. Kill mouse with CO_2 or by cervical dislocation.
2. Wet the left side of mouse with 70% ethanol and remove spleen aseptically.
3. Place spleen in a sterile 60-mm diameter Petri dish containing 10 mL wash medium. Prepare a single cell suspension by mashing spleen between two sterile frosted glass slides.
4. Transfer the suspension to a sterile 15 mL screw top centrifuge tube and allow the tissue debris to settle. Carefully remove the supernatant to a fresh 15-mL centrifuge tube.
5. Centrifuge cells 5 min at 300g at 4°C and discard the supernatant. Resuspend the pellet in mouse erythrocyte lysis solution and proceed according to manufacturer's instructions.
6. Centrifuge erythrocyte-depleted splenocytes and decant supernatant. Resuspend pellet and wash twice with 10 mL of wash medium.
7. Resuspend pellet in 5 mL of complete medium per spleen equivalent and filter the suspension through a 70-µm tissue screen into a new 50-mL tube (*see* **Note 1**).
8. Determine the viable cell density of splenocytes by Trypan Blue exclusion. Dilute a small volume of cells 1:10 in DPBS. Combine diluted cells 1:1 with Trypan Blue and count in a hematocytometer. Cell viability should be ≥90%.
9. At this point, ELISPOT plate wells can be filled and incubated with complete medium in preparation for plating the splenocytes (*see* **step 2, Subheading 3.3.1.**).

10. Make twofold serial dilutions of the splenocytes starting with 1×10^7 cells/mL (1 $\times 10^6$ cells/100 µL/well). Optimal plating densities will vary with the specific immunogen/mouse strain used and should be determined in preliminary experiments. In BALB/c mice immunized with the CTL peptide epitopes used here, splenocytes were plated at a density ranging from 0.63×10^5 to 5×10^5 cells/well.

3.3. ELISPOT Procedure

The R&D Systems ready-to-use murine IFN-γ ELISPOT kit contains a dry 96-well precoated plate, biotinylated detection antibody, streptavidin–alkaline phosphatase conjugate, 5-5-bromo-4-chloro-3-indolyl phosphate/Nitroblue tetrazolium (BCIP/NBT) Chromagen, wash buffer, and recombinant mouse IFN-γ for use as a plate positive control (*see* **Note 2**). The steps described in **Subheadings 3.3.1–3.3.3** outline ELISPOT assay set up and plate development. In some instances, minor variations from instructions provided with the kit are used.

3.3.1. Plating Splenocytes

The steps described in the next section detail the plating of splenocytes and the appropriate controls and treatments used to assess their specific reactivity. Each experimental splenocyte population should be plated with a negative control (medium alone), test peptide, and a positive control (ConA). Assay plate controls should include a background control (medium only) a positive control (reconstituted mouse IFN-γ), and a detection antibody control in which DPBS is substituted for detection antibody.

1. Prepare an assay template to direct sample plating. Allocate duplicate or triplicate wells for each splenocyte/test combination (*see* **Note 3**).
2. Fill all the wells of the 96-well plate with 200 µL of complete medium and incubate for approx 20 min at room temperature. Flick the plate and blot upside down on absorbent paper towels. Plates are pre-blocked; medium incubation serves to condition the wells prior to cell plating.
3. Repeat **step 2**.
4. Plate splenocyte dilutions at 100 µL/well. If the assay design requires a large number of replicate wells to be plated, an Eppendorf Repeater® pipetor or similar device is convenient for this step.
5. As an alternative, a P-200 micropipet can be used. Continuously agitate cells to ensure their homogeneity during plating.
6. Once splenocytes have been plated, move ELISPOT plate to a 37°C 5% CO_2 humidified incubator for 30–45 min to allow cells to settle. During this incubation period, antigenic peptide and ConA solutions are prepared.

3.3.2. Preparation and Addition of Antigenic Peptide and ConA

1. Thaw peptide stocks and dilute to a 2X final concentration in complete medium. Add 100 µL/well. Preliminary experiments performed with peptide-immunized

BALB/c mice established that optimal specific IFN-γ responses were stimulated with peptide concentrations of 1 μg/mL (10^{-6} *M*) and 10 μg/mL (10^{-5} *M*) for RPQASGVYM and TPHPARIGL, respectively. The optimal concentration for different peptides should be experimentally determined.

2. Add 100 μL medium alone to unstimulated splenocyte control wells and to the plate background control wells (*see* **Note 4**).

3. Add ConA to splenocyte positive control wells at a final concentration of at 2.5 μg/mL. This control demonstrates that stimulated splenocytes are of capable of IFN-γ secretion. Seed positive control wells at a density of $\leq 1.25 \times 10^5$ cells/well. Otherwise, spots on the developed membrane will likely appear confluent.

4. After complete addition of all treatments to splenocytes, and the preparation of ELISPOT plate controls as specified in the kit instructions, cover the filled plate with the lid and sit the plate on a piece of aluminum foil of sufficient size to be folded over the edges of the plate. Keep the foil on the plate throughout all incubation steps including the final incubation with Chromagen (**step 8, Subheading 3.3.3.**). This treatment reduces non-specific background staining *(18)*.

5. Incubate the plate in a 37°C humidified CO_2 incubator for 20–36 h (*see* **Note 5**). Do not move the plate during incubation as this may result in smudged spots upon color development.

3.3.2.1. OPTIONAL TECHNIQUE: PREPARATION OF SPLENIC FEEDER CELLS

ELISPOT sensitivity is critically dependent on high cell density. Optimal stimulation of antigen-specific T-cells is best achieved when total cell density approaches 5×10^5 to 1×10^6 cells/well *(19)*. If immune splenocytes are limiting and not available for preparing a twofold dilution series starting with 5×10^5 cells/well, syngeneic spleen cells from unprimed mice can be used as "feeder cells" to increase input cell density. The enhanced sensitivity (increased spot frequency per input cell number) can be experimentally demonstrated by adding peptide-loaded feeder cells to low-density immune splenocytes.

1. Adjust feeder cells to a 5×10^6 cells/mL in complete medium.
2. Add antigenic peptide to feeder cells at 2X final concentration and mix by pipetting. No preincubation is necessary.
3. Plate the peptide/feeder cell mixture at 100 μL/well with immune splenocytes. Feeder cells, without peptide, are used as a negative control.
4. Return plate to incubator as described in **Subheading 3.3.2**.

3.3.3. Plate Harvest and Development

1. Prepare the wash buffer solution by diluting the 10x concentrate in deionized water immediately before harvesting plate. A vacuum-operated plate washer with an attached reservoir is convenient if available but a 500-mL squirt bottle will serve as well. Insure that the washer cannulae are adjusted so as not to touch and damage the plate membranes.

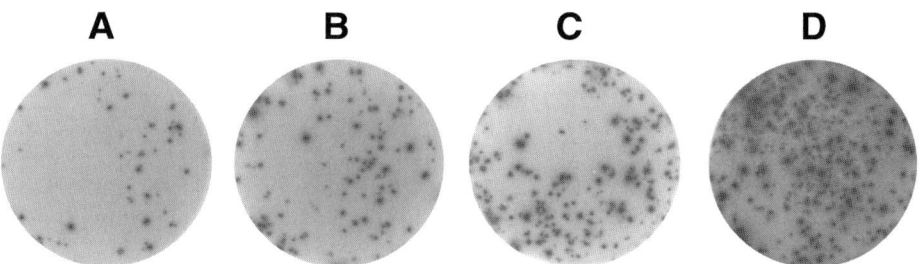

Fig. 1. Detection of antigen-specific CD8+ T-cells by ELISPOT in the spleen of mice immunized with CTL peptide epitope. Representative wells are shown after plate development. Twofold serial dilutions of pooled spleen cells from RPQASGVYM peptide-primed mice were incubated in ELISPOT plate for 36 hrs with peptide. The wells contained 0.63×10^5 (**A**), 1.25×10^5 (**B**), 2.5×10^5 (**C**), and 5×10^5 (**D**) spleen cells/well. Paired medium alone control wells all showed fewer than 15 spots per well (not shown). All results shown were from the same pool of spleen cells.

2. Flick plate to remove the bulk of cells and medium. Fill the wells with wash buffer and aspirate.
3. Repeat **step 2** three times.
4. After the final wash flick the plate and blot upside down on absorbent paper towels to remove as much liquid as possible.
5. Dilute the detection antibody per kit instructions and add 100 µl into each well. Incubate at 2–8°C overnight (*see* **Note 6**).
6. Repeat **steps 2** and **3**.
7. Dilute the streptavidin–alkaline phosphatase conjugate per kit instructions and add 100 µL into each well. Incubate at room temperature for 2 h.
8. Repeat **steps 2** and **3**.
9. Add 100 µL of the BCIP/NBT Chromagen into each well and incubate undisturbed for 1 hr at room temperature. Protect from light.
10. Remove the flexible plastic backing from the plate and fill the wells with deionized water (*see* **Note 7**). Flick the plate to remove the bulk of volume.
11. Repeat **step 10** five times.
12. Invert the plate and blot on absorbent paper towels to remove excess water. Sit the plate upside down on a paper towel and allow it to dry completely at room temperature. Placing the plate at 37°C for 30 min can shorten the drying time. Spots can then be readily visualized (**Figs. 1** and **2**)

3.4. Evaluation of Developed Plates

Developed plates can be evaluated by either manual counting with a stereoscopic microscope or by automated image analysis. The first method is simple and employs an instrument readily accessible in most laboratories. However, it

ELISPOT Stimulation

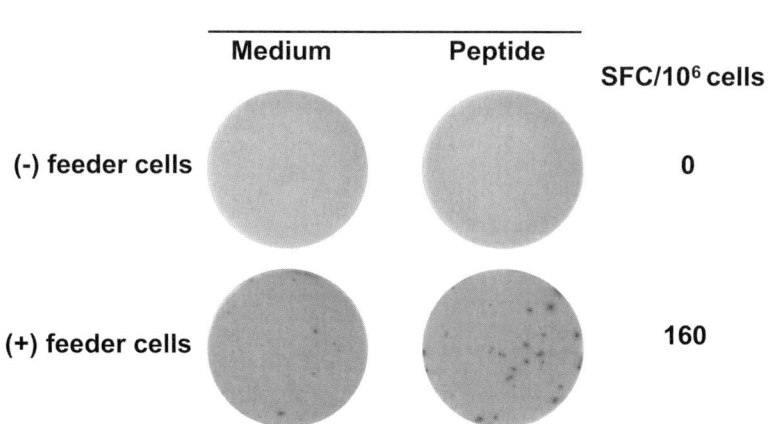

Fig. 2. Addition of feeder cells increases the sensitivity of ELISPOT assay. Pooled spleen cells from TPHPARIGL peptide-primed mice plated at 0.63×10^5 cells/well were incubated in ELISPOT for 36 h in the medium or in the presence of peptide with or without the addition of 5×10^5 feeder cells/well. Plates were developed and spots quantified by automated digital image analysis. Responses are reported as antigen-specific (peptide-medium control) SFC/10^6 immune spleen cells.

is time-consuming, subjective, and under-reports values at high spot frequencies per membrane. Automated measurement is rapid, accurate, and permits high throughput evaluation of many plates.

3.4.1. Stereomicroscopic Counting of Spots

1. Place the dried plate atop a soft paper towel on the stage of the stereomicroscope adjusted to 25X magnification. If microscope lacks an integral light source, membranes can be illuminated using a fluorescent lamp with a flexible stalk.
2. View the ConA-stimulated wells with clearly evident spots to optimize illumination and focus.
3. Count and record the spots. A hand-held push button counter is convenient for this step. Each membrane should be counted in duplicate. Gray-to-black spots will vary somewhat in size and in intensity. However, spots produced by cytokine-secreting cells will appear round with fuzzy or diffuse edges and will show decreasing staining intensity from their center to their periphery. Spots often show asymmetric distribution across the membrane and may abut or overlay each other (*see* **Figs. 1** and **2**). Staining artifacts may be seen, particularly with high cell input numbers per well (*see* **Note 8**). Plate background control wells should be relatively free of spots.

3.4.2. Automated Evaluation of ELISPOT

A variety of automated ELISPOT digital quantification systems as well as commercial optical analysis services are available. Here, quantification of ELISPOT plates performed done using a computer-assisted ImageHub plate reader by a contract service (microanalysis@tcinternet.net; *see* **Note 9**). The resulting spot counts were provided in Excel spreadsheet format. High-resolution photographs of individual membranes were captured and stored as TIFF image files (**Figs. 1** and **2**). Membranes also were removed and affixed to a transparent adhesive backing for compact storage.

3.4.3. Reporting ELISPOT Data

ELISPOT results are commonly reported as IFN-γ spot-forming cells (SFCs) per 10^6 input cells. The number of antigen-specific cells is calculated as the spots per well obtained in the presence of antigen stimulation minus the spots per well obtained in the unstimulated control. A specific response is demonstrated when (1) IFN-γ secretion is detected in response to ConA stimulation; (2) the frequency of SFC detected in presence of antigen is at least two times the frequency of SFC detected in the absence of antigen; and (3) a titratable response is obtained with serially diluted effector cells. When testing serial dilutions of an immune splenocyte population, we report the highest value quantified as the specific response (*see* **Note 10**).

4. Notes

1. Screen mesh filtration of splenocytes prior to plating is essential for preventing macroscopic staining artifacts on membranes.
2. Ready-to-use ELISPOT kits can be purchased from multiple sources. Those who find ready-to-use kits unsuitable for their needs can refer to detailed protocols for preparing murine IFN-γ ELISPOT from individual component reagents *(20,21)*.
3. Triplicate wells were plated here to demonstrate the interwell variability associated with the assay, which generally ranges from negligible up to 15–20% (*see* **Fig. 3**). As a practice we use duplicate wells when evaluating splenocyte/test combinations.
4. In our assays final DMSO concentrations up to 0.5% showed no statistically significant effect on the either antigen-specific or antigen-independent (background) IFN-γ secretion. A 20 mg/mL nonapeptide stock diluted to 10^{-5} *M* (approx 10 μg/mL) final concentration in ELISPOT assay yields a final DMSO concentration of 0.05%. Although experimental design would call for inclusion of DMSO in controls at the same concentration as in peptide wells, operationally it is unnecessary in this system.
5. A commonly cited incubation time is 24 h, which may be based more on operational convenience than biologic mechanism. Power et al. *(19)* reported that murine IFN-γ ELISPOT results obtained with cell-antigen incubation times of 8 and 24 h yielded essential equivalent spot frequencies. IFN-γ secretion by primed murine

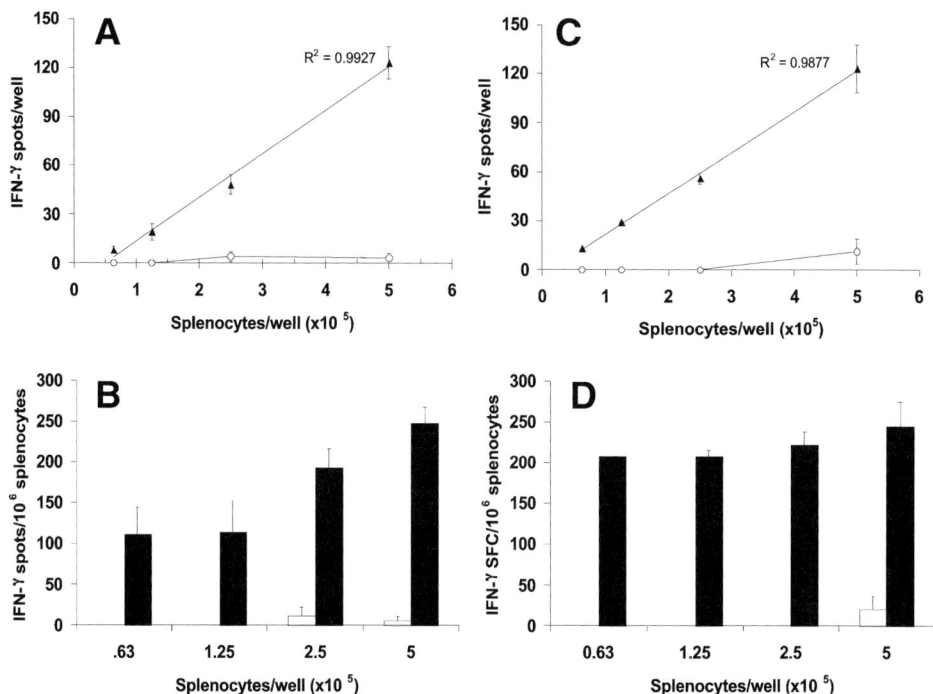

Fig. 3. ELISPOT measurement of antigen-specific CD8[+] T-cells as a function of cell plating density. Twofold serial dilutions of pooled spleen cells from mice immunized with either RPQASGVYM (**A,B**) or TPHPARIGL (**C,D**) were incubated for 36 h in ELISPOT with either medium (open circle) or peptide (black triangle) and the plates were developed. Spots were quantified by automated digital image analysis. The correlation between the splenocytes plated (x-axis) versus the number of spots counted (y-axis) is shown (**A,C**). The frequency of SFC/10[6] spleen cells determined for each dilution plated is shown (**B,D**). Each bar represents the mean ± standard deviation for three replicate wells. Results shown for each individual immune spleen cell population were from the same pool of spleen cells.

CD8[+] T-cells is detected by intracellular cytokine staining 6–8 h after antigenic stimulation (*22*). Although the optimal incubation time should be experimentally determined for each system, 24 h should be sufficient in most cases.

6. The R&D Systems kit instructions specify an overnight incubation. Equally good results are obtained by performing the detection antibody incubation for 2 h at room temperature on an orbital rotator (250 RPM). This shorter incubation time enables assay harvest and complete plate development on the same day.

7. Developed plates will occasionally show a diffuse peripheral darkening of the membranes that hampers visual discrimination of spots. Removal of the flexible

plastic backing from the plate prior to washing out the BCIP/NBT Chromagen solution can help reduce this occurrence.

8. Staining artifacts include small flakes of particulate Chromagen, generalized darkening of membranes, and small punctate spots of low color intensity. We find that background generally increases with longer assay incubation times and with higher cell input numbers. A simple measure to minimize background is to thoroughly wash the plate between all incubations steps and at completion of the BCIP/NBT Chromagen incubation.

9. When spot density is high (approx >100/well) discrimination of spots with the stereomicroscope becomes difficult and automated digital analysis is required to accurately quantify ELISPOT results.

10. Criteria for what constitutes an objective "positive" response are investigator-defined and vary depending of the experimental system. We designate a specific response of ≥30 SFC/10^6 splenocytes as positive. The criteria used in here essentially correspond with those proposed for assessing IFN-γ ELISPOT results from human PBMC stimulated with recall antigens *(23)*.

Acknowledgments

The author thanks Dr. Urban Ramstedt for his helpful suggestions. Point Therapeutics, Inc supported this work.

References

1. Czerkinsky, C. C., Nilsson, L. A., Nygren, H., Ouchterlony, O., and Tarkowski, A. (1983) A solid-phase enzyme-linked immunospot (ELISPOT) assay for enumeration of specific antibody-secreting cells. *J. Immunol. Methods* **16**, 109– 121.
2. Czerkinsky, C., Andersson, G., Ekre, H. P., Nilsson, L. A., Klareskog, L., and Ouchterlony, O. (1988) Reverse ELISPOT assay for clonal analysis of cytokine production: I. Enumeration of gamma-interferon-secreting cells. *J. Immunol. Methods* **110**, 29–36.
3. Tanguay, S., and Killion, J. J. (1994) Direct comparison of ELISPOT and ELISA-based assays for detection of individual cytokine-secreting cells. *Lymphokine Cytokine Res.* **13**, 259–263.
4. Carter, L. L., and Swain, S. L. (1997) Single cell analyses of cytokine production. *Curr. Opin. Immunol.* **2**, 177–182.
5. Altman, J. D., Moss, P. A. H., Goulder, P. J. R., Barouch, D. H., McHeyzer-Williams, M. G., Bell, J. I., et al. (1996) Phenotypic analysis of antigen-specific T-lymphocytes. *Science* **274**, 94–96.
6. Sarawar, S. R., and Doherty, P. C. (1994) Concurrent production of interleukin-2, interleukin-10, and gamma interferon in the regional lymph nodes of mice with influenza pneumonia. *J. Virol.* **68**, 3112–3119.
7. Mo X. Y., Sarawar, S. R., and Doherty, P. C. (1995) Induction of cytokines in mice with parainfluenza pneumonia. *J. Virol.* **69**, 1288–1291.
8. Murata, K., Garcia-Sastre, A., Tsuji, M., Rodrigues, M., Rodriguez, D., Rodriguez, J. R., et al. (1996) Characterization of in vivo primary and secondary CD8[+] T-cell

responses induced by recombinant influenza and vaccinia viruses. *Cell. Immunol.* **173**, 96–107.

9. Murali-Krishna, K., Altman, J. D., Suresh, M., Sourdive, D. J., Zajac, A. J., Miller, J. D., et al. (1998) Counting antigen-specific CD8 T-cells: a reevaluation of bystander activation during viral infection. *Immunity* **8**, 177–187.

10. Oliveira-Ferreira, J., Miyahira, Y., Layton, G. T., Savage, N., Esteban, M., Rodriguez, D., et al. (2000) Immunogenicity of Ty-VLP bearing a CD8+ T-cell epitope of the CS protein of *P. yoelii*: enhanced memory response by boosting with recombinant vaccinia virus. *Vaccine* **18**, 1863–1869.

11. Tobery, T. W., Wang, S., Wang, X. M., Neeper, M.P, Jansen, K. U., McClements, W. L., and Caulfield, M. J. (2001) A simple and efficient method for the monitoring of antigen-specific T-cell responses using peptide pool arrays in a modified ELISpot assay. *J Immunol. Methods* **254**, 59–66.

12. Maecker, B., Sherr, D. H., Vonderheide, R. H., von Bergwelt-Baildon, M. S., Hirano, N., Anderson, K. S., et al. (2003) The shared tumor-associated antigen cytochrome P450 1B1 is recognized by specific cytotoxic T-cells. *Blood* **102**, 3287–3294.

13. Whitton, J. L., Tishon, A., Lewicki, H., Gebhard, J., Cook, T., Salvato, M., et al. (1989) Molecular analyses of a five-amino-acid cytotoxic T-lymphocyte (CTL) epitope: an immunodominant region which induces nonreciprocal CTL cross-reactivity. *J. Virol.* **63**, 4303–4310.

14. Schulz, M., Aichele, P., Schneider, R., Hansen, T. H., Zinkernagel, R. M., and Hengartner H. (1989) Major histocompatibility complex binding and T-cell recognition of a viral nonapeptide containing a minimal tetrapeptide. *Eur. J. Immunol.* **21**, 1181–185.

15. Galvin, M. A., Gilbert, M. J., Riddell, S. R., Greenberg, P. D., and Bevan, M. J. (1993) Alkali hydrolysis of recombinant proteins allows for the rapid identification of class I MHC-restricted CTL epitopes. *J. Immunol.* **151**, 3971–3980.

16. Vitiello, A., Ishioka G, Grey. H. M., Rose, R., Farness, P., LaFond, R., et al. (1995) Development of a lipopeptide-based therapeutic vaccine to treat chronic HBV infection. I. Induction of a primary cytotoxic T-lymphocyte response in humans. *J. Clin .Invest.* 95, 341–349.

17. van der Most, R. G., Sette, A., Oseroff, C., Alexander, J., Murali-Krishna, K., Lau, L. L., et al. (1996) Analysis of cytotoxic T-cell responses to dominant and subdominant epitopes during acute and chronic lymphocytic choriomeningitis virus infection. *J. Immunol.* **157**, 5543–5554.

18. Kalyuzhny, A., and Stark, S. (2001) A simple method to reduce the background and improve well-to-well reproducibility of staining in ELISPOT assays. *J. Immunol. Methods* **257**, 93–97.

19. Power, C. A., Grand, C. L., Ismail, N., Peters, N. C., Yurkowski, D. P., and Bretscher, P. A. (1999) A valid ELISPOT assay for enumeration of ex vivo, antigen-specific, IFN gamma-producing T-cells. *J. Immunol. Methods* **227**, 99–107.

20. Miyahira, Y., Murata, K., Rodriguez, D., Rodriguez , J. R., Esteban, M., Rodrigues, M. M., et al. (1995) Quantification of antigen specific CD8+ T-cells using an ELISPOT assay. *J. Immunol. Methods* **181**, 45–54.

21. Carvalho, L. H., Hafalla, J. C., and Zavala F. (2001) ELISPOT assay to measure antigen-specific murine CD8+ T-cell responses. *J. Immunol. Methods* **252**, 207–218.
22. Slifka, M. K., Rodriguez, F., and Whitton, J. L. (1999) Rapid on/off cycling of cytokine production by virus-specific CD8+ T-cells. *Nature* **401**, 76–79.
23. Currier J. R., Kuta, E. G., Turk, E., Earhart, L. B., Loomis-Price, L., Janetzki, S., et al. (2002) A panel of MHC class I restricted viral peptides for use as a quality control for vaccine trial ELISPOT assays. *J. Immunol Methods* 260, 157–172.

III

ELISPOT ASSAYS IN HUMAN STUDIES

13

Detection of Measles Virus-Specific Interferon-γ-Secreting T-Cells by ELISPOT

Jenna E. Ryan, Inna G. Ovsyannikova, and Gregory A. Poland

Summary

The enzyme-linked immunospot (ELISPOT) assay is a highly sensitive tool used to measure the frequency of antigen-specific T-cells in vitro. Among its many applications, this assay is useful for the characterization of cellular immune responses after immunization against measles and other viral pathogens. A description of the measles virus-specific interferon (IFN)-γ ELISPOT assay optimized in our laboratory is described in detail in this chapter. Procedures for the preparation of measles virus, infection of peripheral blood mononuclear cells with measles virus, and the IFN-γ ELISPOT assay are also outlined. These methods can also be broadly adapted to measure activated T-cells to other viral pathogens and/or pathogen-derived peptides. Therefore, the ELISPOT assay can be used for the design, development, and evaluation of new vaccines.

Key Words: ELISPOT; measles virus; IFN-γ; human PBMC; Vero cells; cryopreservation; $TCID_{50}$.

1. Introduction

Measles is the most transmissible human infectious viral agent known and continues to be a major health problem for children in developing countries. Measles immunization protects against the disease by inducing humoral and cellular immunity. Because both humoral and cell-mediated immunity play essential roles in the immune response to measles, it is important to improve and develop methods to evaluate both arms of the measles virus immune response. Because of the strong immunosuppressive effects of measles and the low precursor frequencies of measles-specific helper T-lymphocytes in peripheral blood, cell-mediated immune responses against measles have been particularly difficult to study. With the materialization of new technologies, assays like enzyme-linked immunospot (ELISPOT) provide an opportunity to study and quantitate antigen-specific T-cells in vitro (*1*).

From: *Methods in Molecular Biology, vol. 302: Handbook of ELISPOT: Methods and Protocols*
Edited by: A. E. Kalyuzhny © Humana Press Inc., Totowa, NJ

We have optimized the use of the ELISPOT assay to determine T-cell activation in response to measles antigen through measurement of interferon (IFN)-γ secretion. This assay allows us to measure functional recall immunity, to determine the measles-specific T-cell frequency in peripheral blood mononuclear cells (PBMCs), and to determine the Th1 cytokine secretion profile in the cells of immunized individuals. Because the number of T-cells from individual subjects can be limited, the opportunity to monitor cytokines at the single cell level is advantageous *(2,3)*. Other advantages of the ELISPOT assay include the ability to measure the functional properties of activated T-cells, to measure the frequency of measles-specific T-cells *(4)*, and to directly determine the measles virus-specific T-cell frequency without additional T-cell in vitro expansion *(5)*. The ELISPOT assay described in this chapter was used to monitor cellular immune responses following measles immunization.

2. Materials

2.1. Collecting and Isolating PBMCs

1. 10 mL Sodium heparin blood collection tubes (Sherwood Medical, St. Louis, MO).
2. 15 mL Sterile polypropylene conical centrifuge tubes (BD Biosciences, San Jose, CA).
3. 50 mL Sterile polypropylene conical centrifuge tubes (BD Biosciences).
4. 1X Phosphate-buffered saline (PBS; Roche Diagnostics Corporation, Indianapolis, IN), prewarmed to 37°C.
5. Ficoll-Paque Plus (Amersham Biosciences, Piscataway, NJ).
6. RPMI culture medium: RPMI 1640 with 25 mM HEPES (Celox Laboratories Inc., St. Paul, MN), 100 U/mL penicillin–100 µg/mL streptomycin (Sigma, St. Louis, MO), and 1 mM sodium pyruvate (Gibco Life Technologies, Gaithersburg, MD), prewarmed to 37°C.
7. Trypan Blue (Sigma).
8. Freezing medium: RPMI 1640 with 25 mM HEPES (Celox Laboratories Inc.), 10% dimethyl sulfoxide (Sigma), and 20% fetal calf serum (FCS; Hyclone, Logan, UT).
9. Cryogenic freezing vials (Nalge Nunc International, Rochester, NY).

2.2. Thawing of Cryopreserved PBMCs

1. 15 mL Sterile polypropylene conical centrifuge tubes.
2. RPMI culture medium supplemented with DNase: RPMI 1640 with 25 mM HEPES (Celox Laboratories Inc.), 10% FCS (Hyclone), 10 µg/mL DNase (Sigma), 100 U/mL penicillin–100 µg/mL streptomycin (Sigma), and 1 mM sodium pyruvate (Gibco), prewarmed to 37°C.
3. RPMI culture medium supplemented with 0.2% bovine serum albumin (BSA): RPMI 1640 with 25mM HEPES (Celox Laboratories Inc.), 0.2% BSA (Sigma),

100 U/mL penicillin–100 µg/mL streptomycin (Sigma), and 1 m*M* sodium pyruvate (Gibco); use at 4°C.
4. Trypan Blue (Sigma).

2.3. Infecting Vero Cells With the Measles Virus

1. Vero cells (American Type Culture Collection [ATCC], Manassas, VA, Number CCL-81).
2. 75 cm² Sterile tissue culture flasks (Corning Incorporated, Corning, NY).
3. 1X PBS (Roche Diagnostics Corporation), prewarmed to 37°C.
4. Measles virus Edmonston vaccine strain (ATCC, Number VR-24).
5. Opti-MEM I Reduced Serum Medium (Gibco).
6. DMEM culture medium supplemented with 5% FCS: Dulbecco's Modified Eagle Medium (Gibco), 5% FCS (Hyclone).

2.4. Harvesting the Measles Virus

1. Opti-MEM I Reduced Serum Medium (Gibco).
2. Cell lifters (Corning Incorporated).
3. 15 mL Sterile polypropylene conical centrifuge tubes (BD Biosciences).
4. 50 mL Sterile polypropylene conical centrifuge tubes (BD Biosciences).
5. Cryogenic tubes (Nalge Nunc International).

2.5. Preparing Control Vero Cell Lysate

1. Vero cells (*see* **Subheading 2.3.**).
2. 75 cm² Sterile tissue culture flasks (Corning Incorporated).
3. 1X PBS (Roche Diagnostics International).
4. Opti-MEM I Reduced Serum Medium (Gibco).
5. 15 mL Sterile polypropylene conical centrifuge tubes (BD Biosciences).
6. 50 mL Sterile polypropylene conical centrifuge tubes (BD Biosciences).
7. Cryogenic tubes (Nalge Nunc International).

2.6. Titrating the Measles Virus

1. 96-Well flat-bottom tissue culture plates (Corning Incorporated).
2. 1.5 mL Sterile centrifuge tubes (Sarstedt Inc., Newton, NC).
3. Vero cells (*see* **Subheading 2.3.**).
4. 10 µL Measles virus stock.
5. DMEM culture medium: DMEM (Gibco) with 5% FCS (Hyclone).

2.7. Infecting PBMCs With the Measles Virus

1. Human IFN-γ ELISpot kit (cat. no. EL285, R&D Systems, Minneapolis, MN).
2. Plain RPMI culture medium: RPMI 1640 with 25 m*M* HEPES (Celox Laboratories, Inc).
3. RPMI culture medium supplemented with 0.2% BSA: RPMI 1640 with 25 m*M* HEPES (Celox Laboratories, Inc.), 0.2% BSA (Sigma), 100 U/mL penicillin-100 µg/mL streptomycin (Sigma), and 1 m*M* sodium pyruvate (Gibco); use at 4°C.

4. Measles virus, multiplicity of infection (MOI) of 0.5, diluted in RPMI culture medium with 0.2% BSA.
5. Vero cell lysate (MOI equivalent of 0.5) diluted in RPMI culture medium with 0.2% BSA.
6. Phytohemagglutinin-P (PHA-P) (Sigma) diluted to 20 μg/mL in RPMI culture medium with 0.2% BSA.
7. RPMI culture medium with 20% pooled human AB serum: RPMI 1640 with 25 mM HEPES (Celox Laboratories, Inc.), 20% pooled human AB serum (PEL-FREEZ Clinical Systems, LLC, Brown Deer, WI), 100 U/mL penicillin-100 μg/mL streptocmycin (Sigma), and 1mM sodium pyruvate (Gibco).

2.8. Detecting Measles Virus-Specific IFN-γ-Secreting T-Cells by ELISPOT

1. Human IFN-γ ELISpot kit (cat. no. EL285, R&D Systems).
2. Dissecting microscope or automated ELISPOT reader.

3. Methods

The following methods describe (1) collecting and isolating PBMCs; (2) thawing of cryopreserved PBMCs; (3) infecting Vero cells with measles virus; (4) harvesting measles virus; (5) preparing control Vero cell lysate; (6) titrating measles virus; (7) infecting PBMCs with measles virus; and (8) detecting measles virus-specific IFN-γ-secreting T-cells by ELISPOT.

3.1. Collecting and Isolating PBMCs

1. Collect 10 mL of whole blood into a 10-mL heparinized blood collection tube (*see* **Note 1**).
2. Invert the tube of blood gently five to six times to ensure thorough mixing of cells.
3. Add the 10 mL of whole blood to one 50-mL conical tube.
4. Dilute the blood 1:1 with 1X PBS prewarmed to 37°C.
5. Mix the diluted blood five to six times by gently inverting the tube.
6. Transfer 10 mL of the blood/PBS mixture to two 15-mL conical tubes.
7. Place a Pasteur pipet into each conical tube and pipet 3.2 mL of Ficoll into the Pasteur pipet so that it slowly drains into the blood/PBS mixture.
8. Centrifuge the conical tubes at 900g for 30 min at room temperature with the centrifuge brake off.
9. Remove the PBMC layer with a Pasteur pipet into one 50-mL conical tube containing 10 mL of prewarmed (37°C) RPMI 1640 culture medium. Ensure that no red blood cells are transferred with the lymphocytes (*see* **Note 2**).
10. Adjust the volume to 50 mL with RPMI 1640 culture medium and wash the cells by gently inverting the tube four to five times.
11. Pellet the cells by centrifugation at 500g for 10 min at room temperature with the brake on.
12. Aspirate the supernatant without disturbing the pellet.

13. Resuspend the cell pellet from one donor in 5 mL of plain RPMI 1640 culture medium.
14. Adjust the volume to 50 mL with plain RPMI 1640 culture medium.
15. Mix thoroughly by inverting the conical tube to form a single cell suspension, ensuring that there are no clumps of cells before counting.
16. Add 25 µL of cell suspension, 75 µL of trypan blue, and 400 µL of plain RPMI culture medium to a small tube. Mix thoroughly.
17. Determine cell number and viability with a hemacytometer and calculate cell concentration (*see* **Note 3**).
18. Pellet the cells by centrifugation at 500*g* for 10 min at room temperature with the brake on.
19. Adjust the cells to a concentration of 1×10^7 cells/mL with freezing medium.
20. Aliquot the cell suspension into prelabeled cryogenic freezing tubes in 1-mL aliquots.
21. Place the cryogenic freezing tubes into a controlled-rate freezing container. Place freezing container in –80°C freezer overnight.
22. Transfer the cryogenic freezing tubes to a liquid nitrogen storage tank for long-term storage (*see* **Note 4**).

3.2. Thawing of Cryopreserved PBMCs

The use of cryopreserved PBMCs in the ELISPOT assay allows for the analysis of a large number of individuals at one time while minimizing the variation associated with multiple assays. Previous studies have shown that cryopreserved PBMCs can be used in the ELISPOT assay without compromising their ability to secrete cytokines *(6–8)*.

1. Remove one vial of PBMCs (cell concentration of 1×10^7 cells/mL) for each sample from the liquid nitrogen storage tank.
2. Rapidly thaw PBMCs by swirling cryogenic freezing vial in a 37°C water bath until a small amount of ice remains.
3. Disinfect the outer surface of the vial with 70% ethanol.
4. Transfer the cells slowly with a 5-mL pipet to a 15-mL polypropylene conical tube without introducing air bubbles.
5. Add 100 µL of RPMI culture medium with DNase to the cryogenic freezing tube to rinse out any remaining cells and transfer to the 15 mL conical tube containing the cells (*see* **Note 5**).
6. Wait 1 min, then double the amount of medium (200 µL) added to the conical tube in a dropwise manner. Gently shake the conical tube to mix the contents.
7. Continue doubling the amount of medium every minute until cells reach a final volume of 10 mL.
8. Centrifuge at 500*g* for 7 min at room temperature to pellet the cells.
9. Aspirate the supernatant and resuspend the cell pellet in 100 µL of RPMI culture medium with DNase by gently tapping the side of the conical tube (*see* **Note 6**).
10. Adjust the volume to 10 mL with RPMI culture medium containing DNase.

11. Incubate in a 37°C water bath for 20 min with intermittent inverting of tube (*see* **Note 7**).
12. Cool the tube of cells on ice for 5 to 7 min.
13. Centrifuge at 500g for 7 min at 4°C to pellet the cells.
14. Aspirate the supernatant and resuspend the cell pellet in 1.6 mL of RPMI culture medium containing 0.2% BSA by gently tapping the bottom of the conical tube (*see* **Note 6**). Keep the cells on ice.
15. Add 25 µL of the cell suspension, 75 µL of trypan blue, and 400 µL of plain RPMI culture medium to a small tube. Mix thoroughly.
16. Determine cell number and viability with a hemacytometer and calculate cell concentration (*see* **Note 3**).
17. Resuspend the cells to a final concentration of 2×10^6 cells/mL with the addition of cold (4°C) RPMI 1640 culture medium containing 0.2% BSA.
18. Keep the cells on ice until the time of plating.

3.3. Infecting Vero Cells With the Measles Virus

The measles virus stocks should be prepared in advance and stored at –80°C.

1. Seed 1.5×10^6 Vero cells in DMEM culture medium containing 5% FCS into each of eight 75-cm^2 tissue culture flasks.
2. Incubate the flasks at 37°C in a 5% CO_2 humidified incubator until 70 to 80% confluent (approx 3 to 4 d).
3. Aspirate off the growth medium.
4. Wash the cells with 5 mL of 1X PBS, prewarmed to 37°C.
5. Infect the Vero cells with 3 mL of measles virus (Edmonston vaccine strain) at a MOI of 0.05–0.1 in Opti-MEM I Reduced Serum Medium per sterile flask (*see* **Note 8**). Ensure that the cells are completely covered with medium.
6. Incubate the flasks with measles-infected Vero cells at 37°C in a 5% CO_2 humidified incubator for a minimum of 2 h and a maximum of 4 h.
7. Remove the measles virus inoculum by aspiration.
8. Add 10 mL of DMEM culture medium containing 5% FCS to each flask.
9. Grow the measles virus at 37°C in a 5% CO_2 humidified incubator for 24–36 h until the cells have formed into an approx 80–90% syncytia. Observe for cytopathic effect both in the morning and afternoon. Increase observation frequency as cells near syncytia.

3.4. Harvesting the Measles Virus

1. Remove the tissue culture flasks from the incubator and aspirate off medium.
2. Add 0.1 mL of Opti-MEM I Reduced Serum Medium to each 75-cm^2 flask.
3. Scrape the cells from the flask surface using a sterile cell lifter.
4. Pool the cells from each flask into a sterile 15-mL conical tube. Do not add more than 6 mL of the cell mixture per conical tube because this will affect the freeze/thaw efficiency.
5. Freeze the conical tubes in liquid N_2 or at –80°C (*see* **Note 9**).

6. Thaw in a 37°C water bath by swirling until the cell mixture has just melted.
7. Repeat the above freeze-thaw cycle one time.
8. Centrifuge at 500g for 10 min at 4°C to pellet the cell debris.
9. Remove the supernatant from each conical tube and pool into one 50-mL conical tube.
10. Mix thoroughly by pipetting without creating air bubbles (*see* **Note 10**).
11. Aliquot 20 µL of the measles virus preparation to a cryogenic tube for titration.
12. Transfer the remaining measles virus preparation into prelabeled cryogenic freezing tubes in 0.3-mL aliquots.
13. Freeze the measles virus stocks at –80°C.

3.5. Preparing Control Vero Cell Lysate

1. Seed 1.5×10^6 Vero cells in DMEM medium containing 5% FCS into each of eight 75-cm^2 tissue culture flasks.
2. Incubate the flasks at 37°C in a 5% CO_2 humidified incubator until cells are 100% confluent.
3. Aspirate off the growth medium.
4. Wash the cells with 5 mL of sterile 1X PBS, prewarmed to 37°C.
5. Add 0.1 mL of Opti-MEM I Reduced Serum Medium to each 75-cm^2 flask.
6. Scrape the cells from the flask surface using a sterile cell lifter.
7. Pool the cells from each flask into a sterile 15-mL conical tube. Do not add more than 6 mL of the cell mixture per conical tube because this will affect freeze/thaw efficiency.
8. Freeze the conical tubes in liquid N_2 or at –80°C (*see* **Note 9**).
9. Agitate the conical tubes in a 37°C water bath by swirling until the cell mixture has just melted.
10. Repeat the above freeze/thaw cycle one time.
11. Centrifuge at 500g for 10 min at 4°C to pellet the cell debris.
12. Remove the supernatant from each centrifuge tube and pool into one 50-mL conical tube. Mix thoroughly.
13. Transfer the Vero cell lysate into prelabeled cryogenic freezing tubes in 0.3 mL aliquots.
14. Freeze the Vero cell lysate stocks in either liquid nitrogen or at –80°C.

3.6. Titrating the Measles Virus

This method uses the Spearman-Karber method of 50% tissue culture infective dose ($TCID_{50}$) titration *(9,10)*. Measles is a replicating virus and syncytia and cytopathic effect are visible on Vero cells after infection. The $TCID_{50}$ represents the virus dose that gives rise to the cytopathic effect in 50% of the measles virus-infected Vero cells.

1. The day before titration of the measles virus, seed 1.4×10^5 Vero cells/mL in DMEM culture medium containing 5% FCS (50 µL/well) in a 96 well flat-bottom tissue culture plate. This assay requires one row (eight wells) per virus dilution to titrate the virus stock and one row for the control.

2. Remove the cryogenic freezing tube from −80°C containing 20 μL of the measles virus stock (from **step 11** in **Subheading 3.4.**) and thaw on ice. Mix the virus stock thoroughly. Keep the virus stock on ice until ready to use.

3. Label seven 1.5-mL sterile centrifuge tubes from 10^{-2} through 10^{-8}.

4. Use 10 μL of the measles virus stock and make a 1:100 dilution of the virus in DMEM culture medium containing 5% FCS in a total volume of 1 mL. Mix thoroughly.

5. Prepare 10-fold serial dilutions of the measles virus by transferring 100 μL of the initial dilution (**step 4**) into 900 μL of DMEM containing 5% FCS. Mix thoroughly. Continue making 10-fold serial dilutions by transferring 100 μL of the previous dilution to the next dilution. Use a new pipet tip for the preparation of each dilution. Keep all tubes on ice.

6. Transfer 50 μL of each virus dilution to each well within a row of cells on the microtiter plate (use eight wells/virus dilution).

7. Transfer 50 μL of DMEM culture medium containing 5% FCS to each of the eight control wells.

8. Incubate the microtiter plate at 37°C in a 5% CO_2 humidified incubator. Examine the cells under a dissecting microscope after 4 d for syncytia formation.

9. Note how many wells possess syncytia formation (i.e., 8/8, 6/8, etc.).

10. Calculate the virus titer using the following formula (*see* **Note 11**):

$$Log_{10} (TCID_{50}/mL) = L + d (s - 0.5) + log_{10} (1/v)$$

Where L = negative log_{10} of the most concentrated virus dilution tested in which all wells are positive; d = log_{10} of dilution factor; s = sum of individual proportions; and v= volume of inoculum (mL/well).

11. Once $TCID_{50}/mL$ is determined, multiply by 0.7 to obtain PFU/mL.

3.7. Infecting PBMCs With the Measles Virus

1. Thirty minutes before adding the cells, add 200 μL of plain RPMI culture medium to all wells of a PVDF-backed microtiter plate pre-coated by the manufacturer (R&D Systems) with monoclonal anti-human IFN-γ.

2. Completely remove medium from all wells by aspiration and pat plate until dry on a paper towel.

3. Add 100 μL of the cell suspension to each of seven wells. Final concentration is 2 × 10^5 cells/well (*see* **Note 12**).

4. Thaw the measles virus and the Vero cell lysate under a cool stream of water, immediately dilute with cold (4°C) RPMI 1640 culture medium containing 0.2% BSA to a MOI of 0.5 (*see* **Notes 13** and **14**).

5. Add 50 μL of the measles virus (MOI 0.5) or Vero cell lysate (MOI equivalent of 0.5) to each of the triplicate wells.

6. Add 50 μL of PHA-P to one well to serve as positive control. Final concentration of PHA-P in each well is 5 μg/mL.

7. Cover plate with aluminum foil and incubate for 2 h at 37°C in a 5% CO_2 humidified incubator (*see* **Note 15**).

8. Add 50 µL of RPMI culture medium containing 20% pooled human AB serum to each well (*see* **Note 16**).
9. Incubate the plate for 42 h at 37°C in a 5% CO_2 humidified incubator (*see* **Note 17**).

3.8. Detecting Measles Virus-Specific IFN-γ-Secreting T-Cells Using ELISPOT

The ELISPOT assay for detection of measles virus-specific IFN-γ-secreting T-cells is performed using a commercially available kit (Human IFN-γ ELISpot, R&D Systems) following the manufacturer's protocol and is briefly outlined in the next section. After development of the ELISPOT plate, the number of IFN-γ-secreting cells is determined with the aid of either a dissecting microscope or an automated ELISPOT reader. However, even with the use of an automated reader, enumeration of the plate involves a degree of subjectivity as the minimum and maximum spot size detected, intensity of the color of the spot detected, and the background balance are all parameters defined by the user *(11)*.

1. Aspirate the wells to remove the cell suspension.
2. Wash the plate four times and pat plate on paper towel until dry.
3. Add 100 µL of diluted biotinylated polyclonal human IFN-γ to each well. Incubate at 4°C overnight.
4. Remove the antibody by aspiration. Wash the plate four times and pat on paper towel until dry.
5. Add 100 µL of diluted streptavidin conjugated to alkaline phosphatase to each well. Incubate for 2 h at room temperature.
6. Remove the conjugate by aspiration and wash plate four times, pat on paper towel until dry.
7. Add 100 µL of BCIP/NBT chromogen to each well. Incubate for 1 hour at room temperature in the dark (*see* **Note 18**).
8. Remove the chromogen solution from each well and rinse microtiter plate with deionized water.
9. Allow the plate to dry completely before determining the number of IFN-γ-secreting cells using either a dissecting microscope or an automated ELISPOT reader.

The sensitivity of the described ELISPOT assay in our laboratory is the detection of an average of 15 measles virus-specific IFN-γ-secreting cells/2×10^5 PBMCs (the range of responses was between 1 and 99 spots).

4. Notes

1. Keep the tubes of whole blood at room temperature until use. Isolate PBMCs within 2 h of collection.
2. The addition of plain RPMI 1640 culture medium prevents the cells from sticking to the bottom and sides of the 50-mL conical tube when transferred. Remove the entire interface with a minimum amount of Ficoll contamination.

3. Cell concentration (number of cells/mL) = average cell count per square × dilution factor × 10^4.
4. While transferring the cryogenic tubes to liquid nitrogen storage tank, place on dry ice to ensure that cells do not begin to thaw.
5. DNase is added to the culture medium to prevent clumping of cells.
6. Do not pipet the cells up and down to resuspend.
7. Invert the conical tube several times every 5 min to mix the cell suspension.
8. The stock of the measles virus inoculum should be at least 1×10^7 virus particles.
9. Freeze the conical tubes overnight if freezing at –80°C before proceeding. The conical tubes must be cryosafe if freezing in liquid nitrogen.
10. The presence of air bubbles will inactivate the measles virus.
11. Example calculation of $TCID_{50}$/mL
    Number of wells with syncytia formation at each dilution of virus:

    10^{-2} 8/8 10^{-3} 8/8 10^{-4} 8/8 10^{-5} 8/8 10^{-6} 6/8 10^{-7} 2/8 10^{-8} 0/8

    $L = 5$

    $D = \log_{10}(10) = 1$

    $s = 8/8 + 6/8 + 2/8 = 2$

    $v = .05$ mL, $1/v = 1/.05$ mL $= 20$, $\log_{10}(20) = 1.3$

    $\log_{10}(TCID_{50}/mL) = 5 + (2-0.5) + 1.3 = 6.3 \times 10^7$

    To calculate PFU/mL, multiply by 0.7, a coefficient calculated from the Poisson distribution.

    $6.3 \times 10^7(0.7) = 4.4 \times 10^7$ PFU/mL

12. Mix the cells before adding to each well to ensure a uniform cell suspension. The cells should be plated previously at various concentrations to determine the optimal cell concentration to use in the assay.
13. Do not freeze-thaw the measles virus multiple times as this destroys virus activity. Immediately after thawing, dilute the virus in cold RPMI culture medium containing 0.2% BSA. Keep the virus on ice until use.
14. The Vero cell lysate is prepared from the same number of Vero cells as the cells that are infected with measles virus; therefore, it is assumed that the Vero cell lysate would have an equivalent number of particles/mL. MOI = number of virus particles/number of cells.
15. Covering the plates with aluminum foil reduces well-to-well variability, as previously reported *(12)*.
16. It is important to test several different lots of human AB serum to ensure that there is no background reactivity.
17. Place the ELISPOT plates in a separate incubator to ensure that cells do not move during the incubation, which could create artificial spots.
18. If background is a problem, reduce the length of incubation with the chromogen solution to 30 min.

Acknowledgments

We thank the members of the Mayo Vaccine Research Group for technical assistance and helpful discussion regarding the development of the assay. We also thank Kim Zabel for her editorial assistance in the preparation of this chapter. This work was supported by NIH grant AI33144.

References

1. Ovsyannikova, I. G., Dhiman, N., Jacobson, R. M., Vierkant, R. A., and Poland, G. A. (2003) Frequency of measles virus-specific CD4+ and CD8+ T-cells in subjects seronegative or highly seropositive for measles vaccine. *Clin. Diag. Lab. Immunol.* 10, 411–416.
2. Bercovici, N., Duffour, M-T., Agrawal, S., Salcedo, M., and Abastado, J-P. (2000) New methods for assessing T-cell responses. *Clin. Diag. Lab. Immunol.* 7, 859–864.
3. Carter, L. L., and Swain, S. L. (1997) Single cell analyses of cytokine production. *Curr. Opin. Immunol.* 9, 177–182.
4. Nanan, R., Rauch, A., Kämpgen, E., Niewiesk, S., and Kreth, H. W. (2000) A novel sensitive approach for frequency analysis of measles virus-specific memory T-lymphocytes in healthy adults with a childhood history of natural measles. *J. Gen. Virol.* 81, 1313–1319.
5. Carvalho, L. H., Hafalla, J. C. R., and Zavala, F. (2001) ELISPOT assay to measure antigen-specific murine CD8+ T-cell responses. *J. Immunol. Methods.* 252, 207–218.
6. Smith, J. G., Liu, X., Kaufhold, R. M., Clair, J., and Caulfield, M. J. (2001) Development and validation of a gamma interferon ELISPOT assay for quantitation of cellular immune responses to varicella-zoster virus. *Clin. Diag. Lab. Immunol.* 8, 871–879.
7. Bailey, T., Stark, S., Grant, A. Hartnett, C., Tsang, M. and Kalyuzhny, A. (2002) A multidonor ELISPOT study of IL-1β, IL-2, IL-4, IL-6, IL-13, IFN-γ, and TNF-α release by cryopreserved human peripheral blood mononuclear cells. *J. Immunol. Methods.* 270, 171–182.
8. Kreher, C. R., Dittrich, M. T., Guerkov, R., Boehm, B. O., Tary-Lehmann, M. (2003) CD4+ and CD8+ cells in cryopreserved human PBMC maintain full functionality in cytokine ELISPOT assays. *J. Immunol. Methods.* 278, 79–93.
9. Karber, G. (1931) Beitrag zur kollektiven behandlung pharmakologischer reihenversuche. *Arch. Exp. Pathol. Pharmakol.* 162, 480–483.
10. Spearman, C. (1908) The method of 'right and wrong cases' ('constant stimuli') without Gauss's formulae. *Br. J. Psychol.* 2, 227–242.
11. Hernandez-Fuentes, M. P., Warrens, A. N., and Lechler, R. I. (2003) Immunologic monitoring. *Immunol. Rev.* 196, 247–264.
12. Kalyuzhny, A., and Stark, S. (2001) A simple method to reduce the background and improve well-to-well reproducibility of staining in ELISPOT assays. *J. Immunol. Methods.* 257, 93–97.

14

Epitope Mapping in Multiple Sclerosis Using the ELISPOT Assay

Clara M. Pelfrey and Ioana R. Moldovan

Summary

Multiple sclerosis (MS) is thought to be an autoimmune disease in which an unknown trigger initiates an immune response against brain proteins. This autoaggressive response causes the breakdown of the myelin sheaths that protect nerve axons, leading to impaired nerve conduction and subsequent neurodegeneration that are characteristic of MS. Many studies have attempted to determine the exact target within the brain. However, there appear to be multiple targets, which may change over time. No single study has examined all targets nor looked at how they can change over the course of the disease and whether these changes are related to the course of disease. We have approached this by using the single-cell resolution capability of the enzyme-linked immunospot assay to examine cytokine reactivity in MS patients in response to a very large set of overlapping peptides that span the two major proteins of myelin: myelin basic protein and proteolipid protein. Our goal was to use the enzyme-linked immunospot assay to perform comprehensive epitope mapping in relapsing-remitting MS patients in a longitudinal study to help define the role of myelin responses in disease progression.

Key Words: Multiple sclerosis; MS; myelin basic protein; proteolipid protein; cytokines; longitudinal study; ELISPOT.

1. Introduction

Multiple sclerosis (MS) is a chronic demyelinating disease, characterized by inflammatory infiltrates of mononuclear cells in the central nervous system (CNS). Relapsing-remitting (RR) MS is defined by a succession of clinical relapses and remissions. Although the etiology and pathogenesis of MS are not fully elucidated, extensive documentation suggests a role for autoimmune cells *(1,2)*. Myelin proteins, such as myelin basic protein (MBP), proteolipid protein (PLP), and myelin oligodendrocyte glycoprotein are considered candidate autoantigens in MS *(3–5)*. Identifying and quantifying autoimmune responses

From: *Methods in Molecular Biology, vol. 302: Handbook of ELISPOT: Methods and Protocols*
Edited by: A. E. Kalyuzhny © Humana Press Inc., Totowa, NJ

in MS has been difficult in the past because of the low frequency of autoanti-gen-specific T-cells, the high number of putative determinants on the autoanti-gens, and the different cytokine signatures of the autoreactive T-cells. An added difficulty includes the chronic myelin damage in the CNS, which releases new epitopes over time, leading to sequential activation of different myelin compo-nents, a process referred to as "epitope spreading" *(6–8)*. A clear link between epitope spreading, as measured by longitudinal myelin-induced cytokine responses, and MS disease progression is still lacking. Here, we describe a method of using the enzyme-linked immunospot (ELISPOT) assay to perform comprehensive epitope mapping in RRMS patients to help define the role of myelin responses in disease progression.

The ELISPOT assay is a solid-phase enzyme linked immunosorbent assay in which cells are stimulated to secrete cytokines. The cytokines are captured on a membrane using anticytokine monoclonal antibodies and detected using cytokine-specific detection antibodies and a chromogenic substrate. Advantages of the ELISPOT assay include: (1) it can be performed without knowing the major histocompatibility complex restriction element for an anti-gen response and (2) it is ideally suited for peptide mapping and determina-tions of very low frequency antigen-specific T-cells (range of 1:10,000 to 1:1,000,000) that are undetectable by other cytokine detection methods, such as intracytoplasmic staining *(9)*. With the development of automated imaging and analysis, the ELISPOT assay is now feasible and extremely useful for peptide mapping.

To define the role of myelin responses in MS disease progression, in our initial studies we used a PLP peptide library consisting of single amino acid overlapping 9-amino-acid-length peptides. Peptides of this length can bind directly on the cell surface to major histocompatibility complex class I and class II molecules of antigen-presenting cells where they can stimulate pep-tide-specific T-cells *(10)*. This approach reveals the total PLP-specific T-cell pool, its fine specificity and the overall clonal sizes of the PLP-peptide reac-tive repertoire.

Findings from studies using a select number of synthetic PLP peptide anti-gens in proliferation assays indicate that T-cell responses to a heterogeneous array of PLP determinants occur in MS patients *(7,11–13)*. Similarly, longitu-dinal analysis of selected myelin protein or peptide responses in MS suggest that chronic immune sensitization to myelin determinants leads to acquired immunity to new self-antigens *(14–17)*. Whereas previous studies focused on the response to whole antigens or just a few selected peptides *(18)*, we have examined reactivity using peptides spanning the entire PLP molecule and, more recently, we have expanded our analysis to include peptides spanning the entire MBP molecule and all its isoforms (21.5 kDa, 18.5 kDa, and 17.2 kDa).

Our goal was to assess the reactivity to PLP and MBP in RRMS patients and healthy controls, and to link myelin-specific reactivity to disease activity, disability, and progression. Disease activity was measured by relapses and gadolinium-enhancing lesions on magnetic resonance imaging (MRI). Disability was assessed by the Multiple Sclerosis Functional Composite (MSFC), a new clinical disability measure recommended by a task force of the National Multiple Sclerosis Society. The MSFC consists of timed tests of walking, arm function, and cognitive function, expressed as a single score along a continuous scale *(19)*. Disease progression was assessed by examining changes in the MSFC over time and the change in brain atrophy by MRI *(19–22)*.

In our initial studies looking at PLP reactivity, we observed very broad epitope recognition in individual patients that was scattered throughout the PLP molecule, showing considerable heterogeneity among MS patients. Frequency measurements showed that the number of PLP peptide-specific interferon (IFN)-γ-producing cells were many times higher in MS patients than in controls, whereas PLP-peptide induced interleukin (IL)-5-producing T-cells occurred in very low frequencies both in MS patients and controls. This first comprehensive assessment of the anti-PLP-Th1/Th2 response in MS showed a greatly increased Th1 effector cell mass in MS patients. Moreover, the highly IFN-γ-polarized, IL-5-negative cytokine profile of the PLP-reactive T-cells suggested that these cells are committed Th1 cells *(23)*. Our follow-up studies showed gender differences in PLP–IFN-γ reactivity such that both MS and control females showed strong Th1 skewing. Th2 cytokine responses suggested that disease and gender are not independent, but rather, interact to influence the cytokine response to myelin. These data suggested a gender bias towards Th1 responses in MS, which may contribute to the female predominance in this disease *(24)*. In a recent cross-sectional study, we addressed the relationship between autoreactivity to myelin antigens and disease progression in MS by comparing the MSFC with immune cytokine responses to both MBP and PLP. MS patients showed a significant correlation between the IFN-γ response to PLP- and MBP peptides and disability. In contrast, in MS patients, there was no correlation between the MSFC and the response to unrelated control antigens or mitogens. These data showed that myelin-specific T-lymphocytes secreting the inflammatory cytokine IFN-γ correlate with functional impairment in MS, supporting an antigen-specific link between the immune response to myelin and disability in MS *(25)*. In our current studies we examine *longitudinal* cytokine responses to PLP and MBP. Our studies show that the temporal evolution of the immune response to self-antigens in MS is more complex than previously expected. Myelin epitope mapping reveals protein-wide bursts of recurrent cytokine responses, which appear to be more strongly associated with MRI activity than with clinical activity in MS (Moldovan et al., unpublished results). By observing both HLA class I and II responses, our current studies reveal the extremely dynamic nature of the anti-myelin cytokine reactivity in the disease, which has direct bearing on the development of emerging therapies for MS.

In summary, this chapter describes our methods for performing myelin peptide mapping using the ELISPOT assay to assess both cross-sectional and longitudinal cytokine responses in RRMS patients and to correlate the cytokine responses with various measures of disease activity and progression.

2. Materials

1. ELISPOT plates (UNIFILTER, low volume no. 7770-0052, regular volume no. 7770-0006, Whatman, Clifton, NJ).
2. ImmunoSpot Series 3 and ImmunoSpot Satellite analyzer (Cellular Technology Ltd., Cleveland, OH) or comparable image analysis equipment and software.
3. Streptavidin-horseradish peroxidase (HRP; DakoCytomation, Carpenteria, CA).
4. Bovine serum albumin (BSA), fraction V, low endotoxin (Sigma, cat. no. A-1933, St. Louis, MO).
5. Primary or capture monoclonal antibodies: anti-IFN-γ (Endogen, cat. no. M-700A, Woburn, MA), anti-IL-2 (BD-Pharmingen, cat. no. 555051, San Diego, CA), anti-IL-5 (eBioscience, cat. no. 14-7052-85, San Diego, CA), anti-tumor necrosis factor (TNF)-α (BD-Pharmingen, cat. no. 554508), anti-IL-10 (eBioscience cat. no. 14-7108-85; *see* **Note 1).**
6. Secondary or detecting biotinylated monoclonal antibodies: anti-IFN-γ (Endogen, cat. no. M-701-B), anti-IL-2 (BD-Pharmingen, cat. no. 555040), anti-IL-5 (eBioscience, cat. no. 13-7059-85), anti-TNF-α (BD-Pharmingen, cat. no. 554511), anti-IL-10 (eBioscience, cat. no. 14-7109-85; *see* **Note 1**).
7. 3-Amino-9-ethylcarbazole, $C_{14}H_{14}N_2$ (AEC) powder (Sigma, cat. no. A-5754).
8. Dimethyl formamide (cat. no. 61032-1000, Fisher/ACROS Chemical, Fair Lawn, NJ).
9. 30% H_2O_2.
10. Tween 20 (polyoxyetylene 20-sorbitan monolaurate).
11. Lymphocyte separation media (LSM) for density gradient separation of lymphocytes (Mediatech, cat. no. 25-072-CV, Herndon, VA).
12. RPMI 1640.
13. Dulbecco's phosphate-buffered saline (dPBS).
14. Antibiotic/antimycotic (Mediatech, cat. no. 30-004-CI).
15. Na-pyruvate,100 mM (Mediatech, cat. no. 25-000-CI).
16. Non-essential amino acids, 100X (Mediatech, cat. no. 25-025-CI).
17. 2 mM L-glutamine (Mediatech).
18. 10% newborn bovine serum (GibcoBRL).
19. Marsh Plates, 96-well polypropylene (Marsh, cat. no. AB-0796). Heatseal film (Marsh, cat. no. AB 0745). Combi Thermo-Sealer (Marsh BioProducts, http://www.marshbio.com/).
20. Sealing mat (Costar, cat. no. 3003, Corning, NY).

21. Millicup, hydrophilic LCR polytetrafluoroethylene, 0.45-μm filters (Millipore, Billerica, MA).

22. Control antigens/mitogens: Tetanus toxoid (Cylex); Diptheria toxoid (Aventis-Pasteur, King of Prussia, PA); Streptokinase (Aventis); Candida extract (Hollister Steer, Spokane, WA); human myelin basic protein (Advanced Immunotech, Westbrook, ME); and PHA-P (L-9017, Sigma). Because native PLP is extremely hydrophobic, we chose to use an MBP-PLP fusion protein known as MP4, which consists of the 21.5-kDa isoform of MBP fused to a genetically engineered form of PLP (deltaPLP4) containing the four major hydrophilic regions of the PLP molecule (kindly provided by Alexion Pharmaceuticals, New Haven, CT; **refs. 26** and **27**).

23. Low retention pipet tips (Fisher, 21-277A).

24. Sterile, needle-nose, polypropylene transfer pipets (Fisher).

25. Peptides were synthesized using the pin method and have >96% purity. Peptides were obtained from Mimotopes (Victoria, Australia). Four hundred seventy two synthetic peptides were used. The peptides spanned the entire PLP and MBP molecules and were 9 amino acids long, with an offset of one amino acid (**Fig. 1A**). Eighty-seven β-galactosidase-derived 9-mer peptides were used as control peptides, and were purchased from Mimotopes (Victoria, Australia). All of the synthetic peptides were reconstituted in the appropriate solvent at a stock concentration of 280 μ*M* (refer to **Subheading 3.4.** for dissolving peptides). They were aliquoted in a volume of 100 μL/well in 96-well polypropylene plates (AB-0796, Marsh Biomedical Products, Rochester, NY), heat-sealed with Easy Peel Heat Sealing Foil (AB-0745, Marsh Biomedical Products), and stored at –20°C until use.

26. AEC stock: AEC, 100 mg in 10 mL of dimethyl formamide in glass container (Note: Dangerous! Use fume hood! Pipet only with glass; *see* **Note 2**).

27. 0.1 *M* acetate buffer: 148 mL of 0.2 *M* acetic acid [11.55 mL of glacial acetic acid/L of H_2O] in 352 mL of 0.2 *M* sodium acetate [27.2 g/L H_2O] bring up to 1 L of H_2O. Adjust pH to 5.0. (sterile filter).

28. 30% H_2O_2. Aliquot and freeze at –20°C. Protect from light.

29. Complete medium: RPMI, 1% antibiotic/antimycotic, 2 m*M* L-glutamine, 25 m*M* HEPES, 1% nonessential amino acids, 1 m*M* Na-pyruvate, 10% newborn calf serum, or fetal calf serum. Sterile filter using 0.22-μm filter.

30. PBS-Tween: 100 μL of Tween-20/L of PBS (0.01%; *see* **Note 3**).

31. PBS-Tween-BSA: 100 μL of Tween-20/L of PBS (0.01%) + 10 g of BSA fraction V (1%) (sterile filter).

32. PBS-BSA: PBS + 1% BSA fraction V (sterile filter).

33. 1:2000 dilution of streptavidin-HRP in PBS-Tween-BSA.

34. Stock solution for red cell lysis. Solution A: 0.16 *M* NH_4Cl (8.3 g/liter) 0.17 *M* Tris, pH 7.65. (Dissolve 20.6 g of Tris base in 900 mL of water, adjust to pH 7.65 with HCl.) Solution B: Make up to 1000 mL.

35. Working solution for red cell lysis: 180 mL of 0.16 *M* NH_4Cl and 20 mL of 0.17 *M* Tris-HCl, pH 7.65. Adjust to pH 7.2 with HCl. Filter with a 0.22-μm filter. Store at room temperature.

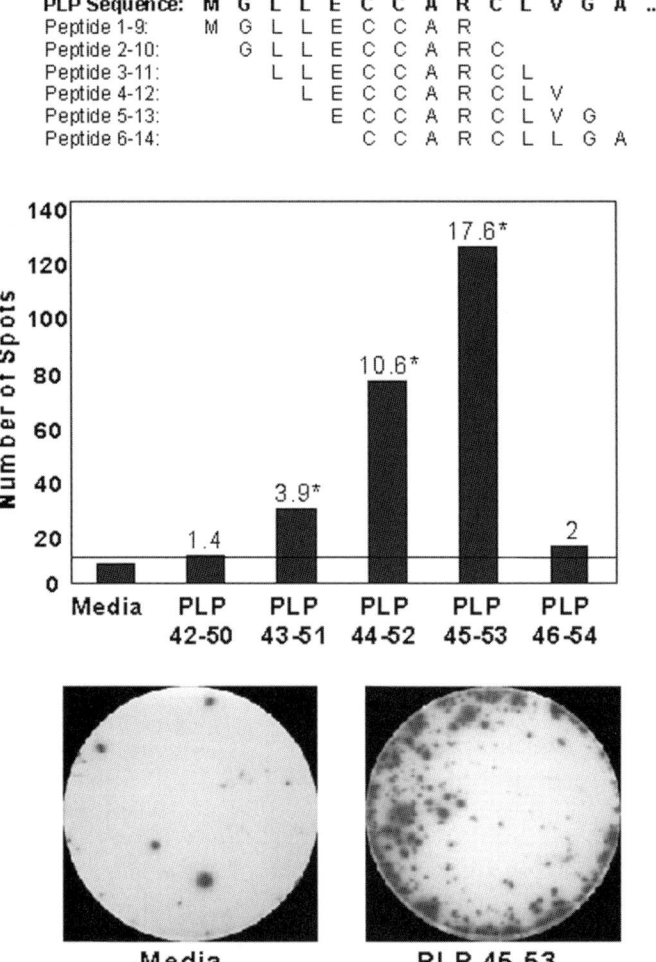

Fig. 1. Determinant mapping of PLP with 9 amino acid length overlapping peptides using the ELISPOT assay. (**A**) Schematic representation of determinant mapping with overlapping peptides. The amino acid sequence of the protein (PLP) is shown on the top. The 9-mer peptides progress along the molecule one amino acid at a time. Thereby, all conceivable determinants are utilized for Class I or Class II binding, and any peptides that bind can activate T-cells. Therefore the complete in vivo primed T-cell repertoire specific for an antigen can be assessed. (**B**) IFN-γ response to multiple adjacent PLP peptides in an MS patient. Each bar represents the number of spots (i.e., IFN-γ-positive cells) obtained per PLP peptide-stimulated well. Numbers at the top of each bar are stimulation indices above the media control. The horizontal line represents the mean + 3SD of the media control well to demonstrate which wells gave significant responses (marked with asterisks). Thus, three adjacent PLP peptides constitute this "positive epitope": PLP 43-51, 44-52, and 45-53. (**C**) Close-up view of individual ELISPOT wells from the analysis in (**B**). The media well is shown on the left and the highest IFN-γ-positive well, PLP 45-53, is shown on the right. (Reproduced from **ref. 23** with permission from The American Association of Immunologists, Inc.)

36. Solutions for dissolving 9-mer peptides (use letter designations):
    (a) 0.1% acetic acid/H_2O (make fresh)
    (b) 100% acetic acid
    (c) 0.1 *M* HEPES, pH 7.4
    (d) acetonitrile, HPLC grade (Sigma)
    (e) 0.1% NH_4OH (make fresh)
    (f) 10% NH_4OH (make fresh)
37. Study Subjects: Twenty patients with clinically definite relapsing-remitting MS and 27 age- and sex-matched controls. Eighty milliliters of blood was collected in sodium heparin for separation of peripheral blood mononuclear cells (PBMCs). At the time of the clinical visit, study subjects were assigned a MSFC score and an Expanded Disability Status Scale score and underwent MRI scans from which lesion number and whole brain atrophy were determined. The patients were either untreated, or on IFN-β1a treatment. All subjects were tested five times (approx every 3 mo) during a 1-yr time span.

3. Methods

3.1. Preparation of ELISPOT Plates and Adding of Human PBMCs (Sterile Steps)

1. Coating: sterile PBS + coating (capture) mAb 100ml/well, overnight at 4°C, covered with plastic wrap (*see* **Notes 4** and **5**).
2. Wash: (Sterile) three times with 200 µL/well PBS (*see* **Note 6**).
3. Blocking: (Sterile) Plates are blocked to prevent non-specific binding of proteins. Add 150 µL PBS + 1% BSA fraction V for a minimum of 2 h at room temperature, or overnight at 4°C (*see* **Note 7**).
4. Wash: (Sterile) three times with 200 µL/well PBS. Blot dry with sterile gauze (*see* **Note 6**).
5. Cells: (Sterile) PBMCs from the study subjects were freshly isolated (*see* **Note 8**) from heparinized blood by density gradient centrifugation, washed one time in PBS and then subjected to a red blood cell lysis step (*see* **Subheading 3.3.**, **Note 9**). After red cell lysis, wash cells two times with complete media, count, adjust cell concentration to 3×10^6/mL (300,000/well, standard volume plates). Add the reagents to the ELISPOT plate in the following order: First, add 100 µL/well media containing nothing/antigen/mitogen (9-mer peptides were added at a final concentration of 7 µ*M*). Then add cells to the plate, *after* antigens, at 100 µL/well. Agitate plate gently before placing in incubator (*see* **Note 10**). Incubate 24 h (for IFN-γ, IL-2, TNF-α) or 48 h (for IL-4, IL-5, IL-10) at 37°C in 5% CO_2/humidified incubator. (For low volume plates, *see* **Note 11.**)

3.2. Completion of ELISPOT Assay and Plate Development (Nonsterile Steps)

1. Discard the contents of the ELISPOT plates. Wash plates three times with PBS and three times with PBS-Tween (0.01%), 200 µL/well. If cells are sticky, last wash can be dH_2O (*see* **Note 12**). Blot-dry the plates on paper towels.

2. Add detection mAb: Dilute Ab in PBS-Tween-BSA and add 100 µL/well. Incubate the plates over night at 4°C, (covered) or incubate 30 min at 37°C.
3. Wash plates three times with PBS-Tween 0.01%, 200 µL/well (*see* **Note 12**).
4. Add enzyme: Dilute streptavidin-HRP 1:2000 in PBS-Tween-BSA and add 100 µL/well for 2 h at room temperature. Do not exceed 2 h incubation.
5. Wash plates three times with 200 µL/well PBS (*see* **Note 12**).
6. Add substrate: (Note: Toxic! Wear gloves). Dilute AEC stock 1:30 in 0.1 *M* acetate buffer and filter through a 0.45-µm low-protein binding hydrophilic-LCR membranes (Millipore) to remove colored precipitates. Substrate should be perfectly clear and colorless after filtration. Just before developing, add 30% H_2O_2 (1:2000, protect H_2O_2 from light). Add 200 µL/well and observe for spot development for a maximum of 1 h at room temperature (*see* **Note 13**).
7. Discard substrate. Wash plates three times with dH_2O, 200 µL/well. The fastest method is to wash the plate under running water to stop the reaction.
8. Allow plates to dry, upright, before imaging the wells (*see* **Note 14**).
9. Acquire images of the wells and save on CD using an automated ImmunoSpot Series 3 (*see* **Fig. 1C**). Enumerate the spots on an ImmunoSpot Satellite analyzer (Cellular Technology Ltd., Cleveland OH) using software specifically designed for the ELISPOT assay (*see* **Note 15**). The cut-off value that is used to decide which responses are positive must be determined empirically. Examples of positive responses can be observed in **Fig. 1B**.

3.3. Separation of PBMCs From Heparinized Blood and Lysis of Red Blood Cells With Tris-Buffered Ammonium Chloride (29)

1. Bring LSM to room temperature. Dispense 17 mL of LSM into a 50-mL polypropylene centrifuge tube. You will need approximately one 50-mL tube for every 15 mL of patient blood.
2. Carefully cover the Vacutainer rubber stopper with an absorbent pad. Slowly pull off stopper, squeezing to trap any blood on the absorbent pad. Pipet blood into a sterile 50-mL polypropylene centrifuge tube.
3. Dilute patient blood 1:1 with dPBS (Ca^{2+}/Mg^{2+} free). Carefully layer 30 mL of diluted blood overtop 17 mL of LSM, without disturbing the interface.
4. Centrifuge the LSM/blood layers for 20 min at 750g at room temperature with the brake OFF.
5. After centrifugation, carefully remove the mononuclear cell-rich interface using a sterile polypropylene transfer pipet with needle tip. Minimize the amount of LSM or plasma you remove while trying to maximize cell recovery.
6. Promptly dilute the cells threefold with dPBS. (Extensive exposure to LSM decreases viable cell recovery.)
7. Centrifuge the cell suspension at 600g for 5 min. Decant supernatant and resuspend pellet in PBS and repeat wash. You can combine all identical samples of the same patient at this time.
8. Decant supernatant and resuspend pellet in 10 mL of red blood cell lysing media. Gently mix well and let sit at room temp for 10 min. Spin at 650g for 5 min.

9. Decant supernatant and resuspend pellet in 30 mL of complete RPMI media. Spin at 650*g* for 5 min.
10. Decant supernatant. Add complete RPMI to 10 mL (final volume).
11. Remove 10 µL and count using trypan blue to assess cell viability.
12. Resuspend cells to appropriate concentration.

3.4. Method for Dissolving 9-mer Peptides From PLP and MBP (see Note 16–18, and Figs. 2–4)

1. First, analyze each peptide according to the amino acid composition to determine whether they are polar (charged) or neutral.
   a) Positively charged amino acids are Arginine (R), Lysine (K), and Histidine (H). Positive or neutral peptides will dissolve sufficiently well in acetic acid (0.1%) followed by the HEPES solution.
   b) Negatively charged amino acids are Aspartic acid (D) and Glutamic acid (E).
2. If the peptide has an overall negative charge, then check for the presence of amino acids that could make them even more difficult to dissolve.
   a) Both Cysteine (C) and Methionine (M) can form disulfide bonds and will cause peptides to polymerize in basic conditions. Thus, they must be kept in acidic conditions.
   b) If 50% or more of the amino acids in a peptide are hydrophobic, then the whole peptide is designated hydrophobic: (1) Strong hydrophobicity amino acids are Isoleucine (I), Leucine (L), Phenylalanine (F), and Valine (V); (2) Intermediate hydrophobicity amino acids are Tryptophan (W), Tyrosine (Y), and Proline (P); (3) low-hydrophobicity amino acids constitute the remaining amino acids.
3. Designate each overall peptide as: (1) positive (follow scheme shown in **Fig. 2**); (2) negatively charged with Cysteine or Methionine (follow scheme shown in **Fig. 3**); (3) negative (follow scheme shown in **Fig. 4**; *see* **Note 17**).
4. Use borosilicate glass tubes (5 mL) covered with parafilm for dissolving peptides. Peptides are delivered in plastic tubes. Add 1 mL of initial diluent (*see* **Figs. 2–4**) to plastic tube. Transfer contents to glass tube. Vortex 1 min at the highest speed and hold it on the edge of the tube. This replaces sonication. Do not allow peptide to foam. If it foams, add 10 µL of acetonitrile. Work at room temperature with sterile materials. Our final concentration of the 9-mer peptide in the ELISPOT well was 7 µ*M*. Our stock peptide solution was prepared at 280 µ*M* with a total volume of 3.72 mL per peptide (1 micromole each). After dissolving, peptides can be placed at 4°C for 5 d. It is best to aliquot them for long-term storage as soon as possible: freeze at –20°C (*see* **Note 18**). Once peptides have been thawed, store at 4°C for up to 4 wk.

4. Notes

1. The selection of primary and secondary antibodies that are listed here do not imply that these are the absolute best antibodies for the assay. Importantly, however, the antibody pairs (primary + secondary) are tested by the manufacturer *together* and should be used as a *matched pair* for the ELISPOT. We do not recommend mixing and matching primary and secondary antibodies from different sources.

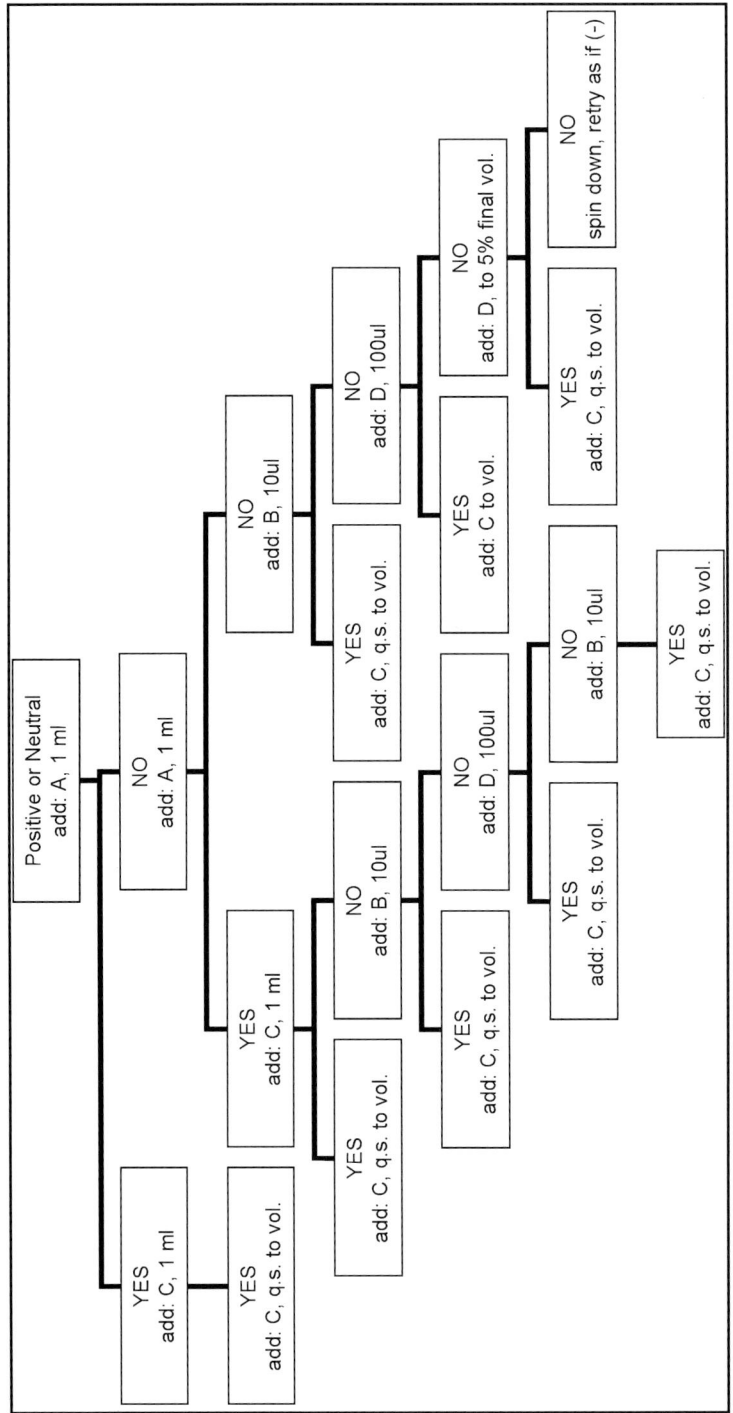

Fig. 2. Schematic flow chart showing the procedure for dissolving positively charged or neutral peptides. "Yes" and "No" refer to whether the peptide has dissolved or not, respectively. The letter designations refer to the flowing solutions: (**A**) 0.1% acetic acid/H_2O (make fresh). (**B**) 100% acetic acid. (**C**) 0.1 M HEPES, pH 7.4. (**D**) Acetonitrile, HPLC grade (Sigma). (**E**) 0.1% NH_4OH (make fresh). (**F**) 10% NH_4OH (make fresh).

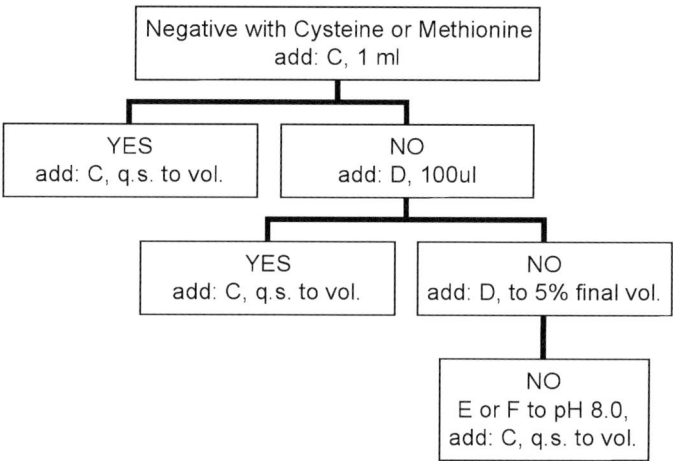

Fig. 3. Schematic flow chart showing the procedure for dissolving negatively charged peptides that *also* contain Cysteine or Methionine. "Yes" and "No" refer to whether the peptide has dissolved or not, respectively. The letter designations refer to the solutions listed in the legend to **Fig. 2.**

2. Do not use AEC tablets. The tablets do not dissolve sufficiently and leave a colored residue that cannot be filtered out. The resulting substrate leaves residue in the ELISPOT plates.
3. Some protocols recommend up to 0.05% Tween-20. We have found that this amount sometimes reduces the hydrophobicity of the polyvinylidene diflouride membranes so much that the ELISPOT plates leak, completely ruining the experiment.
4. The coating (capture) and detection antibody concentrations must be determined for each new commercial source of antibody and for each new lot that is produced. We find that the coating and detection antibodies generally work best in the range of 1 to 5 µg/mL. However, different lots/sources require titration in an ELISPOT assay. We perform a grid titration such that during coating the primary antibody is titrated *horizontally in rows* across the plate. The ELISPOT is conducted with the same number of PBMCs added to every well and with low-dose anti-CD3 or PHA as a stimulator of PBMC. After discarding the cells, the detection antibody is titrated *vertically in columns* down the plate. After developing, each combination of antibodies is observed visually using ImmunoSpot image analysis (*see* **Note 15**) to determine the optimal signal to noise ratio. Use the least amount of antibodies that gives you the optimal signal to noise ratio.
5. When wrapping the plate in plastic wrap, make sure that it sits perfectly horizontally in order to coat the bottom of the wells evenly. Plates can be prepared up to 3 d ahead of time with antibody coating solution and stored at 4°C. If the plate is kept for longer than 3 d, it should be washed with sterile PBS and blocked with PBS + 1% BSA fraction V. The blocking solution can be left on the plate for up to 7 d

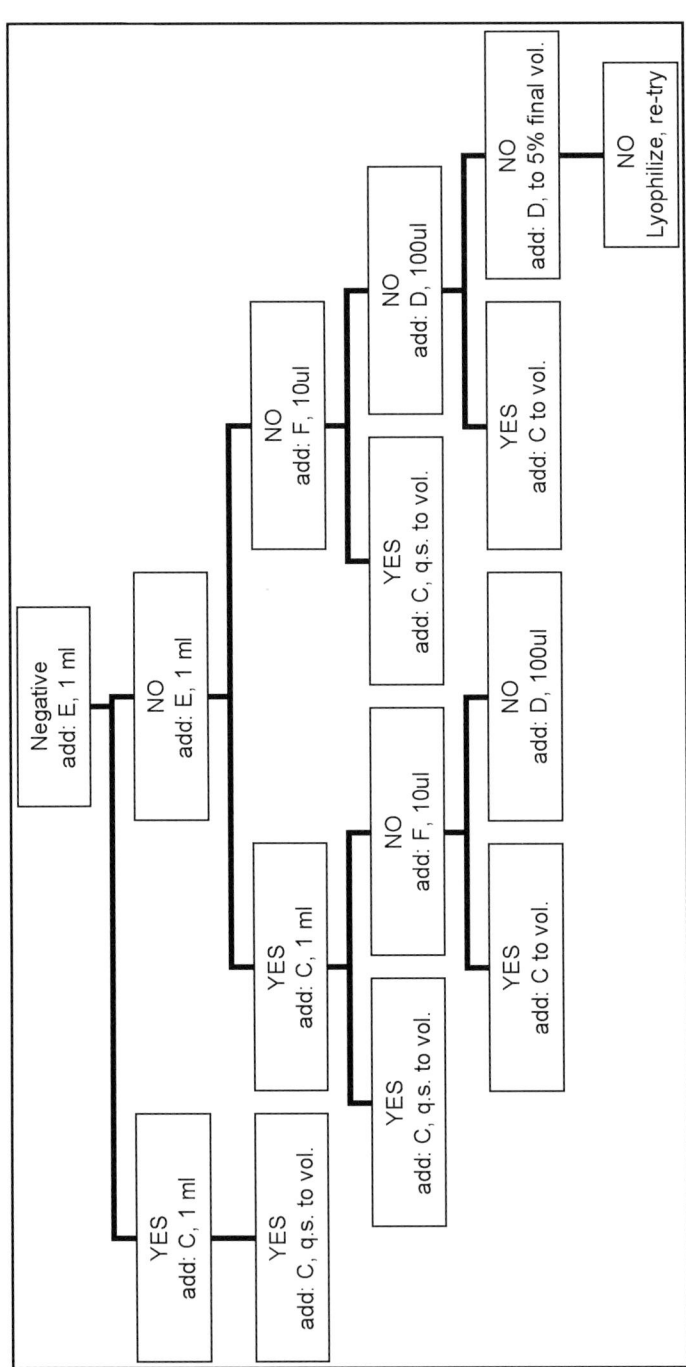

Fig. 4. Schematic flow chart showing the procedure for dissolving negatively charged peptides. "Yes" and "No" refer to whether the peptide has dissolved or not, respectively. The letter designations refer to the solutions listed in the legend to **Fig. 2**.

before washing and using the plate for an assay. We have not tested longer periods of time. Be sure to seal the plate against evaporation during storage at 4°C.

6. The plate can be turned upside-down and "flicked" to empty the coating solution, however, remember to do this into a container in the hood and be careful to prevent splashing back into the plate (i.e., keep sterile). We have found that a rapid method for sterile washing can be done with a 12-channel pipet or a 96-well Transtar 96™ (Costar). Washing following the blocking steps involves BSA, which has a tendency to form bubbles with repeated pipetting. Pipetting very gently and watching to avoid bubble formation can prevent this. Dry the lid with sterile gauze before adding the cells to the plate.

7. Blocked plates can be kept sealed at 4°C for up to 1 wk. Then they can be washed and used as normal. We have not used plates that were blocked for longer than 1wk.

8. We have done some minimal testing using freshly isolated PBMCs vs cryopreserved PBMCs. In our hands, the detection of IFN-γ was not dramatically affected by freezing the lymphocytes, however, IL-10 was very dramatically affected (decreased). Other researchers have examined the effects of cryopreservation on the ELISPOT more thoroughly *(28)*.

9. Although many laboratories do not perform a red blood cell lysis step at this point, we have found that this has several advantages. First, our PBMCs are more pure and therefore easier to count. This increases the accuracy when plating the cells in the ELISPOT, which is important when you are measuring frequencies of very rare events and comparing between individuals or between treatments. Second, removing the red blood cells helps prevent excessive washing, which can lead to cell loss and diminished viability. Third, contaminating red blood cells can increase background responses by stimulating lymphocytes and cause artifacts in the ELISPOT wells when they are imaged.

10. This agitation step (gentle tapping on all four sides of the plate) is meant to evenly distribute cells; however, be careful when agitating the plate to avoid cross-contamination between wells.

11. With human studies, the volume of blood that can be obtained from any study subject may be limiting. To reduce the number of cells per well that are plated, we used "low volume plates" which have somewhat conical-shaped sides and a smaller surface area of membrane in the bottom. This allows one to use 7.5×10^4 cells/mL, which is 75,000 cells/well in a volume of 50 μL/well for low volume plates. The other steps that can be scaled down correspondingly include the coating Ab step (use 50 μL/well) and the detecting Ab step (use 50 μL/well). The obvious advantages here include the use of less blood and less antibody. In our hands, disadvantages include the fact that the quality of the low volume plates is highly variable and may lead to unacceptable assay-to-assay variability.

12. An automatic washer can be used at this point. For consistency and "hands-free" washing of up to 25 plates at a time, we use the SkanWasher 300 (Molecular Devices, Sunnyvale, CA).

13. Observing plates for spot development after adding the substrate requires some patience to master. It is tempting to wait too long until you observe big, dark spots

developing, however, often this means that the background color is also increasing. Increased background can sometimes obscure the fainter spots and the substrate may form precipitates in a ring pattern around the edge of the well, which can also obscure spots around the rim of the well. We recommend using a magnifying glass with a built-in light source (most drugstores have these) to get a close-up view into the wells as they develop.

14. We dry plates over night at room temperature in the *upright position* to prevent any remaining substrate that is caught beneath the well from soaking back through the membrane and obscuring the spots. We also lightly cover the open, drying plates with paper towels to prevent dust from settling in the plates (which can be misinterpreted as spots by the image analysis).

15. What follows is a brief description of the imaging and analysis process. Digitized images are analyzed for the presence of areas in which color density exceeds background by a factor calculated from comparing control wells to experimental wells. After separating spots that touch or partially overlap, additional criteria of spot size and circularity are applied to gate out noise caused by spontaneous substrate precipitation and nonspecific antibody binding. Objects that do not meet these criteria are ignored and areas that meet them are recognized as spots, counted and highlighted. Our lab has tested multiple different imagers and software programs that are designed to analyze ELISPOT assays. In our hands, the imaging and software capabilities from Cellular Technologies Limited, Inc (CTL, Inc.) are superior to the competition in most respects.

16. We obtained extremely useful literature entitled "T-cell epitope mapping with PIN peptides" and "Guide to handling and storing peptides" from Mimotopes (http://www.mimotopes.com/peptides/lit.html). PLP is an extremely hydrophobic molecule and has poor aqueous solubility. Many of the 9-mer peptides were strongly hydrophobic and required multiple solutions to solubilize them. An advantage to using different diluents is that each peptide can be custom-dissolved to achieve the maximum solubility. Our intention was to maximize the ability of PMBCs to bind and respond to each different peptide. However, this method is quite tedious and has the disadvantage that peptides cannot be pooled if they are dissolved in different (i.e., incompatible) solutes.

17. Negatively charged peptides that are particularly hydrophobic may not be easily dissolved even using the scheme described in **Fig. 4**. In our hands, occasional peptides like this must be lyophilized and subjected to the scheme in **Fig. 3**. The acetonitrile sometimes allows dissolving of difficult peptides.

18. Peptides are aliquoted (100 μL/well) into specialized polypropylene 96-well plates that are low-protein binding. The position of the peptides in the plates should exactly match the position that they will be tested when they are added to the ELISPOT assay. This saves lots of time and prevents mix-ups when you are doing mapping with many peptides. To prevent evaporation, plates are heat sealed and then frozen for long-term storage (*see* **Subheading 2., step 25**). Do not use a "frost-free" freezer for long-term storage of peptides. The freeze-thaw cycles that are meant to prevent frost build-up are damaging to the peptides.

Acknowledgments

We thank Drs. Paul Lehmann and Magdalena Tary-Lehmann for their expertise, knowledge and training in all aspects of the ELISPOT assay and the development of a revolutionary new system for automated image analysis.

References

1. Raine, C. S., and Scheinberg, L. C.. (1988) On the immunopathology of plaque development and repair in multiple sclerosis. *J. Neuroimmunol.* **20,** 189–201.
2. Martin, R., McFarland, H. F., and McFarlin, D. E. (1992) Immunological aspects of demyelinating diseases. *Annu. Rev. Immunol.* **10,** 153–187.
3. Allegretta, M., Nicklas, J. A., Sriram, S., and Albertini, R. J. (1990) T-cells responsive to myelin basic protein in patients with multiple sclerosis. *Science* **247,** 718–721.
4. Johnson, D., Hafler, D. A., Fallis, R. J., Lees, M. B., Brady, R. O., Quarles, R. H., and Weiner, H. L. (1986) Cell-mediated immunity to myelin-associated glycoprotein, proteolipid protein, and myelin basic protein in multiple sclerosis. *J. Neuroimmunol.* **13,** 99–108.
5. Kerlero, d. R., Hoffman, M., Mendel, I., Yust, I., Kaye, J., Bakimer, R., et al.. (1997) Predominance of the autoimmune response to myelin oligodendrocyte glycoprotein (MOG) in multiple sclerosis: reactivity to the extracellular domain of MOG is directed against three main regions. *Eur. J. Immunol.* **27,** 3059–3069.
6. Lehmann, P. V., Forsthuber, T., Miller, A., and Sercarz, E. E. (1992) Spreading of T-cell autoimmunity to cryptic determinants of an autoantigen. *Nature* **358,** 155–157.
7. Tuohy, V. K., Yu, M., Weinstock-Guttman, B., and Kinkel, R. P. (1997) Diversity and plasticity of self recognition during the development of multiple sclerosis. *J. Clin. Invest* **99,** 1682–1690.
8. Vanderlugt, C. J., and Miller, S. D. (1996) Epitope spreading. *Curr .Opin. Immunol.* **8,** 831–836.
9. Helms, T., Boehm, B. O., Asaad, R. J., Trezza, R. P., Lehmann, P. V., and Tary-Lehmann, M. (2000) Direct visualization of cytokine-producing recall antigen-specific CD4 memory T-cells in healthy individuals and HIV patients. *J. Immunol.* **164,** 3723–3732.
10. Adorini, L., Muller, S., Cardinaux, F., Lehmann, P. V., Falcioni, F., and Nagy, Z. A. (1988) In vivo competition between self peptides and foreign antigens in T- cell activation. *Nature* **334,** 623–625.
11. Tuohy, V. K., Yu, M., Yin, L., Kawczak, J. A., and Kinkel, R. P. (1999) Spontaneous regression of primary autoreactivity during chronic progression of experimental autoimmune encephalomyelitis and multiple sclerosis. *J. Exp. Med.* **189,** 1033–1042.
12. Pelfrey, C. M., Trotter, J. L., Tranquill, L. R., and McFarland, H. F. (1993) Identification of a novel T-cell epitope of human proteolipid protein (residues 4060) recognized by proliferative and cytolytic CD4+ T-cells from multiple sclerosis patients. *J. Neuroimmunol.* **46,** 33–42.
13. Pelfrey, C. M., Trotter, J. L., Tranquill, L. R., and McFarland, H. F. (1994) Identification of a second T-cell epitope of human proteolipid protein (residues 89106)

recognized by proliferative and cytolytic CD4+ T-cells from multiple sclerosis patients. *J. Neuroimmunol.* **53,** 153–161.

14. Calabresi, P. A., Fields, N. S., Farnon, E. C., Frank, J. A., Bash, C. N., Kawanashi, T., et al. (1998) ELI-spot of Th-1 cytokine secreting PBMC's in multiple sclerosis: correlation with MRI lesions. *J. Neuroimmunol.* **85,** 212–219.

15. Pender, M. P., Csurhes, P. A., Greer, J. M., Mowat, P. D., Henderson, R. D., Cameron, K. D., et al. 2000. Surges of increased T-cell reactivity to an encephalitogenic region of myelin proteolipid protein occur more often in patients with multiple sclerosis than in healthy subjects. *J. Immunol.* **165,** 5322–5331.

16. Hellings, N., Gelin, G., Medaer, R., Bruckers, L., Palmers, Y., Raus, J., et al. (2002) Longitudinal study of antimyelin T-cell reactivity in relapsing- remitting multiple sclerosis: association with clinical and MRI activity. *J. Neuroimmunol.* **126,** 143–160.

17. Arbour, N., Holz, A., Sipe, J. C., Naniche, D., Romine, J. S., Zyroff, J., et al. (2003) A new approach for evaluating antigen-specific T-cell responses to myelin antigens during the course of multiple sclerosis. *J. Neuroimmunol.* **137,** 197–209.

18. Olsson, T., Zhi, W. W., Hojeberg, B., Kostulas, V., Jiang, Y. P., Anderson, G., et al (1990) Autoreactive T-lymphocytes in multiple sclerosis determined by antigen-induced secretion of interferon-gamma. *J. Clin. Invest* **86,** 981–985.

19. Rudick, R. A., Cutter, G., Baier, M., Fisher, E., Dougherty, D., Weinstock-Guttman, B., et al. (2001) Use of the Multiple Sclerosis Functional Composite to predict disability in relapsing MS. *Neurology* **56,** 1324–1330.

20. Fisher, E., Rudick, R. A., Cutter, G., Baier, M., Miller, D., Weinstock-Guttman, B., et al. (2000) Relationship between brain atrophy and disability: an 8-year follow-up study of multiple sclerosis patients. *Mult. Scler.* **6,** 373–377.

21. Rudick, R. A., Fisher, E., Lee, J. C., Duda, J. T., and Simon, J. (2000) Brain atrophy in relapsing multiple sclerosis: relationship to relapses, EDSS, and treatment with interferon beta-1a. *Mult. Scler.* **6,** 365–372.

22. Rudick, R. A., Fisher, E., Lee, J. C., Simon, J., and Jacobs, L. (1999) Use of the brain parenchymal fraction to measure whole brain atrophy in relapsing-remitting MS. Multiple Sclerosis Collaborative Research Group. *Neurology* **53,** 1698–1704.

23. Pelfrey, C. M., Rudick, R. A., Cotleur, A. C., Lee, J. C., Tary-Lehmann, M., and Lehmann, P. V. (2000) Quantification of self-recognition in multiple sclerosis by single-cell analysis of cytokine production. *J. Immunol.* **165,** 1641–1651.

24. Pelfrey, C. M., Cotleur, A. C., Lee, J. C., and Rudick, R. A. (2002) Sex differences in cytokine responses to myelin peptides in multiple sclerosis. *J. Neuroimmunol.* **130,** 211–223.

25. Moldovan, I. R., Rudick, R A., Cotleur, A. C., Born, S., Lee, J. C., Karafa, M., and Pelfrey, C. M. 2003. Interferon gamma responses to myelin peptides in multiple sclerosis correlate with a new clinical measure of disease progression. *J. Neuroimmunol.* **141,** 132–140.

26. Elliott, E. A., McFarland, H. I., Nye, S. H., Cofiell, R., Wilson, T. M. Wilkins, J. A., et al. 1996. Treatment of experimental encephalomyelitis with a novel chimeric

fusion protein of myelin basic protein and proteolipid protein. *J. Clin. Invest* **98,** 1602–1612.

27. Nye, S. H., Pelfrey, C. M., Burkwit, J. J., Voskuhl, R. R., Lenardo, M. J., and Mueller, J. P. (1995) Purification of immunologically active recombinant 21.5 kDa isoform of human myelin basic protein. *Mol. Immunol.* **32,** 1131–1141.
28. Kreher, C. R., Dittrich, M. T., Guerkov, R., Boehm, B. O., and Tary-Lehmann, M. (2003) CD4+ and CD8+ cells in cryopreserved human PBMC maintain full functionality in cytokine ELISPOT assays. *J. Immunol. Methods* **278,** 79–93.
29. Mishell, B. B., and Shiigi, S. M. (eds.) *Selected Methods in Cellular Immunology.* W.H. Freeman and Company, New York, p. 23.

15

Use of Interferon-γ ELISPOT in Monitoring Immune Responses in Humans

Mark Matijevic and Robert G. Urban

Summary

The interferon (IFN)-γ enzyme-linked immunospot (ELISPOT) assay has become a useful tool for immunologists seeking to quantify immune responses on a per-cell basis. The assay is sensitive and allows for the enumeration of low-frequency T-cells. Many have applied this assay to clinical trials as a way to measure biological activity in a patient cohort. It is critical that each laboratory attempting to use the assay in their facility perform rigorous development and qualification work to establish an assay that suits their particular needs. This chapter serves as a demonstration of two practical and slightly different approaches to using the ELISPOT assay to monitor immune activity in the human periphery: (1) assays using whole samples of peripheral blood mononuclear cells with and without the use of additional antigen presenting cells and (2) assays using enriched T-cell populations. Detailed protocols and procedures will be covered, as well as a demonstration of results obtained from three separate applications.

Key Words: ELISPOT; PBMCs; clinical trial; immune response; IFN-γ.

1. Introduction

The interferon (IFN)-γ enzyme-linked immunospot (ELISPOT) assay has become a useful tool for immunologists seeking to quantify immune responses on a per-cell basis. The assay is sensitive and allows for the enumeration of low-frequency T-cells (1–9). The ability to detect enhanced T-lymphocyte responses is a critical step in evaluating the immunological efficacy of a vaccine (10–13). The methods available to test peripheral blood samples for the presence and/or number of antigen specific T-lymphocytes have been steadily improving. These include using flow cytometry to evaluate intracellular cytokines, tetramer assays, and ELISPOT. It is critical that each laboratory attempting to use the assay in their facility perform rigorous development and qualification work to establish an assay that suits their particular needs (14–16).

From: *Methods in Molecular Biology, vol. 302: Handbook of ELISPOT: Methods and Protocols*
Edited by: A. E. Kalyuzhny © Humana Press Inc., Totowa, NJ

Induction of potent lymphocyte effector function is thought to be central to the activity of many biological therapeutics. The cytokine IFN-γ has become a useful marker for measuring the induction of cellular immune responses *(17)*. Several assays have been developed to quantify IFN-γ secreted by activated lymphocytes in response to antigen. The ELISPOT assay is one that can detect low frequency, antigen specific T-lymphocytes in the periphery.

Depending on the sensitivity of the antigen used in the assay, there are multiple ELISPOT formats. Herein, two methods with slightly different protocols are covered. They consist of: (1) direct IFN-γ ELISPOT assay using whole peripheral blood mononuclear cells (PBMCs; with and without antigen-presenting cells; APCs) and (2) direct IFN-γ ELISPOT assay using T-cell subsets (i.e., CD8$^+$ T-cells) isolated from PBMCs. Each assay may be applied to different settings. For example, a whole PBMC assay may be required when cell material is limited, human leukocyte antigen (HLA) type is unknown and/or the antigen is known to be highly immunogenic. Alternatively, if a trial is evaluating immune responses and has collected an abundant supply of PBMCs, T-cell subsets may be used to increase assay sensitivity. Finally, if the antigen contained within the therapeutic is known to be weakly immunogenic, in vitro expansion in vitro of low-frequency effector T-cells in a PBMC population may be required to detect and demonstrate immunological activity. There are multiple protocols covering in vitro expansion in vitro of T-cells with peptides, but they will not be covered in this chapter.

2. Materials

1. Gloves (Microflex Corporation; Reno, NV).
2. Dupont Tyvek sleeves (VWR; Aurora, CO).
3. Sterile heparinized Vacutainer tubes (Fisher Scientific; Pittsburgh, PA).
4. Microcentrifuge tubes (National Scientific; Claremount, CA).
5. Dulbecco's phosphate-buffered saline (DPBS; JRH Biosciences; Lenexa, KS).
6. 15 mL Polypropylene centrifuge tubes (Corning Incorporated; Corning, NY).
7. 50 mL Polypropylene centrifuge tubes (Corning Incorporated; Corning, NY).
8. Ficoll-Paque PLUS (Amersham Biosciences; Uppsala, Sweden).
9. Red blood cell lysing solution (Sigma Chemical Company; St. Louis, MO).
10. Trypan blue (Invitrogen Corporation; Carlsbad, CA).
11. Freeze medium. Filter sterilize 9 mL of fetal bovine serum (JRH Biosciences; Lenexa, KS] and 1 mL of dimethyl sulfoxide (Malinckrodt Baker Incorporated; Paris, KY); store at –20°C, expires 1 yr after day of preparation.
12. Cryotube 1.8 mL OF SI EXT, starfoot round (Nunc; Roskilde, Denmark).
13. Styrofoam rack (Sarstedt Incorporated; Newton, NC).
14. PBMC wash medium with DNase. Filter sterilize 500 mL of RPMI 1640 (JRH Biosciences; Lenexa, KS), 5 mL of Penicillin–Streptomycin (Invitrogen Corporation; Carlsbad, CA), 5 mL of HEPES buffer (Invitrogen Corporation; Carlsbad, CA), and 15,000 units of DNase (Sigma Chemical Company; St. Louis, MO); store at 2–8 °C, expires 2 mo after day of preparation.

15. PBMC wash medium without DNase. Filter sterilize 500 mL of RPMI 1640 (JRH Biosciences; Lenexa, KS), 5 mL of Penicillin–Streptomycin (Invitrogen Corporation; Carlsbad, CA), and 5 mL of HEPES buffer (Invitrogen Corporation; Carlsbad, CA); store at 2–8°C, expires 2 mo after day of preparation.

16. PBMC medium. Filter sterilize 180 mL of RPMI 1640 (JRH Biosciences; Lenexa, KS), 2 mL of Penicillin–Streptomycin (Invitrogen Corporation; Carlsbad, CA), 2 mL of HEPES buffer (Invitrogen Corporation; Carlsbad, CA), 2 mL of L-Glutamine (Invitrogen Corporation; Carlsbad, CA), 200 µL of 2-Mercaptoethanol (Invitrogen Corporation; Carlsbad, CA), and 20 mL of Human AB serum (C-Six Diagnostics; Germantown, WI); store at 2–8°C, expires 2 wk after day of preparation.

17. Ten percent RPMI. Filter sterilize 450 mL of RPMI 1640 (JRH Biosciences; Lenexa, KS), 5 mL of Penicillin–Streptomycin (Invitrogen Corporation; Carlsbad, CA), 5 mL of HEPES buffer (Invitrogen Corporation; Carlsbad, CA), 5 mL of L-glutamine (Invitrogen Corporation; Carlsbad, CA), 500 µl of 2-Mercaptoethanol (Invitrogen Corporation; Carlsbad, CA), and 50 mL of fetal bovine serum (JRH Biosciences; Lanexa, KS); store at 2–8°C, expires 2 mo after day of preparation.

18. T2 cell line (American Type Culture Collection; Manassas, VA).

19. Phytohemagglutinin (PHA; Sigma Chemical Company; St. Louis, MO).

20. A2-restricted influenza matrix peptide (Multiple Peptide Systems; San Diego, CA).

21. Cytomegalovirus, Epstein–Barr virus, and influenza (CEF) peptide pool (NIH AIDS Research & Reference Reagent Program; Rockville, MD).

22. Fourteen milliliters Polypropylene round-bottom tubes (Becton Dickinson Labware; Franklin Lakes, NJ).

23. Human IFN-γ ELISPOT kits (R&D Systems, Minneapolis, MN).

24. Nunc-Immuno wash 8 (Nunc; Roskilde, Denmark).

25. CD8$^+$ T-cell columns (R&D Systems, Minneapolis, MN).

3. Methods

Note: When working with human blood or blood products, Biosafety Level 2 containment practices must be followed. The methods below describe the process of isolating PBMCs from heparinized whole blood samples.

3.1. Isolating PBMCs From Heparinized Blood Sample

1. Intravenous blood sample is collected into sterile vacutainer tubes containing sodium heparin and then stored at room temperature until processing (*see* **Note 1**).

2. Centrifuge tubes for 15 min at 230g at room temperature, **with no brake**.

3. Remove stopper from tube and carefully aspirate plasma from sample using a pipet aid and serological pipet (*see* **Note 2**). Aliquot the plasma into sterile microcentrifuge tubes if needed for additional assays or discard into a biohazardous waste container containing appropriate disinfectant solution (i.e., bleach).

4. Dilute the remaining blood sample 1:2 with DPBS and transfer the sample to a 50-mL centrifuge tube.

5. Using a pipet aid and serological pipet, slowly and carefully layer the diluted blood sample onto ficoll hypaque in a sterile 50-mL centrifuge tube (*see* **Note 3**). This

should be performed at a ratio **no higher** than 1:2; Ficoll hypaque:diluted blood sample (i.e., 10 mL of Ficoll with 20 mL of diluted blood sample layered on top of it).

6. Centrifuge tubes for 30 min at 600g at room temperature, **with no brake**.
7. Using a pipet-aid and serological pipet, carefully remove the PBMC layer (buffy coat) from the centrifuge tube and transfer it to a sterile centrifuge tube (*see* **Note 4**).
8. Add DPBS to the PBMCs at a ratio of 1:3; or 1 mL of PBMC:3 mL of DPBS.
9. Centrifuge tubes for 10 min at 275g at room temperature, **with brake**.
10. Visually inspect the cell pellet and if significant red blood cell contamination exists, perform red blood cell lysing procedure (*see* **Subheading 3.1., step 11**). If minimal or no red blood cell contamination exists, then resuspend the cell pellet with 10 mL of DPBS and repeat **step 9**.
11. Resuspend the cell pellet in 3 mL of red blood cell lysing buffer, mix, and incubate for 3 min at room temperature. Add DPBS to the cell suspension to a total volume of 20 mL. Centrifuge for 10 min at 275g at room temperature, **with brake**. Pour off the supernatant and resuspend the cell pellet in 10 mL of DPBS and centrifuge for 10 min at 275g at room temperature, **with brake** (*see* **Note 5**).
12. Resuspend the cell pellet with 10 mL of DPBS and count viable PBMCs using trypan blue (*see* **Note 6**).
13. Repeat **step 9**.
14. Resuspend the cell pellet with freeze media to a concentration of 5×10^6/mL and keep on ice (*see* **Note 7**).
15. Quickly aliquot 1 mL of PBMCs (5×10^6) into sterile cryovials (each kept on ice until proceeding to the next step; *see* **Note 8**).
16. Transfer aliquots of PBMC from ice into styrofoam rack (*see* **Note 9**).
17. Place Styrofoam rack into –80°C storage for 24 h.
18. Transfer PBMC aliquots to liquid nitrogen vapor phase for long-term storage.

3.2. Preparing Cryopreserved PBMC Samples for Future Use in ELISPOT

The process below describes the methods involving preparation of cryopreserved PBMC for use in ELISPOT. Because of inter-assay variability, it is good practice to test (i.e., ELISPOT) all time points from a given individual or clinical trial subject on the same day. PBMC samples can be stored in liquid nitrogen vapor phase for extended periods of time or until an entire sample set is collected. This will minimize the effect of assay variability on data analysis.

1. Remove sufficient number of cryovials from liquid nitrogen storage to perform all ELISPOT assays for that day.
2. Transfer cryovials to a water bath set at 37°C and thaw PBMCs to the point where only a few ice crystals remain in each vial.
3. Using a pipet aid and serological pipet, quickly transfer the contents of one vial to a centrifuge tube containing 9 mL of PBMC wash medium (*see* **Note 10**).

4. Centrifuge samples for 10 min at 275g at room temperature, **with brake**.
5. Resuspend the cell pellet with 10 mL of PBMC wash medium and repeat **step 4**. In the event that visible clumping is present in the PBMC sample, perform an additional wash step with wash medium containing DNase (*see* **Note 11**).
6. Resuspend the cell pellet in PBMC medium and count viable PBMCs using trypan blue.
7. Adjust PBMCs to final per mL concentration for appropriate task (i.e., direct assay, in vitro expansion, T-cell subset purification; *see* **Note 12**).

3.3. Direct ELISPOT Using Whole PBMCs in HLA-A2 Individuals (With APCs)

The process below describes an ELISPOT assay that can be applied to samples with a known HLA type. This example is specifically for HLA-A2 individuals. Individuals with alternate HLA haplotypes (i.e., A1, A3, A11, and A24) may be screened using this method with an alternate, HLA-matched APC line and peptides. (i.e., K562; **ref. 20**). A demonstration of the antigen-specific control data using this method, captured from an institutional review board (IRB)-approved Phase I clinical trial, is illustrated in **Fig. 1**. The trial was sponsored by ZYCOS, Inc., and involved the safety screening of a potential biotherapeutic for human papillomavirus (HPV) associated anal high-grade squamous intra-epithelial neoplasia (HSIL). ELISPOT reagents (i.e., matched antibody pairs, plates, buffers, etc.) and prefabricated kits are commercially available through several vendors. Each investigator should select reagents that suit their particular needs and requirements.

1. Collect cultured T2 cell line *(18)* into a sterile centrifuge tube using a pipet-aid and serological pipet (*see* **Note 13**).
2. Centrifuge tubes for 10 min at 275g at room temperature, **with brake**.
3. Resuspend the cell pellet in 10 mL of PBMC medium and count viable cells in trypan blue.
4. Adjust the T2 cell population to a final concentration of 2×10^6/mL in PBMC medium (*see* **Note 14**).
5. Label the appropriate number of polypropylene tubes for each T2 test group, that is, T2 without peptide, T2 with a mitogen control—PHA, T2 with an antigenic control—A2-restricted influenza matrix peptide (Flu), and T2 with test peptide(s). Peptides should be acquired from vendor at >**90% purity**.
6. Dispense the appropriate volume of T2 into labeled tubes.
7. Add mitogen control (PHA; *see* **Note 15**) to the T2 at a 2X concentration of 10 μg/mL (final concentration after plating T2 cells with PBMCs in the ELISPOT plate will be 5 μg/mL) and antigenic peptides to the T2 cells at a 2X concentration of 150 μg/mL (final concentration after plating T2 cells with PBMC in the ELISPOT plate will be 75 μg/mL; *see* **Note 16**).

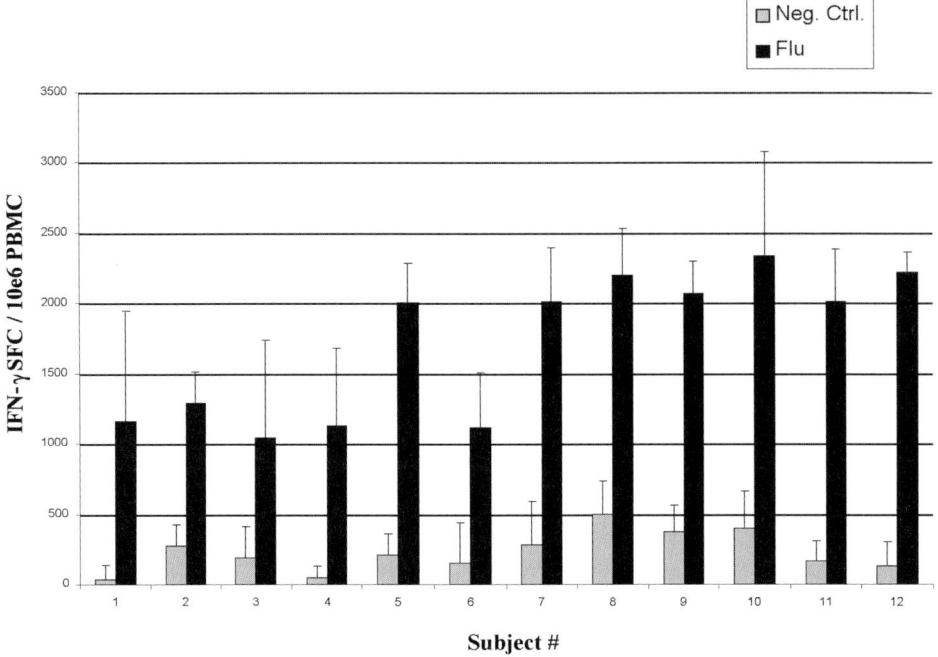

Fig. 1. IFN-γ ELISPOT: whole PBMCs with APCs assay. The mean responses of 12 subjects (indicated on the *x*-axis) in a phase I clinical trial (HPV-associated anal HSIL therapeutic) to the assay antigen control (Flu). The number of separate trial time points used to calculate the mean of each subject was six. Mean ± SD. IFN-γ SFC/10^6 PBMCs is shown on the *y*-axis. The negative control (Neg. Ctrl.) represents PBMCs plated with T2 without peptide. The experimental data are PBMCs plated with T2 with Flu peptide.

8. Gently vortex each T2 population for 10 s and place the tubes at a 30-degree angle on a tray (i.e., the lid of a pipet tip box) and place into a 37°C incubator with 5% CO_2 (do not tighten the caps of the tubes; **keep them loose**).

9. Incubate the T2 cell populations for a total of 4 h using gentle vortexing each hour.

10. Prepare PBMC populations (*see* **Subheading 2.**) to a final concentration of 2 × 10^6/mL in PBMC medium (*see* **Note 14**). This procedure should be performed close to the end of the T2 incubation so as not to let the PBMC "rest" too long prior to adding them to the ELISPOT plate (maximum "rest" time for PBMC should be 0.5 h).

11. Remove ELISPOT plates from 2 to 8°C storage and equilibrate to room temperature (ELISPOT plates are pre-coated with capture antibodies by manufacturer of the ELISPOT kits).

12. Add 100 μL of PBMC medium to all wells of the ELISPOT plate and incubate for 20 min at room temperature.

13. At the end of the incubation, remove the PBMC medium from the plates by "flicking" it into a biohazardous waste container and then blot the plates on a dry paper towel.
14. Add 100 μL (2×10^5) of the T2 cell populations to the appropriate wells (in triplicate if possible to account for assay variability). Immediately after adding the T2 cells to the ELISPOT plate, add 100 μL of PBMCs (2×10^5) to the appropriate wells.
15. Mix the contents of the ELISPOT plate by gently tapping on the sides of the plate for approx 10 s.
16. Place ELISPOT plates in a 37°C incubator with 5% CO_2 and let incubate for 24 h.
17. Remove ELISPOT plates from the incubator and wash five times with a 1X wash buffer prepared from the 10X wash concentrate supplied in the kit. Take care not to make contact with the membrane at the bottom of each well. After completing the wash, blot the plates on a dry paper towel (*see* **Note 17**).
18. Add 100 μL of diluted detection antibodies to all wells and incubate the plates at 2–8°C overnight.
19. Remove ELISPOT plates from the refrigerator and wash five times (*see* **Subheading 3.3., step 17**).
20. Add 100 μL of diluted strepavodin AP to all wells and incubate the plates for 2 h at room temperature.
21. Wash plates five times (*see* **Subheading 3.3., step 17**).
22. Add 100 μL of the chromogenic substrate to all wells and incubate the plates for one hour at room temperature, **in the dark**.
23. Wash plates five times with distilled water. Blot the plates on a dry paper towel, remove the bottom casing and let air-dry.
24. Count spots using a dissecting microscope or preferably, an automated spot counter with appropriate software (*see* **Note 18**).

3.4. Direct ELISPOT Using Whole PBMCs (Without Additional APCs)

The process below describes an ELISPOT assay that uses whole PBMC samples without using additional APCs. When additional APCs are not available, the APCs that are present in a PBMC population may be sufficient to stimulate antigen-specific responses. A demonstration of the control data using this method, captured from an IRB-approved Phase I clinical trial, is illustrated in **Fig. 2**. The trial was sponsored by ZYCOS Inc. and involved the safety screening of a potential bio-therapeutic for various cancer types. ELISPOT reagents (i.e., matched antibody pairs, plates, buffers) and prefabricated kits are commercially available through several vendors. Each investigator should select reagents and/or kits that suit their particular needs and requirements.

1. Prepare PBMC populations (*see* **Subheading 2.**) to a final concentration of 4×10^6/mL in PBMC medium (*see* **Note 14**). This procedure should be performed so as not to let the PBMCs "rest" too long prior to adding to the ELISPOT plate (maximum "rest" time for PBMCs should be 0.5 h).

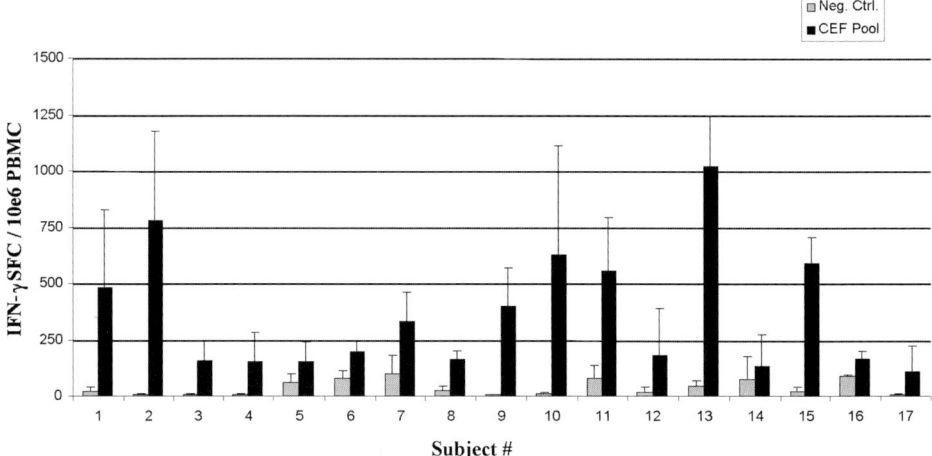

Fig. 2. IFN-γ ELISPOT: whole PBMCs assay. The mean responses of 17 subjects (indicated on the *x*-axis) in a phase I clinical trial (cancer therapeutic) to the assay antigen control (CEF). The number of separate trial timepoints used to calculate the means varied among the subjects (range: 2–8). Mean ± SD. IFN-γ SFC/10^6 PBMC is shown on the *y*-axis. The negative control (Neg. Ctrl.) represents PBMC plated with PBMC media alone. The experimental data are PBMC plated with CEF peptide pool in PBMC media.

2. Remove ELISPOT plates from 2 to 8°C storage and equilibrate to room temperature (ELISPOT plates are precoated with capture antibodies by manufacturer of the ELISPOT kits).

3. Add 100 µL of PBMC medium to all wells of the ELISPOT plate and incubate for 20 min at room temperature.

4. Prepare 2X solutions of mitogen/antigen, for addition to the PBMCs in sterile polypropylene tubes. Prepare each 2X solution in PBMC medium. These should include; PBMC medium alone, PBMC medium with a mitogen, that is PHA (*see* **Note 15**) at a 2X concentration of 10 µg/mL, PBMC medium with antigenic control peptides, that is, CEF pool *(19)* at a 2X concentration of 6 µg/mL and PBMC medium with individual test peptides or pools of test peptides at a 2X concentration of 100 µg/mL. Final concentrations of each solution will be: PHA, 5 µg/mL; CEF pool, 3 µg/mL; and test peptides, 50 µg/mL (*see* **Note 16**).

5. Gently vortex 2X solutions for 10 s.

6. At the end of the incubation, remove the PBMC medium from the plates by "flicking" it into a biohazardous waste container and then blot the plates on a dry paper towel.

7. Add 100 µL of the 2X solution to the appropriate wells (in triplicate if possible to account for assay variability). Immediately after adding the 2X solution to the ELISPOT plate, add 100 µL of PBMCs (4×10^5) to the appropriate wells.

8. Mix the contents of the ELISPOT plate by gently tapping on the sides of the plate for approx 10 s.

9. Place ELISPOT plates in a 37°C incubator with 5% CO_2 and let incubate for 24 h.

10. Remove ELISPOT plates from the incubator and wash five times with a 1X wash buffer prepared from the 10X wash concentrate supplied in the kit. Take care not to make contact with the membrane at the bottom of each well. After completing the wash, blot the plates on a dry paper towel (*see* **Note 17**).

11. Add 100 μL of diluted detection antibodies to all wells and incubate the plates at 2–8°C overnight.

12. Remove ELISPOT plates from the refrigerator and wash five times (*see* **Subheading 3.4., step 10**).

13. Add 100 μL of diluted strepavodin AP to all wells and incubate the plates for 2 h at room temperature.

14. Wash plates five times (*see* **Subheading 3.4., step 10**).

15. Add 100 μL of the chromogenic substrate to all wells and incubate the plates for one hour at room temperature, **in the dark**.

16. Wash plates five times with distilled water. Blot the plates on a dry paper towel, remove the bottom casing and let air-dry.

17. Count spots using a dissecting microscope or preferably, an automated spot counter with appropriate software (*see* **Note 18**).

3.5. Direct ELISPOT Using T-Cell Subsets (i.e., CD8⁺ T-Cells) Isolated From Whole PBMCs (With Additional APCs)

The method described here can be used when an abundant PBMC sample is available. Isolating specific cell subsets and making use of an antigen presenting cell will result in a more sensitive assay. A demonstration of the antigen-specific control data using this method, captured from an IRB-approved-randomized Phase II clinical trial, is illustrated in **Fig. 3**. The trial was sponsored by ZYCOS, Inc., and involved the safety and efficacy screening of a potential bio-therapeutic for HPV-associated cervical HSIL. ELISPOT reagents (i.e., matched antibody pairs, plates, buffers, etc.) and prefabricated kits are commercially available through several vendors. Each investigator should select reagents that suit their particular needs and requirements.

1. Collect cultured T2 cells (*18*) into a sterile centrifuge tube (*see* **Note 13**).

2. Centrifuge tubes for 10 min at 275*g* at room temperature, **with brake**.

3. Resuspend the cell pellet in 10 mL of PBMC medium and count viable cells in trypan blue.

4. Adjust the T2 cell population to a final concentration of 1×10^6/mL in PBMC medium (*see* **Note 14**).

5. Label the appropriate number of polypropylene tubes for each T2 test group, that is, T2 without peptide, T2 with a mitogen control—PHA, T2 with an antigenic control—CEF pool (*19*), and T2 with test peptide(s).

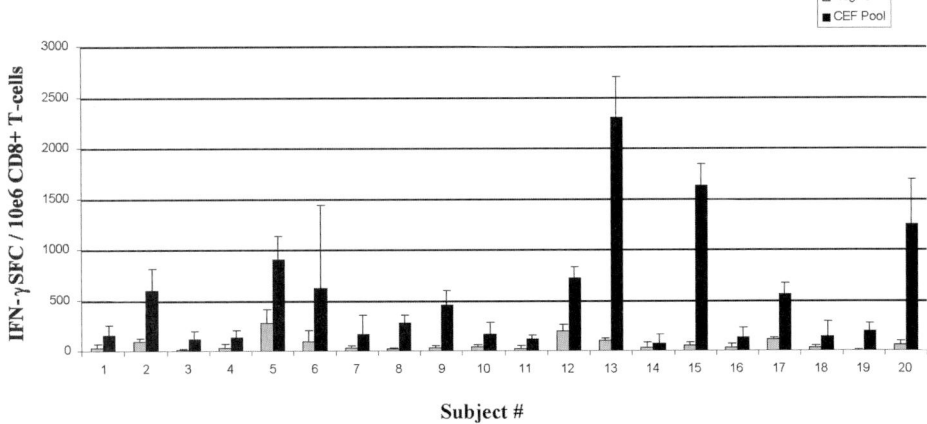

Fig. 3. IFN-γ ELISPOT: CD8$^+$ T-Cells with APC assay. The mean responses of 20 subjects (indicated on the *x*-axis) from a phase II clinical trial (HPV-associated cervical HSIL therapeutic) to the assay antigen control (CEF). The number of trial timepoints used to calculate the mean of each subject was three. Mean ± SD. IFN-γ SFC/10^6 CD8$^+$ T-cells is shown on the *y*-axis. The negative control (Neg. Ctrl.) represents CD8$^+$ T-cells plated with T2 without peptide. The experimental data are CD8$^+$ T-cells plated with T2 with the CEF peptide pool.

6. Dispense the appropriate volume of T2 into labeled tubes using a pipet-aid and serological pipet.
7. Add mitogen control (PHA; *see* **Note 15**) to the T2 at a 2X concentration of 10 μg/mL (final concentration after plating T2 cells with PBMCs in the ELISPOT plate will be 5 μg/mL) and antigenic peptides to the T2 cells at a 2X concentration of 6 μg/mL of CEF pool and 100 μg/mL of test peptide(s) (final concentration after plating T2 cells with PBMCs in the ELISPOT plate will be 3 and 50 μg/mL, respectively; *see* **Note 16**).
8. Gently vortex each T2 population for 10 s and place the tubes at a 30-degree angle on a tray (i.e., the lid of a pipet tip box) and place into a 37°C incubator with 5% CO_2 (do not tighten the caps of the tubes; **keep them loose**).
9. Incubate the T2 cell populations for a total of 4 h, with gentle vortexing each hour.
10. Begin preparation of CD8$^+$ T-cells from whole PBMC population prepared as in **Subheading 3.2., steps 1–5**; *see* **Note 19**).
11. After the second wash, resuspend the PBMC cell pellet in 1 mL of the 1X column wash buffer.
12. Add the contents of one vial (1 mL) of antibody to the resuspended cell population (total cell volume = 2 mL).
13. Gently vortex for 7 s and then incubate the cells at room temperature for 15 min.

14. Add 8 mL of the 1X column wash buffer to the cell population and centrifuge the tube for 10 min at 275g at room temperature, **with brake**.

15. Poor off the supernatant into a biological waste container, resuspend the cell pellet in 10 mL of 1X column wash buffer and centrifuge the tube for 10 min at 1100 RPM at room temperature, **with brake**.

16. Poor off the supernatant and resuspend the cell pellet in 2 mL of 1X column wash buffer.

17. All columns and column reagents should be equilibrated to room temperature prior to usage. Place columns on a stand or rack, place a waste receptacle (i.e., centrifuge tube) underneath each column, remove caps from columns and rinse each column with 10 mL of 1X column wash buffer. Collect all wash buffer flushed through columns into a waste receptacle.

18. Discard waste receptacle and replace it with a sterile 15-mL centrifuge tube that will collect eluted cells.

19. Add the cell population (2 mL) to the individual column and incubate at room temperature for 10 min.

20. After completing the incubation, wash each column with 10 mL of 1X column wash buffer.

21. Centrifuge eluted cell population for 10 min at 275g at room temperature, **with brake**.

22. Poor off the supernatant, resuspend the cell pellet in 1 mL of PBMC medium and count viable cells in trypan blue.

23. Adjust CD8$^+$ T-cell populations to a final concentration of 5×10^5/mL in PBMC medium (*see* **Note 14**). This procedure should be performed towards the end of the T2 incubation so as not to let the CD8$^+$ T-cells "rest" too long prior to adding to the ELISPOT plate (maximum "rest" time for CD8$^+$ T-cells should be 0.5 h).

24. Remove ELISPOT plates from 2 to 8°C storage and equilibrate to room temperature (ELISPOT plates are pre-coated with capture antibodies by manufacturer of the ELISPOT kits).

25. Add 100 μL of PBMC medium to all wells of the ELISPOT plate and incubate for 20 min at room temperature.

26. At the end of the incubation, remove the PBMC medium from the plates by "flicking" it into a biohazardous waste container and then blot the plates on a dry paper towel.

27. Add the 100 μL (1×10^5) of the T2 cell population to the appropriate wells (in triplicate if possible to account for assay variability). Immediately after adding the T2 cells to the ELISPOT plate, add 100 μL of CD8$^+$ T-cells (5×10^4) to the appropriate wells.

28. Mix the contents of the ELISPOT plate by gently tapping on the sides of the plate for approx 10 s.

29. Place ELISPOT plates in a 37°C incubator with 5% CO_2 and let incubate for 24 h.

30. Remove ELISPOT plates from the incubator and wash five times with a 1X wash buffer prepared from the 10X wash concentrate supplied in the kit. Take care not to make contact with the membrane at the bottom of each well. After completing the wash, blot the plates on a dry paper towel (*see* **Note 17**).

31. Add 100 µL of diluted detection antibodies to all wells and incubate the plates at 2–8°C overnight.
32. Remove ELISPOT plates from the refrigerator and wash five times (*see* **Subheading 3.5., step 30**).
33. Add 100 µL of diluted strepavodin AP to all wells and incubate the plates for 2 hours at room temperature.
34. Wash plates five times (*see* **Subheading 3.5., step 30**).
35. Add 100 µL of the chromogenic substrate to all wells and incubate the plates for one hour at room temperature, **in the dark**.
36. Wash plates five times with distilled water. Blot the plates on a dry paper towel, remove the bottom casing and let air-dry.
37. Count spots using a dissecting microscope or preferably, an automated spot counter with appropriate software (*see* **Note 18**).

3.6. Data Analysis

ELISPOT plates may be evaluated by visual inspection aided by a dissecting microscope or an automated plate reader using counting software. There are several automated instruments that are commercially available. All data reported in this chapter were captured using a ZEISS instrument (Carl Zeiss Vision, Germany) with KS ELISPOT 4.0 software, by Zellnet Consulting Inc. (New York, NY). The analysis was performed in a blinded fashion in that the analyst was only informed of the negative and positive control locations on each plate. Parameters were set on the ELISPOT instrument based on the results captured in the negative and positive control wells within each subject. All results were normalized values reported as spot-forming cells (SFCs) per million PBMCs or CD8$^+$ T-cells.

Other factors to consider when performing analysis include, but are not limited to, spot size, spot clarity, spot clustering, and spot confluency (*see* **Note 20**). Each of these may affect or contribute to results and analysis and should be evaluated in assay development/qualification stages.

4. Notes

1. Whole blood should be kept at room temperature until processing. Do not refrigerate. If collected at an alternate site, which requires shipment, samples should be shipped at ambient temperature to desired laboratory. Samples should not be at room temperature for more than 24–30 h after draw.
2. Do not disturb white blood cells and red blood cells during this step. Avoid any contamination of plasma sample with cells.
3. Tilt centrifuge tube containing ficoll to a 45-degree angle and slowly dispense diluted blood sample down the side of the tube. The diluted blood will "rest" on top of the ficoll. The tube should not contain more than 40 mL of total volume.
4. Be sure not to aspirate red blood cells into PBMC sample when removing the buffy coat.

5. This second wash ensures that all red blood cell lysis solution is removed from the sample.

6. When counting cells, it is important to only count viable cells and make the appropriate cell concentrations based on these numbers.

7. We determined the optimal cell concentration for freezing to be 5×10^6/mL and placed no more than 1.5 mL in a cryovial.

8. It is important to minimize the amount of time PBMC samples stay in freeze medium prior to placement to –80°C storage.

9. Styrofoam racks designed to hold cryovials are commercially available and their purpose is to ensure a slow freezing process.

10. Be sure not to let freshly processed or thawed PBMC stand in freeze medium for extended periods of time. It may cause cell lysing and as result, cell clumping may take place.

11. Cells have a tendency to lyse and as a result, free DNA may stick to adjacent cells and cause clumping. Using wash medium containing DNase will help minimize clumping. Exercising care and efficiency when freezing and thawing PBMCs should keep cell lysing to a minimum in a normal PBMC population.

12. At this point, PBMCs are ready to be used in a multitude of laboratory procedures: direct ELISPOT assays, T-cell subset enrichment, or in vitro-culturing procedures.

13. It is important to minimize the amount of time that T2 cells are in culture. It was found that culturing for no more than 4 wk before use in assays is important in keeping nonspecific responses at a minimum. 10% RPMI is the culture medium used to maintain the T2 cell line.

14. Optimal cell concentrations for each assay should be developed prior to performing assays with clinical samples. It was found that adding more than 5×10^5 cells, total, per well may exceed the limit in a standard 96-well ELISPOT plate.

15. A mitogen control, such as the polyclonal T-cell stimulator PHA, is used to demonstrate basic cell and assay function. The optimal final assay concentration was determined to be 5 μg/mL.

16. Peptide concentration may vary between peptides and between different assays. It is critical to determine optimal concentration prior to setting up assays with clinical samples.

17. Washing can be performed one of several ways; manually by a multichannel pipet, manually by an automatic multichannel washer or by an automated plate washer. Take care not to make contact with the membrane at the bottom of each well. After completing the wash, blot the plates on a dry paper towel. We used a NUNC 8-channel washer fitted to a carboy with 1X wash buffer, a vacuum pump and a waste receptacle containing a disinfectant.

18. A description of options for use in plate analysis is outlined in the data analysis section of the chapter. Briefly, a dissecting microscope may be used to obtain initial results but it is preferable to count spots using one of many available, automated spot analyzers. This will ensure a high level of accuracy, precision and reproducibility.

19. CD8⁺ T-cell enrichment columns were used to isolate CD8⁺ T-cells from whole PBMC samples. Magnetic beads are also available for T-cell subset isolation but

columns were selected for their ease of use, consistency and high CD8⁺ T-cell recovery.

20. If confluency (complete color development in a well which inhibits accurate quantification of spots) occurs within a well, a repeat of the assay may be required to ensure more accurate results. Typically, it is restricted to the mitogen control wells. One way to prevent confluency from taking place is to add fewer cells in the mitogen control wells then the other antigen-control or test wells. Optimal cell concentrations should be determined in assay development stages.

References

1. Pittet, M. J., Zippelius, A., Speiser, D. E., Assenmacher, M., Guillaume, P., Valmori, D., Lienard, D., Lejeune, F., Cerottini, J.-C., and Romero, P. (2001) Ex vivo IFN-γ secretion by circulating CD8 T-lymphocytes: implications of a novel approach for T-cell monitoring in infectious and malignant diseases. *J. Immunol.* **166**, 7634–7640.

2. Schmittel, A., Keilholz, U., Bauer, S., Kuhne, U., Stevanovich, S., Thiel, E., et al. (2001) Application of the IFN-γ ELISPOT assay to quantify T-cell responses against proteins. *J. Immunol. Methods.* **247**, 17–24.

3. Schmittel, A., Keilholz, U., and Scheibenbogen, C. (1997) Evaluation of the IFN-γ ELISPOT-assay for quantification of peptide specific T-lymphocytes from peripheral blood. *J. Immunol. Methods.* **210**, 167–174.

4. Di Fabio, S., Mbawuike, I. N., Kiyono, H., Fujihashi, K., Couch, R. B., and McGhee, J. R. (1994) Quantitation of human influenza virus-specific cytotoxic T-lymphocytes: correlation of cytotoxicity and increased numbers of IFN-γ producing CD8⁺ T-cells. *Int. Immunol.* **6**, 11–19.

5. McCutcheon, M., Wehner, N., Wensky, A., Kushner, M., Doan, S., Hsiao, L., et al. (1997) A sensitive ELISPOT assay to detect low frequency human T-lymphocytes. *J. Immunol. Methods.* **210**, 149–166.

6. Larsson, M., Jin, X., Ramratnam, B., Ogg, G. S., Engelmayer, J., Demoitie, M. A., et al. (1999) A recombinant vaccinia virus-based ELISPOT assay detects high frequencies of Pol-specific CD8 T-cells in HIV-1-positive individuals. *AIDS.* **13**, 767–777.

7. Kumar, A., Weiss, W., Tine, J. A., Hoffman, S. L., and Rogers, W. O. (2001) ELISPOT assay for detection of peptide specific IFN-γ secreting cells in rhesus macaques. *J. Immunol. Methods.* **247**, 49–60.

8. Corne, P., Huguet, M. F., Briant, L., Segondy, M., Reynes, J., and Vendrell, J. P. (1999) Detection and enumeration of HIV-1 producing cells by ELISPOT (enzyme-linked immunospot) assay. *J. Acquir. Immune Defic. Syndr. Hum. Retrovirol.* **20**, 442–447

9. Griffioen, M., Borghi, M., Schrier, P. I., and Osanto, S. (2001) Detection and quantification of CD8⁺ T-cells specific for HLA-A*0201-binding melanoma and viral peptides by the IFN-γ ELISPOT assay. *Int. J. Cancer.* **93**, 549–555.

10. Clay, T. M., Hobeika, A. C., Mosca, P. J., Lyerly, H. K., and Morse, M. A. (2001) Assays for monitoring cellular immune responses to active immunotherapy of cancer. *Clin. Cancer Res.* **7**, 1127–1135.

11. Klencke, B., Matijevic, M., Urban, R. G., Lathey, J. L., Hedley, M. L., Berry, M., et al. (2002) Encapsulated plasmid DNA treatment for human papillomavirus 16-associated anal dysplasia: a phase I study of ZYC101. *Clin. Cancer Res.* **8**, 1028–1037.

12. Pass, H. A., Schwarz, S. L., Wunderlich, J. R., and Rosenberg, S. A. (1998) Immunization of patients with melanoma peptide vaccines: immunologic assessment using the ELISPOT assay. *Cancer J. Sci. Am.* **4**, 316–323.

13. Lewis, J. J., Janetzki, S., Schaed, S., Panageas, K. S., Wang, S., Williams, L., et al. (2000) Evaluation of CD8(+) T-cell frequencies by the Elispot assay in healthy individuals and in patients with metastatic melanoma immunized with tyrosinase peptide. *Int. J. Cancer.* **87**, 391–398.

14. Mwau, M., McMichael, A. J., and Hanke, T. (2002) Design and validation of an enzyme-linked immunospot assay for use in clinical trials of candidate HIV vaccines. *AIDS Res. Human Retroviruses.* **18**, 611–618.

15. Smith, J. G., Liu, X., Kaufhold, R. M., Clair, J., and Caulfield, M. J. (2001) Development and validation of a gamma interferon ELISPOT assay for quantification of cellular immune responses to varicella-zoster virus. *Clin. Diagn. Lab. Immunol.* **8**, 871–879.

16. Lathey, J. L., Sathiyaseelan, J., Matijevic, M., and Hedley, M. L. (2003) Validation of pretrial ELISPOT measurements for predicting assay performance during a clinical trial. *BioProcess Int.* **September**, 34–41.

17. Flynn, K., Beltz, G., Altman, J., Ahmed, R., Woodland, D., and Doherty, P. (1998) Virus-specific CD8+ T-cells in primary and secondary influenza pneumonia. *Immunity.* **8**, 683–691.

18. Anderson, K., Cresswell, P., Gammon, M., Hermes, J., Williamson, A., and Zweerink, H. (1991) Endogenously synthesized peptide with an endoplasmic reticulum signal sequence sensitizes antigen processing mutant cells to class I-restricted, cell-mediated lysis. *J. Exp. Med.* **174**, 489–492.

19. Currier, J. R., Kuta, E. G., Turk, E., Earhart, L. B., Loomis-Price, L., Janetzki, S., et al. (2002) A panel of MHC class I restricted viral peptides for use as a quality control for vaccine trial ELISPOT assays. *J. Immunol. Methods.* **260**, 157–172.

20. Britten, C. M., Meyer, R. G., Kreer, T., Drexler, I., Wolfel, T., and Herr, W. (2002) The use of HLA-A*0201-transfected K562 as standard antigen-presenting cells for CD8+ T-lymphocytes in IFN-γ ELISPOT assays. *J. Immunol. Methods.* **259**, 95–110.

16

ELISPOT Determination of Interferon-γ T-Cell Frequencies in Patients With Autoimmune Sensorineural Hearing Loss

C. Arturo Solares and Vincent K. Tuohy

Summary

Autoimmune sensorineural hearing loss (ASNHL) is the most common cause of sudden hearing loss in adults. Although the etiopathogenesis of this disease is unclear, it is widely believed that antibody and/or T-cell responses directed against inner ear-specific proteins may mediate ASNHL. Using the enzyme-linked immunospot (ELISPOT) assay, we have recently found that many patients with ASNHL have increased frequencies of peripheral blood T-cells capable of producing interferon (IFN)-γ in response to a homogenate of human inner ear tissue. Our studies may ultimately lead to the identification of inner ear-specific autoimmune targets in ASNHL, and our ELISPOT approach may be particularly useful in supporting the diagnosis of this disease entity. In the current chapter we detail how to use the ELISPOT assay for measuring frequencies of IFN-γ-producing T-cells in patients with ASNHL.

Key Words: Autoimmunity; sensorineural hearing loss; deafness, T-cell, Th1/Th2, cytokines; ELISPOT.

1. Introduction

Autoimmune sensorineural hearing loss (ASNHL) typically is characterized by bilateral, rapidly progressive hearing loss that responds therapeutically to corticosteroid treatment. Although its name implies an autoimmune etiopathogenesis, data implicating self-recognition events in the development and progression of ASNHL have been limited predominantly to circumstantial evidence *(1)*. Clinically, this evidence involves a therapeutic response to corticosteroid treatment, whereas immunologically, this evidence includes detection of serum antibody to a 680kDa protein derived from inner ear tis-

From: *Methods in Molecular Biology, vol. 302: Handbook of ELISPOT: Methods and Protocols*
Edited by: A. E. Kalyuzhny © Humana Press Inc., Totowa, NJ

sues *(2)* and/or recall responses of peripheral blood mononuclear cells (PBMCs) to crude inner ear homogenate *(3–6)*. The significance of ASNHL, as opposed to other forms of hearing loss, resides in its potential for medical intervention *(1)*.

Currently, ASNHL is diagnosed by clinical criteria that often implicate patients with confounding systemic disorders that may contribute to the development of inner ear autoimmunity (e.g., systemic lupus erythematosus, rheumatoid arthritis). However, the existence of patients with ASNHL that have no evidence of any systemic abnormalities implies that inner ear-specific self-recognition events may be playing a decisive role in most if not all ASNHL patients *(1)*. Several assays have demonstrated that sensitized T-cells recognize autoantigens in ASNHL *(3–5)*. Other authors have focused on the presence of crossreacting antibodies in sera from patients by indirect immunofluorescence or immunoperoxidase techniques on temporal bone sections of human patients *(7,8)*. More recent approaches include Western blot assays using bovine inner ear tissue as the substrate to detect antibodies against to the 68-kDa protein believed to be heat shock protein 70 *(2,9,10)*.

Despite the different approaches used in the development of laboratory methods for ASNHL, laboratory testing to date has played only a supportive role in the diagnosis of ASNHL. No test has shown sufficient sensitivity and specificity for its detection *(11)*. We have successfully used the enzyme-linked immunospot (ELISPOT) assay to determine the frequency of interferon (IFN)-γ-producing T-cells responding to inner ear antigens in PBMCs from patients with ASNHL *(12)*. By analyzing the production of IFN-γ using ELISPOT, we have provided evidence implicating inner ear specific IFN-γ-producing T-cells in the pathogenesis of ASNHL. Further support for the role of autoreactive T-cells in the pathogenesis of ASNHL comes from our murine studies in which we showed that CD4$^+$ T-cells specific for inner ear peptides derived from cochlin and β-tectorin are capable of mediating experimental autoimmune hearing loss *(13)*. Our initial work with the ELISPOT was performed using human inner ear homogenate as a recall antigen (**Fig. 1**). We recently have initiated more refined studies using recombinant human inner ear-specific proteins.

The current chapter provides a list of materials required, a step-by-step description of the ELISPOT methodology as it refers to patients with ASNHL and some relevant notes highlighting potential pitfalls are provided.

2. Materials

Materials required include: (1) laboratory equipment, (2) human specimens, (3) media and solutions, and (4) immunologic reagents.

ASNHL
High IFNγ
Producers

ASNHL
Low IFNγ
Producers

Normal
Subjects

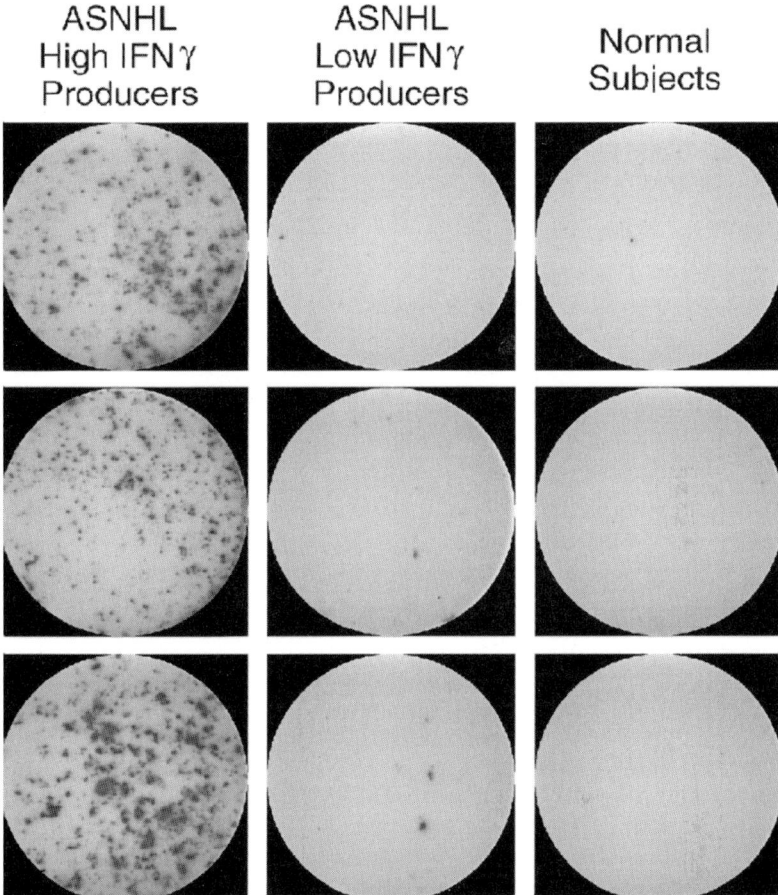

Fig. 1. IFN-γ ELISPOTs in response to inner ear antigens. PBMCs from three patients with ASNHL showed a high frequency of IFN-γ-producing T-cells 24 h after activation with 50 μg/mL human inner ear homogenate (left column). In contrast, ELISPOTS from three representative nonresponder ASNHL study subjects (middle column) showed IFN-γ-producing T-cell frequencies that were indistinguishable from those generated by PBMCs from three representative age- and sex-matched normal control subjects (right column; reprinted from **ref.** *12* with permission from Elsevier).

2.1. Laboratory Equipment

1. Dremel Moto-Tool tissue homogenizer (Dremel, Racine, WI).
2. Optical density spectrophotometer.
3. Centrifuge.

4. 7% CO_2/humidified incubator.
5. Series-1 Immunospot Analyzer (Cellular Technology, Cleveland, OH).

2.2. Human Specimens

1. Membranous tissue from human vestibular and/or cochlear structures.
2. Blood from patients (50 mL) in heparinized 10-mL tubes.

2.3. Media and Solutions

1. PBMC separation medium (Ficoll-Hypaque, Pharmacia Biotech, Uppsala, Sweden).
2. Phosphate-buffered saline (PBS) solution.
3. Hanks balanced salt solution (HBSS; Gibco Invitrogen, Carlsbad, CA).
4. Serum-free HL-1 medium (Hycor, Irvine, CA).
5. 2 m*M* L-glutamine (Gibco Invitrogen).
6. 100 U/mL Penicillin (Gibco Invitrogen).
7. 100 µg/mL Streptomycin (Gibco Invitrogen)
8. HEPES buffer (Gibco Invitrogen).
9. Tween (Sigma, St. Louis, MO).
10. 1 *M* Acetate buffer.
11. 30% H_2O_2.

2.4. Immunologic Reagents

1. Mouse capture anti-human IFN-γ (Endogen, Cambridge, MA).
2. ELISPOT plates (Polyfiltronics, Rockland, MA).
3. 1% BSA fraction V (Sigma).
4. Mouse anti-human CD3 (Pharmingen, San Diego, CA).
5. Phytohemagglutinin-P (Sigma).
6. Biotin-labeled mouse anti-human IFN-γ (Endogen).
7. Peroxidase-conjugated streptavidin (Dako, Carpenteria, CA).
8. 3-amino-9-ethylcarbazole (AEC).

3. Methods

The methods described below outline: (1) the preparation of human inner ear homogenate, (2) the preparation of peripheral blood leukocytes, (3) a description of the ELISPOT assay, and (4) the analysis of ELISPOT assays.

3.1. Preparation of Human Inner Ear Homogenate

To test the reactivity of the PBMCs to human inner ear tissues, membranous tissue from vestibular and/or cochlear structures is obtained from surgical patients undergoing labyrinthectomies for disease other than neoplasm or autoimmune disorders. The tissue collected at the time of surgery is processed as follows:

1. Homogenize the tissue using a Dremel Moto-Tool (*see* **Note 1**).
2. Ultracentrifuge the homogenate at 20,000*g* for 60 min.

3. Decant the supernatant and filter it through a 0.45-μm filter.
4. Determine the protein concentration by a modified Lowry method using optical density values at wavelengths of 215, 225, and 280 nm.

3.2. Preparation of PBMCs

1. Collect 50 mL of blood in heparinized 10-mL tubes (*see* **Note 2**)
2. Bring Ficoll-Hypaque lymphocyte separation media to room temperature. Dispense 17 mL of Ficoll-Hypaque into a 50-mL polypropylene centrifuge tube. One 50-mL tube is used for every 15 mL of patient blood.
3. Pipet blood into a sterile 50-mL polypropylene centrifuge tube.
4. Dilute patient blood 1:1 with PBS and carefully layer 30 mL of diluted blood over 17 mL of Ficoll–Hypaque, being careful not to disturb the interface.
5. Centrifuge Ficoll-Hypaque/blood layers for 30 min at 700*g* at 20°C with the brake off.
6. Collect the mononuclear cell interface using a sterile polypropylene transfer pipet with needle tip. Minimize the amount of Ficoll-Hypaque or plasma removed while maximizing the cell recovery.
7. Promptly dilute the cells 3X in HBSS. Avoid extensive exposure to Ficoll–Hypaque as it decreases cell viability.
8. Centrifuge cell suspension at 600*g* for 5 min. Decant the supernatant and resuspend the cells in HBSS. Repeat wash two more times.
9. Resuspend at 3×10^6 cells/mL in culture medium (90 mL serum-free HL-1 medium supplemented with 0.5 mL of 2 mM L-glutamine, 0.25 mL of 100 U/mL penicillin, 0.25 mL of 100 μg/mL streptomycin, 2 mL of 30 m*M* HEPES buffer, and 5 mL of autologous serum).

3.3. ELISPOT Assay

1. Dilute cytokine-capturing mouse anti-human IFN-γ capture antibody in sterile PBS so that 100 μL/well may be dispensed.
2. Precoat ELISPOT plates with cytokine-capturing mouse anti-human IFN-γ capture antibody at 0.4 μg/well by incubation at 4°C for 12 h. Seal the plates in plastic wrap. Plates can be prepared up to 1 wk in advance if necessary.
3. Wash wells with 200 μL/well of PBS. Repeat three times.
4. Dispense 200 μL of PBS containing 1% BSA fraction V to prevent nonspecific binding. Incubate for a minimum of 2 h at room temperature.
5. Repeat sterile wash as outlined in **step 3**. Do not remove the last wash until you are ready to add media (plates must not be allowed to dry).
6. Remove final PBS wash from the plate. Add 100 μL of PBMCs at a concentration of 3×10^6 cells/mL of supplemented medium to each well. Supplemented medium is added so that each well contains 3×10^5 cells in a final volume of 200 μL.
7. Add human inner ear homogenate at various final concentrations ranging from 0.5 to 50 μg/mL.
8. Add media alone to PBMCs as a negative control.

9. For positive controls, wells are treated with either mouse anti-human CD3 or phytohemagglutinin-Pat 5 µg/mL.
10. Incubate for 24 h at 37°C in a 7% CO_2/humidified incubator.
11. Decant cells and media and rinse plates with double-distilled deionized water (ddH_2O). Cells are then washed away using three washes of PBS followed by three washes of PBS/0.05% Tween/1% bovine serum albumin (BSA).
12. Dilute secondary biotin-labeled mouse anti-human IFN-γ antibody (Endogen, Cambridge, MA) in PBS/0.05% Tween/1% BSA. Add 50 µL/well of secondary antibody and incubate overnight at 4°C.
13. Wash three times with PBS-Tween.
14. Dilute peroxidase-conjugated streptavidin (Dako, Carpenteria, CA) 1:2000 in PBS/Tween/BSA. Add 100 µL/well and incubate at room temperature for 2 h.
15. Wash three times with 200 µL/well of PBS.
16. Dilute AEC 1:30 in 0.1 *M* acetate buffer, filter the AEC/acetate buffer solution through a 0.45-µm pore size filter prior to removing colored precipitates. Add 200 µL/well of a 1:2000 dilution of 30% H_2O_2 and observe for spot development for up to 1 h.
17. Wash three times with 200 µL/well of ddH_2O to stop the reaction. This can be done under running ddH_2O.
18. Dry plates without lids in the upright position at room temperature for 12 h.

3.4. ELISPOT Analysis

ELISPOT detection is performed using an automated Series-1 Immunospot Analyzer (Cellular Technology, Cleveland, OH) with proprietary software designed to distinguish real spots from artifact (*see* **Note 3**). Digitized images of the wells are analyzed for concentrated spots of red color in which the density exceeds background by a factor individually calculated per plate based on the positive and negative control wells. Parameters for automated spot counting are established so that spots that touch or overlap can be separated and spot size criteria and circularity are used to exclude noise caused by nonspecific antibody binding.

4. Notes

1. Tissue homogenization. It is important to assure that one has a completely homogeneous solution. At the time of surgery, the inner ear membranes are collected in either 0.9% sodium chloride solution or PBS in a volume of approx 30–50 mL per sample. Avoid using excessively large volumes of solution. Subsequently, the tissues are thoroughly homogenized, either with a Dremel Moto-Tool or alternatively with a mortar and pestle. Centrifugation eliminates unwanted particles from the solution and subsequent filtering eliminates bacterial contamination. Bring the solution to a final concentration of 1–2 mg/mL and store at 4°C for up to 1 mo.
2. Preparation of PBMCs. Make sure the blood is collected in heparinized tubes. Collect at least 50 mL per patient; if possible, collect 80 mL. When performing the

technique for the first time, the cellular yields may be lower and additional amounts of blood may be helpful. The critical steps in the isolation of PBMCs involve the use of Ficoll–Hypaque. When adding blood to the Ficoll–Hypaque layer, do it slowly to prevent disruption of the interface. Also, after centrifugation, when collecting the cells from the interface be extremely careful to collect all the cells in the interface while minimizing the amount of plasma or lymphocyte separation media (Ficoll–Hypaque). Lastly, minimize the exposure of the cells to Ficoll–Hypaque. Immediately after collecting the PBMCs from the interface, the cells should be placed in HBSS or similar media. Extended exposure of PBMC to Ficoll–Hypaque causes cell lysis leading to a decrease in cell yield.

3. ELISPOT assay. We perform our ELISPOTS based on a previously published protocol by Pelfrey and colleagues *(14)*. A few notes on our personal experience with the technique will suffice. A critical step, in our experience, is the addition of 1% BSA fraction V. If this step is not done properly, or not done at all, significant nonspecific binding will be observed and the results will be at best, suboptimal. The other important caveat is to be extremely diligent when adding different volumes of antigen. Make sure the appropriate concentration and volume are added to each well. Multi-channel and/or pipets can be extremely helpful to achieve this goal. Other helpful hints in performing ELISPOT can be found elsewhere in this book.

References

1. Solares, C. A., Hughes, G. B., and Tuohy, V.K. (2003) Autoimmune sensorineural hearing loss: An immunologic perspective. *J. Neuroimmunol.* **138**, 1–7.
2. Moscicki, R. A., San Martin, J. E., Quintero, C. H., Rauch, S. D., Nadol, J. B., Jr., and Bloch, K. J. (1994) Serum antibody to inner ear proteins in patients with progressive hearing loss. *JAMA* **272**, 611–616.
3. McCabe, B. F., and Mccormick, K. J. (1984) Tests for autoimmune disease in otology. *Amer. J. Otol.* **5**, 447–449.
4. Hughes, G. B., Barna, B. P., Calabrese, L. H., Kinney, S. E., and Nalepa, N. L. (1986) Predictive value in laboratory tests in autoimmune inner ear disease: preliminary report. *Laryngoscope* **96**, 502–505.
5. Hughes, G. B., Moscicki, R, Barna, B. P., and San Martin, J. E. (1994) Laboratory diagnosis of immune inner ear disease. *Am. J. Otol.* **15**, 198–202.
6. Billings, P. (2004) Experimental autoimmune hearing loss. *J. Clin. Invest.* **113**, 1114–1117.
7. Arnold, W., Pfaltz, C. R., and Alternatt, H. J. (1985) Serum antibodies against inner ear tissues in the blood of patients with sensorineural hearing disorders. *Acta Otolaryngol.* **99**, 437–444.
8. Arnold, W., and Pfaltz, R. (1987) Critical evaluation of the immunofluorescence microscopic test for identification of serum antibodies against human inner ear tissue. *Acta Otolaryngol.* **103**, 373–378.
9. Billings, P. B., Keithley, E. M., and Harris, J. P. (1995) Evidence linking the 68 kilodalton antigen identified in progressive sensorineural hearing loss patient sera with heat shock protein 70. *Ann. Otol. Rhinol. Laryngol.* **104**, 181–188.

10 Hirose, K., Wener, M. H., and Duckert, L. G. (1999) Utility of laboratory testing in autoimmune inner ear disease. *Laryngoscope* **109**, 1749–1754.

11. Barna, B. P., and Hughes, G. B. (1997) Autoimmune inner ear disease—a real entity? *Clin. Lab. Med.* **17**, 581–594.

12. Lorenz, R. R., Solares, C. A., Williams, P., Sikora, J., Pelfrey, C. M., Hughes, G. B., et al. (2002) Interferon-gamma production to inner ear antigens by T-cells from patients with autoimmune sensorineural hearing loss. *J. Neuroimmunol.* **130**, 173–178.

13. Solares, C. A., Edling, A. E., Johnson, J. M., Baek, M.-J., Hirose, K., Hughes, G. B., et al. (2004) Murine autoimmune hearing loss mediated by CD4+ T-cells specific for inner ear peptides. *J. Clin. Invest.* **113**, 1210–1217.

14. Pelfrey, C. M., Rudick, R. A., Cotleur, A. C., Lee, J. C., Tary-Lehmann, M., and Lehmann, P. V. (2000) Quantification of self-recognition in multiple sclerosis by single-cell analysis of cytokine production. *J. Immunol.* **165**, 1641–1651.

IV

MULTIPLEX AND MODIFIED ELISPOT ASSAY FORMATS

17

Dual-Color ELISPOT Assay for Analyzing Cytokine Balance

Yoshihiro Okamoto and Mikio Nishida

Summary

A dual-color enzyme-linked immunospot (ELISPOT) assay enabled us to analyze three kinds of cytokine-secreting cells simultaneously. T helper (Th) cells can be subdivided into at least two distinct functional subsets based on their cytokine secretion profiles. The first type of clones (Th1) produces interleukin (IL)-2 and interferon (IFN)-γ but not IL-4 or IL-5. The second type of clones (Th2) produces IL-4 and IL-5 but not IL-2 or IFN-γ. Furthermore, the presence of the third type (Th0) cell, which is a precursor of Th1 or Th2 cells, has been demonstrated to produce both Th1- and Th2-type cytokines. The dual-color ELISPOT assay is developed to differentiate these three subtypes of Th cells in an identical well. In the system, the red spots corresponding to IL-2-secreting cells (Th1) were developed with horseradish peroxidase and amino-ethyl-carbazole/H_2O_2. The light blue spots corresponding to IL-4-secreting cells (Th2) were developed with alkaline phosphatase and Vector blue (chromogenic substrate for alkaline phosphatase). The mixed colored (indigo) spots corresponding to both kinds of cytokine-secreting cells (Th0 cells) were developed with both chromogenic substrates. With this system, we could detect the IL-2- and/or IL-4-secreting cells simultaneously in a murine spleen cell or human peripheral mononuclear cell preparation.

Key Words: Dual-color enzyme-linked immunospot assay; interleukin-2; interleukin-4; cytokine balance; mouse; human.

1. Introduction

The enzyme-linked immunospot (ELISPOT) assay is an efficiently sensitive technique for the enumeration of single cells secreting cytokines *(1)*. Variations of the ELISPOT assay have been developed by some investigators, including our group *(2–9)*. Recently, we developed a dual-color ELISPOT assay *(4)*, which was named "Stardust Assay," by improving an ordinary ELISPOT assay. This new method enabled us to analyze three kinds of cytokine-secreting cells simultaneously.

From: *Methods in Molecular Biology, vol. 302: Handbook of ELISPOT: Methods and Protocols*
Edited by: A. E. Kalyuzhny © Humana Press Inc., Totowa, NJ

T helper (Th) cells can be subdivided into at least two distinct functional subsets based on their cytokine secretion profiles *(10)*. The first type of clones (Th1) produces interleukin (IL)-2 and interferon (IFN)-γ but not IL-4 or IL-5. The second type of clones (Th2) produces IL-4 and IL-5 but not IL-2 or IFN-γ. Furthermore, the presence of the third type (Th0) cell, which is a precursor of Th1 or Th2 cells, has been demonstrated to produce both Th1- and Th2-type cytokines *(11,12)*.

The dual-color ELISPOT assay is developed to differentiate these three subtypes of Th-cells in an identical well. In the system, the red spots, which correspond to IL-2-secreting cells (Th1-cells), were developed with horseradish peroxidase and amino-ethyl-carbazole (AEC)/H_2O_2. The light blue spots, which correspond to IL-4-secreting cells (Th2-cells), were developed with alkaline phosphatase and Vector blue (chromogenic substrate for alkaline phosphatase, Vector Laboratories, CA, USA). The mixed colored (indigo) spots, which correspond to both kinds of cytokine-secreting cells (Th0-cells), were developed with both chromogenic substrates (**Fig. 1**). A photographic profile of different colored spots resembles "Stardust." Thus we call this technique "Stardust Assay." With this system, we could detect the IL-2- and/or IL-4-secreting cells simultaneously in a murine spleen cell preparation (**Fig. 2A,B**). In the present article, the dual color ELISPOT assay enabling simultaneous detection for plural numbers of cytokines will be described.

2. Materials

2.1. Reagents and Buffers

1. Capture (first) antibody: anti-mouse IL-2 monoclonal antibody (Genzyme, Cambridge, MA) and anti-mouse IL-4 monoclonal antibody (clone BVD4-1D11: Pharmingen, San Diego, CA).
2. Detection (second) antibody: rabbit polyclonal antibody for mouse IL-2 (Bectone Dickinson, Bedford, MA) and biotinylated monoclonal antibody for mouse IL-4 (clone BVD6-24G2: Pharmingen, San Diego, CA).
3. Streptavidin-conjugated alkaline phosphatase (GIBCO BRL Co. Ltd., NY).
4. Horseradish peroxidase-conjugated F(ab'2) fragment donkey anti-rabbit IgG(H+L) (Jackson ImmunoResearch Laboratories, Inc., West Grove, PA).
5. 30% Hydrogen peroxide (H_2O_2).
6. Bovine serum albumin (BSA, globulin free).
7. Cell culture medium (e.g., RPMI 1640 containing 10% heat-inactivated fetal bovine serum [FBS]).
8. Phosphate-buffered saline (PBS): Dissolve 80 g of NaCl, 2.0 g of KCl, 11.5 g of Na_2HPO_4, and 2.0 g of KH_2PO_4 in 900 mL of distilled water (dH_2O). Check pH and adjust to 7.4 with 1 M NaOH if necessary. Make volume up to 1 L with dH_2O. Store at room temperature. Dilute 1 in 10 with dH_2O for use.

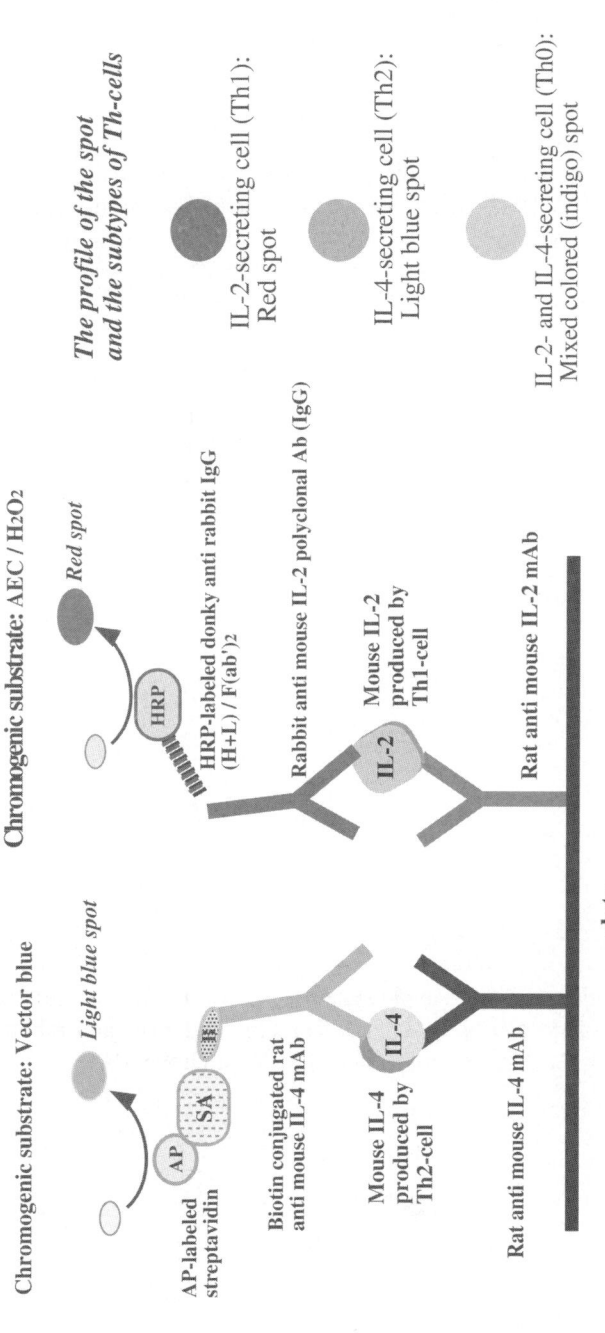

Fig. 1 The schematic representation of a dual-color detection (Stardust assay). Abbreviations: mAb, monoclonal antibody; AP, alkaline phosphatase; HRP, horseradish peroxidase; SA, streptavidin; B, biotin. (Reprinted from **ref. 4** with permission from Elsevier.)

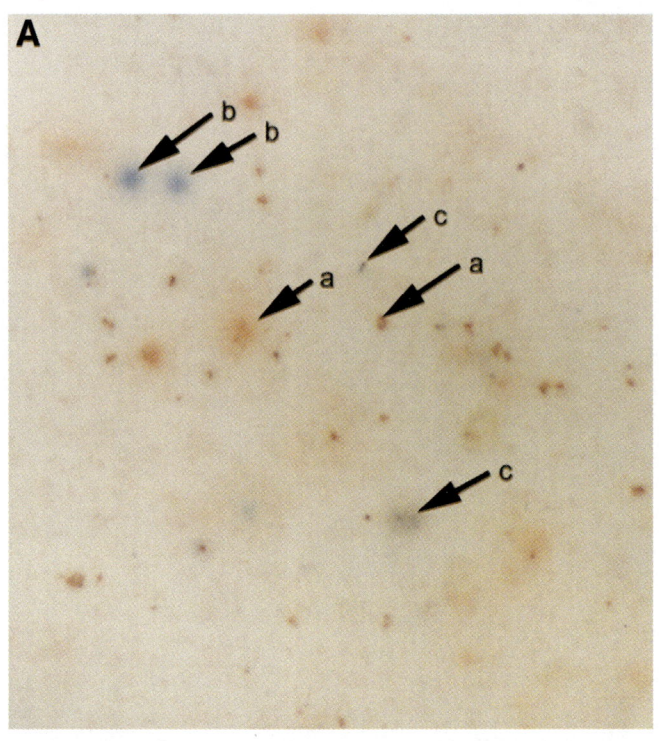

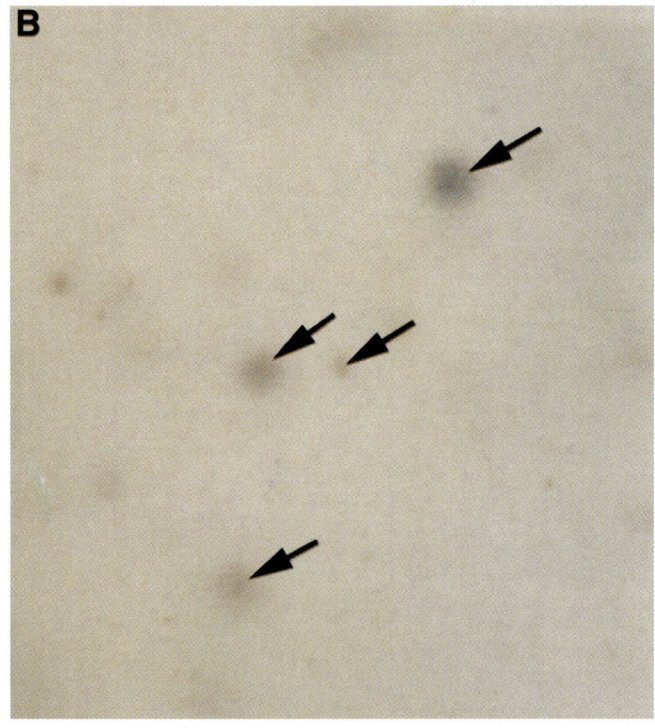

9. 0.05% Tween-20 in PBS (PBS-T): Add 0.5 mL of Tween-20 to 1 L of PBS from above.
10. Blocking solution: 5% BSA in PBS.
11. Substrate solution: 3-amino-9-ethylcarbazole (AEC) substrate kit (Vector Laboratories, CA) and Vector Blue substrate kit (Vector Laboratories, CA).

2.2. Equipment

1. 37°C CO_2 incubator: It is important that the incubator is absolutely leveled to prevent cells from rolling to one side of the well.
2. Microscope (magnification; ×10–40)
3. 96-Well nitrocellulose-backed plate (Millipore Multiscreen HA plate, Millipore, MA).
4. Plastic plate seal (Sumitomo Bakelite Co., Ltd., Tokyo, Japan).

3. Methods

1. Prepare first antibody mixture including an anti-mouse IL-2 monoclonal antibody (5 µg/mL) and an anti-mouse IL-4 monoclonal antibody (5 µg/mL) in PBS. Coat the wells of a 96-well nitrocellulose-backed plate with 100 µl of the antibody mixture per well (*see* **Notes 1** and **2**).
2. Seal the plate with plastic plate seal to prevent evaporation. Incubate overnight at 4°C.
3. Wash the plate three times with PBS-T (*see* **Note 3**).
4. Add 300 µL of blocking solution (5% BSA/PBS).
5. Incubate for 2 h at room temperature.
6. Wash the plate three times with sterile PBS.
7. Prepare cell suspension at different concentrations, for example, 1×10^5 cells/mL, 2×10^4 cells/mL, and 4×10^3 cells/mL. Add 100 µL of each cell suspension per well, in triplicate (*see* **Notes 4** and **5**).
8. Incubate at 37°C in 5% CO_2 for 18 h (*see* **Note 5**).
9. Wash the plate five times with PBS-T.
10. Add 100 µL of the detection antibody mixture including a rabbit polyclonal antibody for mouse IL-2 (2 µg/mL) and a biotinylated monoclonal antibody for mouse IL-4 (2 µg/mL) in PBS-T containing 1% BSA per well.

Fig. 2 (A). Typical profile of the dual color ELISPOT assay. **(A)** Crude spleen cells of normal BALB/c mice were stimulated with 1 µg/mL Concanavalin A for 18 h. After the stimulation, the cells were added to wells coated with the mixture of anti-IL-2 and IL-4 antibody, and subsequently spots were developed by the enzyme-substrate system shown in **Fig. 1**. Red spots corresponding to IL-2-secreting cells are indicated by *arrow a*, light blue spots corresponding to IL-4-secreting cells are indicated by *arrow b*, and the indigo spots corresponding to the Th0 type cells are indicated by *arrow c* (×40). **(B).** The ideal spots of Th0 cells. The mixed colored spots (indigo) are shown as the ideal profile of the spots corresponding to a Th0 cell (arrows; ×40). (Reprinted from **ref. 4** with permission from Elsevier.)

11. Seal the plate to prevent evaporation. Incubate overnight at 4°C.
12. Wash the plate five times with PBS-T.
13. Add 100 µL of the mixture including a horseradish peroxidase-conjugated F(ab')$_2$ fragment donkey anti-rabbit IgG(H+L) (diluted 1: 5000) and a streptavidin-conjugated alkaline phosphatase (diluted 1:2000) per well.
14. Seal the plate to prevent evaporation. Incubate for 2 h at room temperature.
15. Wash the plate five times with PBS-T.
16. Expose wells to 100 µl of AEC/H$_2$O$_2$ substrate solution (Vector Laboratories, Inc., Burlingame, CA) and examine for red spots to identify IL-2. These reactions developed for 5–7 min at room temperature.
17. Wash with PBS several times to eliminate AEC/H$_2$O$_2$ substrate solution.
18. Next, 100 µL of the Vector blue substrate solution (Vector Laboratories, Inc., Burlingame, CA) was added to each well, yielding light blue spots within 10–20 min to stain IL-4. The mixed-colored (indigo) spots correspond to both kinds of cytokine-secreting cells (*see* **Note 6**).
19. Wash the plate several times with dH$_2$O.
20. The developed plate is dried and count the number of spots in each well under low magnification (approx ×40) with a microscope. A typical profile of the dual color ELISPOT assay is shown in **Fig. 2** (*see* **Notes 7– 9**).

Recent studies revealed that the balance of cytokines secreted by different types of cells affected the state and progression of various diseases, including infectious, allergic and autoimmune disorders *(13)*. The present procedure provides a useful tool for quantitatively analyzing micro-levels of dynamic immune responses. Practically we analyzed the changes in cytokine balance in collagen-induced arthritic (CIA) mice as an animal experimental model of human rheumatoid arthritis using the dual-color ELISPOT assay. We could obtain the valuable results that, at the prearthritic phase Th1 cells, were dominant, and after the onset of clinical arthritis, there was a shift from a Th1-dominant to a Th2-dominant state (**Fig. 3**; **ref.** *14*).

Furthermore, we have optimized a human dual-color ELISPOT assay system with replacing antibodies for murine cytokines to those for human, and evaluated the cytokine balance in a patient with juvenile rheumatoid arthritis (JRA). It was demonstrated that the frequency of both IL-2- and IL-4-secreting cells in the peripheral mononuclear cells of the patient with JRA was markedly higher than those of healthy individuals. The ratio of Th1/Th2 of the patient was lower than that of healthy subjects (**Fig. 4**; **ref.** *15*).

In summary, the dual-color ELISPOT assay (Stardust assay) is an excellent method to monitor the cytokine balance in diseases and should be one of the most powerful tools for not only animal experiments but also clinical investigation.

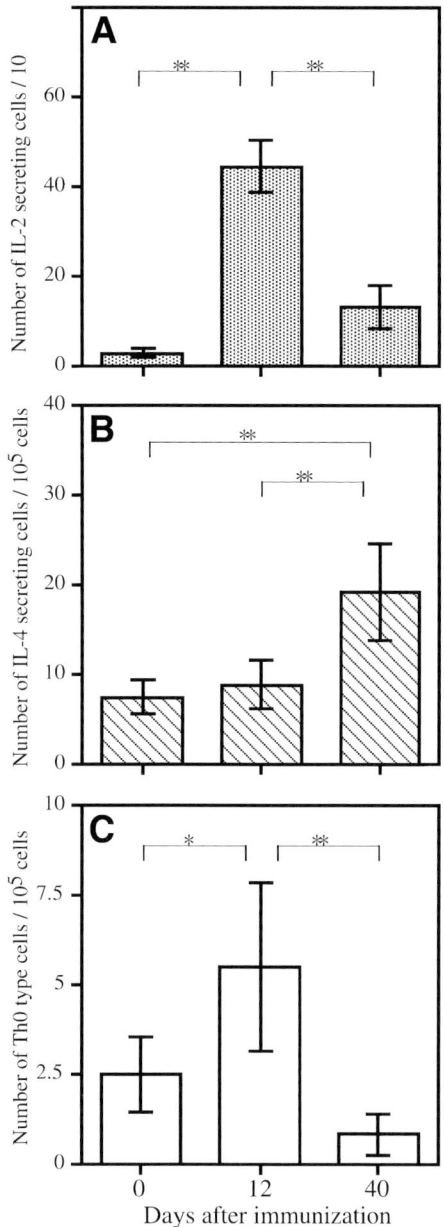

Fig. 3. The population change of Th cells in spleen of mice with CIA. Dual-color ELISPOT assay was conducted on spleen cells obtained from mice with CIA at different times after collagen type II (CII)-immunization. Cells were incubated with CII (50 µg/mL) in a well coated with the mixture of anti-IL-2 antibody and anti-IL-4 antibody for 18 h and, subsequently, spots were developed as described in Materials and Methods. **(A)** The frequency of IL-2 secreting cells (Th1 cell), **(B)** The frequency of IL-4 secreting cells (Th2), **(C)** The frequency of cells (Th0) secreting both cytokines. The results are expressed as mean ± SD of six assay wells. Significant differences were determined by Kruskal–Wallis nonparametric one-way analysis of variance and Scheffé's F test. $**p < 0.01$, $*p < 0.05$. (Reprinted from **ref. _14_** with permission from Mary Ann Liebert, Inc.)

4. Notes

1. Keep reagents and assay plate sterile during **steps 1** to **8**.
2. Higher concentration of coating antibody may give better results. However, optimal concentration (usually 2–10 µg/mL) should be examined in preliminary experiments.

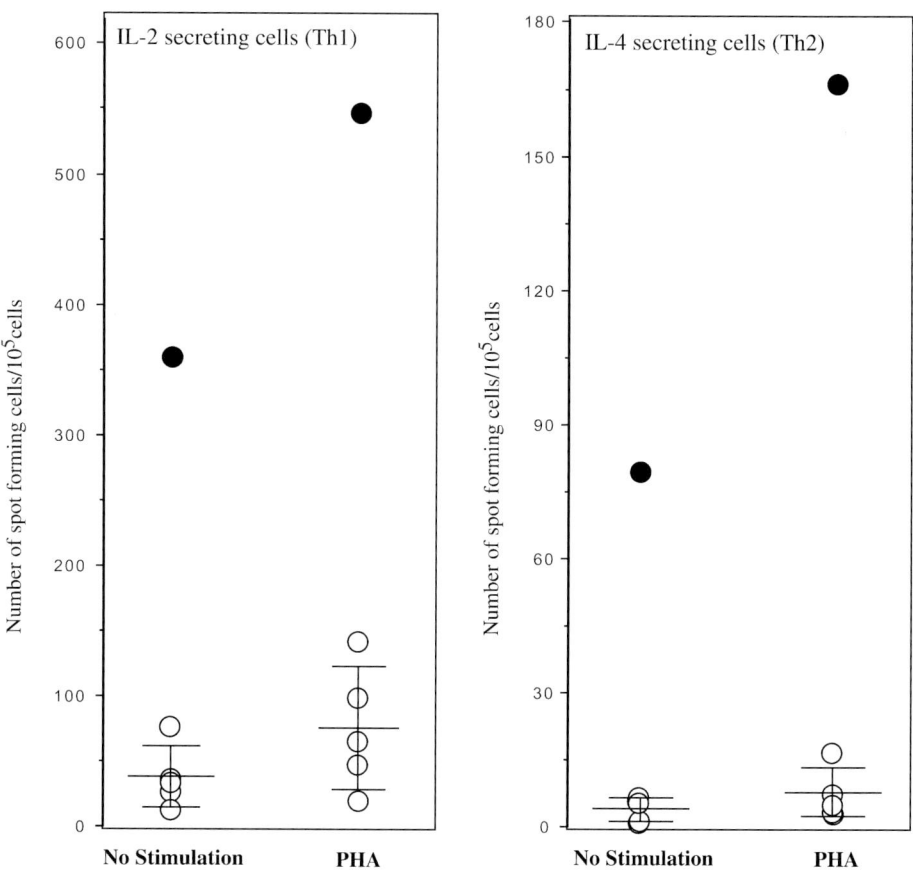

Fig. 4. Frequency of IL-2- and IL-4-secreting cells in a patient with JRA. PBMCs of each healthy volunteer (24 ± 2 yr old, range 22–33, two males and three females) and one patient with JRA (4 yr old, male) were prepared. The cells were washed with RPMI 1640 medium. Subsequently the cells were stimulated with 2 µg/mL phytohemaglutinin for 18 h. Nonstimulated cells were employed as a control. After the stimulation, the cells were applied to the dual-color ELISPOT assay. Open circles represent healthy individuals; closed circles represent the patients with JRA. Each circle represents the mean values of at least six assay wells. Each horizontal bar represents the mean values ± SD of five healthy individuals. (Reprinted from **ref. *15*** with permission from Elsevier.)

3. For each wash, fill wells with approx 300 µL of PBS(-T), soak for at least 1 min per wash, and invert plate to discard a washing solution.
4. Various types of cell specimens are applied to this assay (e.g., spleen, lymph nodes, bone marrow, or a cell-fraction purified from various sources). The cell suspension is prepared by washing cells extensively with incomplete medium, then resus-

pending the cells in medium containing 10% heat inactivated FBS. The cell specimen should be kept on ice until use. The viability of the cells should be assessed by trypan blue dye exclusion test before use to identify the number of living cells.

5. The optimal cell concentration and time of incubation will differ in individual experiment. The cell specimen is sequentially diluted to detect the appropriate number of spots in a well, and the conditions to produce 10–200 spots per well should be used to count the total number of cytokine-secreting cells per sample. It is difficult to count the number of spots precisely when more than 200 spots per well were developed.

6. To detect the cells secreting both kinds of cytokines precisely, a reference profile of the double-stained spots (indigo) should be provided in the plate. To obtain the ideal spots corresponding to the cells secreting both kinds of cytokines, the plate was incubated with biotinylated monoclonal antibody for IL-4 and followed by the two kinds of chromogenic system; the mixture of the horseradish peroxidase-labeled and the alkaline phosphatase-labeled streptavidin was added to the well after the incubation with the biotinylated monoclonal antibody for IL-4. By this procedure, we obtained a typical profile corresponding to the cells secreting both cytokines (*see* **Fig. 2B**).

7. The color depth or the size of spots depends on the amount of secreted cytokines. The strong and well-defined spots should be counted; any small or faint spots are likely to be artifacts and should not be counted.

8. The developed spots would be kept for several weeks if the plates are stored at 4°C under light protection.

9. To confirm specificity of the assay, the experiment using the wells coated with an irrelevant antibody (e.g., anti-IL-6 antibody coated well is used) as a negative control should be included.

References

1. Sedgwick, J. D., and Czerkinsky, C. (1992) Detection of cell-surface molecules, secreted products of single cells and cellular proliferation by enzyme immunoassay. *J. Immunol. Methods* **150**, 159–175.
2. Shirai, A., Sierra, V., Kelly, C. I., and Klinman, D. M. (1994) Individual cells simultaneously produce both IL-4 and IL-6 in vivo. *Cytokine* **6**, 329–336.
3. Okamoto, Y., Murakami, H., and Nishida, M. (1997) Detection of interleukin 6-producing cells among various organs in normal mice with an improved enzyme-linked immunospot (ELISPOT) assay. *Endocr. J.* **44**, 349–355.
4. Okamoto, Y., Abe, T., Niwa, T., Mizuhashi, S., and Nishida, M. (1998) Development of a dual color enzyme-linked immunospot assay for simultaneous detection of murine T helper type 1- and T helper type 2-cells. *Immunopharmacology* **39**, 107–116.
5. Herr, W., Linn, B., Leister, N., Wandel, E., Meyer zum Buschenfelde, K. H., and Wolfel, T. (1997) The use of computer-assisted video image analysis for the quantification of CD8+ T-lymphocytes producing tumor necrosis factor alpha spots in response to peptide antigens. *J. Immunol. Methods* **203**, 141–152.

6. Ronnelid, J. and Klareskog, L. (1997) A comparison between ELISPOT methods for the detection of cytokine producing cells: greater sensitivity and specificity using ELISA plates as compared to nitrocellulose membranes. *J. Immunol. Methods* **200**, 17–26.

7. Vaquerano, J. E., Peng, M., Chang, J. W., Zhou, Y. M., and Leong, S. P. (1998) Digital quantification of the enzyme-linked immunospot (ELISPOT). *Biotechniques* **25**, 830–834, 836.

8. Kalyuzhny, A. and Stark, S. (2001) A simple method to reduce the background and improve well-to-well reproducibility of staining in ELISPOT assays. *J. Immunol. Methods* **257**, 93–97.

9. Snyder, J. E., Bowers, W. J., Livingstone, A. M., Lee, F. E., Federoff, H. J., and Mosmann, T. R. (2003) Measuring the frequency of mouse and human cytotoxic T-cells by the Lysispot assay: independent regulation of cytokine secretion and short-term killing. *Nat. Med.* **9**, 231–235.

10. Mosmann, T. R., Cherwinski, H., Bond, M. W., Giedlin, M. A., and Coffman, R. L. (1986) Two types of murine helper T-cell clone. I. Definition according to profiles of lymphokine activities and secreted proteins. *J. Immunol.* **136**, 2348–2357.

11. Firestein, G. S., Roeder, W. D., Laxer, J. A., Townsend, K. S., Weaver, C. T., Hom, J. T., et al. (1989) A new murine CD4+ T-cell subset with an unrestricted cytokine profile. *J. Immunol.* **143**, 518–525.

12. Street, N. E., Schumacher, J. H., Fong, T. A., Bass, H., Fiorentino, D. F., Leverah, J. A., et al. (1990) Heterogeneity of mouse helper T-cells. Evidence from bulk cultures and limiting dilution cloning for precursors of Th1 and Th2 cells. *J. Immunol.* **144**, 1629–1639.

13. Abbas, A. K., Murphy, K. M., and Sher, A. (1996) Functional diversity of helper T-lymphocytes. *Nature* **383**, 787–793.

14. Okamoto, Y., Gotoh, Y., Tokui, H., Mizuno, A., Kobayashi, Y., and Nishida, M. (2000) Characterization of the cytokine network at a single cell level in mice with collagen-induced arthritis using a dual color ELISPOT assay. *J. Interferon Cytokine Res.* **20**, 55–61.

15. Okamoto, Y., Gotoh, Y., Shiraishi, H., and Nishida, M. (2004) A human dual color enzyme-linked immunospot (ELISPOT) assay for simultaneous detection of interleukin 2 (IL-2)- and interleukin 4 (IL-4)-secreting cells. *Int. Immunopharm.* **4**, 149–156.

18

Simultaneous Detection of Multiple Cytokines in ELISPOT Assays

Sarah Palzer, Tanya Bailey, Chris Hartnett, Angela Grant, Monica Tsang, and Alexander E. Kalyuzhny

Summary

Living in the era of multiplex detection systems, it appears attractive to develop enzyme-linked immunospot (ELISPOT) assays for the detection of more than one cytokine released by the same cell. However, despite technical simplicity in building such an assay, several factors have to be considered when designing multiplex ELISPOT assays. We have used four capture antibodies (hIFN-γ, hIL-2, hIL-4, and hTNF-α) either in combination or individually to coat polyvinylidene diflouride membrane-backed Millipore 96-well plates. Several cell stimulations were also used, including Concanavalin A, Phorbol Myristate Acetate (PMA) and calcium ionophore (CaI), phytohemagglutinin, CD3e, and lipopolysaccharide. Biotinylated antibodies were used either individually or combined together to detect secreted cytokines. We have found that when plates were coated with all four capture antibodies and captured cytokines were detected using either one detection antibody or all four detection antibodies combined together, fewer spots could be seen when compared with a plate coated with a single capture antibody followed by using its matched detection antibody counterpart. Interestingly, negative interferences between antibodies were less profound when detection antibodies rather than capture antibodies were mixed together.

Key Words: ELISPOT assay; multiple cytokines; multiple antibodies; double-color assays; IFN-γ; IL-2; IL-4; TNF-α; PVDF membrane; Millipore; capture antibodies; detection antibodies; BCIP/NBT, Phorbol Myristate Acetate (PMA), calcium ionophore (CaI), PHA, Con A, LPS, CD3E, PBMC, cell stimulants.

1. Introduction

Conventional enzyme-linked immunospot (ELISPOT) assays are designed in such a way that they can detect release of only a single cytokine. However, when performing an ELISPOT experiment, it is tempting to collect as much data as possible about the cytokine-secreting capacity of cells and this brings

From: *Methods in Molecular Biology, vol. 302: Handbook of ELISPOT: Methods and Protocols*
Edited by: A. E. Kalyuzhny © Humana Press Inc., Totowa, NJ

forth a question as to whether it is possible to detect simultaneous release of multiple cytokines by the same cells plated in a single well in the ELISPOT plate. Technically speaking, setting up such a multiplex ELISPOT assay appears to be simple: plate is coated with multiple capture antibodies and a mixture of multiple detection antibodies is added into each well followed by either developing or adding reporter tags of different colors. However, in spite of technical simplicity, several factors have to be considered when preparing and performing multiplex ELISPOT assays.

1.1. Maintaining Efficient Antibody Concentration

When coating a well with multiple monospecific capture antibodies, there is a possibility that antibodies may either compete for binding to a solid phase (filter membrane) or bind to each other *(1–3)* This, in turn, may reduce the binding efficiency of monospecific antibodies in a polyspecific antibody cocktail. There also is a concern that immobilization (by capture antibodies) of secreted cytokines of one type will exclude them from the feedback regulatory loop and affect release of other cytokines from the same cells.

1.2. Cell Stimulation

Induction of either TH_1 or TH_2 responses may require different cell-stimulation strategies and, therefore, it could be technically difficult or impossible to induce both TH_1 and TH_2 responses of the same cell. This limits the application of a multiplex ELISPOT assay to detect release of either TH_1 or TH_2 cytokines.

1.3. Cell Concentration

It is known to ELISPOT users that choosing the optimal cell concentration is of critical importance since it determines the quality of resulting spots and background staining. To generate detectable spots the number of plated cells may vary over a wide range of concentrations for various cytokines. For example, 1000 cells per well may be enough to detect tumor necrosis factor (TNF)-α secretion, whereas for the detection of interleukin (IL)-13 as many as 200,000 cells per well may be needed because of a very low frequency of cells secreting IL-13. However, plating 200,000 cells per well will be excessive for the detection of TNF-α because of the high frequency of TNF-α-secreting cells and will result in overdeveloped and, therefore, nonanalyzable wells. Thus, it may not be possible to choose a "one-fits-all" cell concentration to measure release of multiple cytokines.

1.4. Detection of Multiple Colors

Detection of two and more cytokines released by the same cells can be accomplished by using multicolor reporter tags. Recognition of individual overlapping colors in the multicolor mix is a prerequisite to unambiguous detection of multiple cytokines released by the same cell. It appears that when using fluorescent tags *(4)* individual colors (including overlapped ones) can be detected by simply switching between filter cubes on the microscope, whereas the detection of overlapped colors produced by enzyme-converted chromogens such as 3-amino-9-ethylcarbazole, $C_{14}H_{14}N_2$ and 5-5-bromo-4-chloro-3-indolyl phosphate/Nitroblue tetrazolium (BCIP/NBT; **refs. 5–7**) may be quite challenging and require expensive real-color image analysis systems.

We have conducted a study to evaluate performance of a multiplex ELISPOT assay. Our approach was to use a single-color detection system to compare the sensitivity of mono-specific ELISPOT assays vs poly-specific ones, and compared the quality of spots and intensity of background. Even though we have shown that monospecific ELISPOT assays are superior to multiplex ELISPOT assays, the information presented in this chapter will still prove helpful to those who dedicate themselves to designing multi-cytokine ELISPOTs.

2. Materials

1. Capture antibodies from R&D Systems, Inc.; Ms × hIFN-γ, Gt × hIL-2, Gt × hIL-4, and Ms × hTNF-α.
2. Biotinylated detection antibodies from R&D Systems, Inc.; Btn Gt × hIFN-γ, Btn Gt × hIL-2, Btn Gt × hIL-4, and Btn Gt × hTNF-α.
3. Ficoll (Amersham Biosciences).
4. Human Leukopack (Memorial Blood Centers of Minnesota).
5. Hemacytometer (Hausser Scientific).
6. Trypan Blue Dye (Gibco BRL).
7. RPMI complete which contains 1 L RPMI 1640 (Gibco BRL), 10% Fetal calf serum (Sigma), 2 g of sodium bicarbonate (Gibco BRL), 1.19 g of HEPES (Sigma), 1 mL of 50 mg/mL of Gentamycin (Gibco BRL), and 3.5 μL of 2-Mercaptoethanol (Sigma; *see* **Notes 1** and **2**).
8. Cell stimulants used; lipopolysaccharide (Sigma), CaI (Sigma), phytohemagglutinin (Sigma), CD3ε (R & D Systems, Inc.), PMA (Sigma), and Concanavalin A (Sigma).
9. Block buffer (R&D Systems, Inc.).
10. PBS (pH 7.4).
11. Human Red Blood Cell Lysis solution: 4.15 g of NH_4Cl, 18.61 mg of ethylenediamine tetraacetic acid, and 0.42 g of $NaHCO_3$, pH to 7.4.
12. ELISPOT wash buffer (R&D Systems, Inc.).
13. Centrifuge.
14. Biosafety cabinet.
15. Humidified cell incubator set at 37°C and 5% CO_2.
16. Biotinylated antibody and streptavidin diluent (R&D Systems, Inc.).
17. Streptavidin conjugated to alkaline phosphatase (R&D Systems, Inc.).

18. Ready-to-use BCIP/NBT chromogen solution (R&D Systems, Inc.).
19. 96-Well PVDF filter microplates (Millipore, cat. no. MAIPNOB).
20. 50 mL Conical centrifuge tubes (Falcon).

3. Methods

The methods described below outline (1) coating and blocking of the membrane-backed ELISPOT plates, (2) cell preparation, (3) addition of detection antibodies, and (4) color development of spots.

3.1. Plate Coating

In this study, four different capture antibodies were used either in combination or alone. The antibodies were chosen from a panel of monoclonal and polyclonal antibodies, then the antibody that produced the highest quantity and quality of spots with the least amount of background were chosen from that panel and used for future experiments. These antibodies include Ms × hIFN-γ, Gt × hIL-2, Gt × hIL-4, and Ms × hTNF-α. Each antibody was diluted in PBS to reach concentrations that were determined as optimal in preliminary experiments. All wells were coated adding 100 µL of antibody solution. Then the plates were kept at 4°C for 18–24 h.

3.1.1. Plate Washing

The following day plates were removed and washed three times with a manual eight channel vacuum manifold (three times with PBS; *see* **Notes 3** and **4**).

3.1.2. Blocking

After washing, plates were blocked with 200 µL of block buffer per well for 1.5 to 5 h. Then the block buffer was removed, and 200 µL of RPMI complete was added into each well to condition wells for cell cultures.

3.2. Cell Preparation

The following methods describe how to separate cells from a leukopack and then how to plate them into an ELISPOT plate. Cells were prepared from a leukopack (a blood pack containing concentrated human peripheral blood mononuclear cells [PBMCs]). Blood was layered over Ficoll and then spun at 500*g* for 30 min (*see* **Note 5**). The plasma layer was discarded, and the PBMC layer was transferred into a clean 50-mL centrifuge tube. Then sterile PBS was added to the tube and then spun for 5 min at 500*g*. The supernatant was discarded, and pellet was broken up by repetitive pipetting. After that, 10 mL of cold human red blood cell lyse was added to the tube, mixed with cells, and then incubated for 5 min. Then the tube was filled with PBS and spun again for 5 min at 500*g*. The supernatant was discarded, and 50 mL of RPMI complete culture media was added into the tube and cells were counted as described below.

3.2.1. Counting of PBMCs

Cells were counted in a 1:2 mix of cells and a Trypan blue dye. Of that mixture, 10 μL was pipetted into each side of a hemacytometer (*see* **Notes 6** and **7**) and cells were counted under routine laboratory microscope using a 10X or 20X lens and phase contrast. After counting cells as described in the Hemacytometer's insert appropriate dilutions of cells were made to culture in the ELISPOT plates (*see* **Note 8**).

3.2.2. Cell Dilutions

For this experiment, cells were diluted to 1×10^6 cells/mL and 1×10^5 cells/mL. Cell suspension (100 μL) was added into the wells so that resulting cell concentrations were 1×10^5 and 1×10^4 cells per well. It is important to determine the appropriate cell concentrations for each experiment depending on the targeted cytokine and stimulants used to induce its release (*see* **Note 8**). For the cytokines measured in our experiment, the range of 1×10^4 to 1×10^5 cells per well produced quantifiable spot numbers. Once the cell dilutions were made, the RPMI complete was aspirated from the ELISPOT plates, and nonstimulated cells were added into the plate (*see* **Note 9**). To induce cytokine release, stimulants were mixed with PBMCs in sterile tubes, and then PBMCs (100 μL/well) were added into the ELISPOT plates.

3.2.3. Plate Map

The plate map used in this experiment was as follows. Columns 1–6 contained capture antibodies for all analytes; column 7 contained only interferon (IFN)-γ capture antibody; column 8 contained only IL-2 capture antibody; column 9 contained only IL-4 capture antibody; and column 10 contained only TNF-α capture antibody. Columns 11 and 12 were used for controls. Row A was used for IFN-γ (1×10^6 cells/mL); row B, IFN-γ (1×10^5 cells/mL); row C, IL-2 (1×10^6 cells/mL); row D, IL-2 (1×10^5 cells/mL), row E, IL-4 (1×10^6 cells/mL); row F, IL-4 (1×10^5 cells/mL); row G, TNF-α (1×10^6 cells/mL); and row H, TNF-α (1×10^5 cells/mL). Detection antibodies were added as follows: columns 1 and 2, all four biotinylated antibodies; column 3, IFN-γ; column 4, IL-2; column 5, IL-4; column 6, TNF-α, and in columns 7 through 10, all four detection antibodies were added. In rows A and B there were stimulated cells at 1×10^6 cells/mL; in rows C and D, stimulated cells at 1×10^5 cells/mL; in rows E and F, nonstimulated cells at 1×10^6 cells/ml; in row G, a "no detection antibody" control to determine background staining; row H, a "no cell control" group containing media only without any cells. Five different cytokine-inducing treatments were tested in this study.

3.2.4. Incubating Cells

After the cells are plated, the bottom of the plate was covered with aluminum foil then placed in a humidified incubator at 37°C and 5% CO_2 *(8)*. The plate was then incubated for 16–24 h (*see* **Notes 10–13**).

3.3. Detection Antibodies

This section describes the addition of biotinylated detection antibodies into the plates.

3.3.1. Plate Washing

After incubating the plate overnight the cells were removed by washing the plate four times with ELISPOT wash buffer (*see* **Note 14**).

3.3.2. Adding Detection Antibodies

The detection antibodies were diluted in biotinylated antibody diluent. When several antibodies were mixed in combination they were mixed in their respective concentrations, which was determined when each analyte was developed alone. Some mixtures only contained one antibody at its respective dilution. Diluted antibody solution (100 µL) was added to the 96-well plates according to the plate map as noted in the plate map section. The detection antibodies are incubated on the plates overnight (18–24 h) at 4° C.

3.4. Color Development

The next steps describe the color development of spots using BCIP/NBT chromogen.

3.4.1. Plate Washing

After finishing the incubation with detection antibodies plates were washed three times with ELISPOT Wash Buffer to remove excess detection antibodies. In addition, plates are taped out onto paper towel to remove excess wash buffer.

3.4.2. Streptavidin Conjugated to Alkaline Phosphatase

Streptavidin conjugated to alkaline phosphatase was mixed in streptavidin dilution buffer to a concentration of 0.15 µg/mL. Then 100 µL of this solution was added to all wells in the plates and then plates were incubated for 2 h at room temperature on a rocking plate.

3.4.3. Plate Washing

After finishing the incubation with Streptavidin conjugated to alkaline phosphatase, plates were washed as described in **Subheading 3.4.1.**

3.4.4. Color Development

After plates were washed and tapped out on paper towel, 100 µL of BCIP/NBT color chromogen was added into each well. Then the plates were incubated from 30 min to 1 h (until dark spots are seen on the bottom of the plate under the microscope) in the dark at room temperature. Then the plates were washed three times with deionized water. Plates were dried on a hot plate and spots were analyzed under the microscope using a 4X lens.

3.5. Muliplex ELISPOT Assay Sensitivity

3.5.1. Quality of Spot and Background Staining in Multiplex ELISPOT Assays

Tables 1–5 illustrate the visual appearance of spots and background staining for each type of stimulation. Results are given for 1×10^5 cells/mL for all stimulations when all four cytokines were detected. For mono-specific ELISPOT assays (i.e., IFN-γ capture antibody and IFN-γ detection antibody) results are given for 1×10^6 cells/mL.

3.5.2. Quantification of Multiplex ELISPOT Assays

The numbers in **Tables 6–10** represent average values of two or more wells. In the series, the first antibody listed is the capture antibody and the second antibody listed is the detection antibody. When the term "All" is used it designates that all four capture or detection antibodies were used in combination.

Our results indicate that polyspecific multiplex ELISPOT assays are qualitatively (**Tables 1–5**) and quantitatively (**Tables 6–10**) different from monospecific ones. Because it is not clear which factors (or combination of factors) determine performance of multiplex ELISPOT assay, its optimization appears to be a challenging task not only to beginners but to experienced ELISPOT developers as well. It is possible that when several capture monospecific antibodies are mixed together and incubated for a long period of time, they may bind to each other and, thus, inhibit each others' activity. Interestingly, reciprocal inhibiting effects of antibodies were less profound when detection antibodies rather than capture antibodies were mixed together (**Tables 1–10**). If multiplex ELISPOT assays are not producing satisfactory results, an alternative method for measuring the release of multiple cytokines by the cells of the same donor would be an array type assay, in which cells are cultured in wells coated with various mono-specific capture antibodies followed by using matched mono-specific detection antibodies.

Table 1
Human PBMCs Stimulated With 3 µg/mL CD3ε

Capture antibodies	Detection antibodies				
	All four antibodies	IFN-γ	IL-2	IL-4	TNF-α
All four antibodies in a mix	Spots are fuzzy; the background is high.	No spots are present; non-cellular debris are sticking to membranes.	Spots are sharp and background is low.	No spots are present; the background is low.	There are many fuzzy spots and also cells sticking to the membrane.
IFN-γ	Spots are nonspecific; the background is low.	Few spots due to non-responding donor; medium background.	NA	NA	NA
IL-2	Nonspecific spots; the background is low.	NA	Few sharp spots; the background is low.	NA	NA
IL-4	Nonspecific spots; the background is low.	NA	NA	No spots; the background is low.	NA
TNF-α	Non-specific spots; the background is low.	NA	NA	NA	Sharp spots ("crawling" from the center towards periphery); the background is medium.

Table 2
Human PBMCs Stimulated With 1 µg/mL of LPS

Capture antibodies	Detection antibodies				
	All four antibodies	IFN-γ	IL-2	IL-4	TNF-α
All four antibodies	Spots are sharp; the background is high.	Few spots due to non-responding donor; the background is low.	No spots; the background is low.	No spots; the background is low.	There are many fuzzy spots and also cells sticking to the membrane.
IFN-γ	No spots; the background is medium; there are stained cells sticking to the membrane.	No spots; the background is low.	NA	NA	NA
IL-2	No spots; the background is low; there are stained cells sticking to membrane.	NA	No spots; the background is low.	NA	NA
IL-4	No spots; there are stained cells sticking to the membrane.	NA	NA	No spots; the background is low.	NA
TNF-α	Sharp spots; the background is low.	NA	NA	NA	Sharp spots; the background is low.

antibodies	antibodies	IFN-γ	IL-2	IL-4	TNF-α
All four antibodies	Sharp spots; the background is high.	No spots;- the background is medium.	No spots; the background is low.	No spots; the background is low.	Sharp spots; the background is medium.
IFN-γ	No spots; the background is medium; there are stained cells sticking to the membrane.	No spots; the background is low.	NA	NA	NA
IL-2	No spots; the background is low; there are stained cells sticking to the membrane.	NA	Spots are fuzzy; the background is low.	NA	NA
IL-4	No spots; the background is low; there are cells sticking to the membrane.	NA	NA	No spots; the background is low.	NA
TNF-α	Fuzzy spots; the background is low.	NA	NA	NA	Sharp spots; the background is low.

Table 4
Human PBMCs Stimulated with 0.5 and 0.05 µg/mL of CaI and PMA, respectively

Capture antibodies	All four antibodies	Detection antibodies			
		IFN-γ	IL-2	IL-4	TNF-α
All four antibodies	Fuzzy spots; the background is low; there are stained cells sticking to the membrane.	No spots; the background is low; there are stained cells sticking to the membrane.	Fuzzy spots; the background is medium.	Sharp spots; the background is low; there are stained cells sticking to the membrane.	Fuzzy spots; the backgrounds is low; there are stained cells sticking to the membrane.
IFN-γ	Few sharp spots; the background is medium.	Few sharp spots; the background is low.	NA	NA	NA
IL-2	Few sharp spots; the background is medium.	NA	Sharp spots; the background is low.	NA	NA
IL-4	No spots; the background is low. there are stained cells sticking to the membrane.	NA	NA	Sharp spots; the background is medium.	NA
TNF-α	Sharp spots; the background is low.	NA	NA	NA	Sharp spots; spots are stronger than in TNF-α: All group; the background is medium.

Table 5
Human PBMCs Stimulated With 4 µg/mL of Con A

Capture antibodies	Detection antibodies				
	All four antibodies	IFN-γ	IL-2	IL-4	TNF-α
All four antibodies	Fuzzy spots; the background is high.	No spots; the background is low; there are stained cells sticking to the membrane.	No spots; the background is low; there are stained cells sticking to the membrane.	No spots; there are stained cells sticking to the membrane.	Fuzzy spots; the background is medium.
IFN-γ	No spots; the background is medium.	Few sharp spots; the background is low; there are stained cells sticking to the membrane.	NA	NA	NA
IL-2	No spots; the background is medium.	NA	Sharp spots; the background is low.	NA	NA
IL-4	No spots; the background is medium.	NA	NA	Sharp spots; the background is low.	NA
TNF-α	Sharp spots; Spots are stronger than in All:All group; the background is medium.	NA	NA	NA	Sharp spots; wells are overdeveloped and staining is more intense than in TNF-alpha: All group; the background is medium.

Table 6
Spot Count for Human PBMCs Stimulated With 3 μg/ml CD3ε

All:All	All:IFN-γ	All:IL-2	All:IL-4	All:TNF-α
189	36	27	33	185
	IFN-γ:All	IL-2:All	IL-4:All	TNF-α:All
	Too numerous to count	Too numerous to count	Too numerous to count	591
	IFN-γ:IFN-γ	IL-2:IL-2	IL-4:IL-4	TNF-α:TNF-α
	19	Too numerous to count	0	304

Table 7
Spot Count for Human PBMCs Stimulated With 1 μg/mL of LPS

All:All	All:IFN-γ	All:IL-2	All:IL-4	All:TNF-α
215	43	27	0	311
	IFN-γ:All	IL-2:All	IL-4:All	TNF-α:All
	649	170	0	652
	IFN-γ:IFN-γ	IL-2:IL-2	IL-4:IL-4	TNF-α:TNF-α
	28	16	16	394

Table 8
Spot Count for Human PBMCs Stimulated With 3 μg/mL of PHA

All:All	All:IFN-γ	All:IL-2	All:IL-4	All:TNF-α
263	24	6	34	367
	IFN-γ:All	IL-2:All	IL-4:All	TNF-α:All
	40	28	44	767
	IFN-γ:IFN-γ	IL-2:IL-2	IL-4:IL-4	TNF-α:TNF-α
	19	26	23	715

Table 9
Spot Count for Human PBMCs Stimulated With 0.5 and 0.05 µg/mL
of Cal and PMA respectively

All:All	All:IFN-γ	All:IL-2	All:IL-4	All:TNF-α
432	18	20	68	469
	IFN-γ:All	IL-2:All	IL-4:All	TNF-α:All
	193	38	114	726
	IFN-γ:IFN-γ	IL-2:IL-2	IL-4:IL-4	TNF-α:TNF-α
	26	54	51	691

Table 10
Human PBMCs Stimulated With 4 µg/mL of Con A

All:All	All:IFN-γ	All:IL-2	All:IL-4	All:TNF-α
270	26	24	0	339
	IFN-γ:All	IL-2:All	IL-4:All	TNF-α:All
	387	172	100	747
	IFN-γ:IFN-γ	IL-2:IL-2	IL-4:IL-4	TNF-α:TNF-α
	70	0	43	661

4. Notes

1. Sterilize RPMI complete culture medium and reagents that will be used to separate out the white blood cells through 0.2-µm sterile filter to allow their long-term storage.
2. When using fetal calf serum it is important to heat inactivate the serum at 56°C for 30 min. After the heat inactivation the serum should be filtered.
3. To get consistent development of spots it is important to make sure that wash buffer does not come into contact with the wells in the ELISPOT plate until after the cells have been finished incubating overnight. Otherwise wash buffer may cause cell death and at times may result in a completely blank or under-developed wells. This may require a different washing device for washing coated plates, which requires PBS vs all other washes that require ELISPOT wash buffer. This becomes even more relevant when multiple users are running multiple assays on different time courses.
4. When using an automated or a hand held plate washer, it is important to make sure its prongs do not puncture the membranes on the bottom of the plates.
5. When layering Ficoll make sure that the blood does not mix with the Ficoll to gain the best separation and highest yield of PBMCs.

6. When using a hemacytometer do not overfill the chambers; this may result in inaccurate counts.
7. While counting on a hemacytometer, find the middle square that contains several smaller squares (25 squares) and count cells in five of them. Then take the average and with that number multiply by 25 (total number of squares in that area) multiplied by 2 (dilution factor) multiplied by 10,000 (to get cells/mL). Use this number when making serial dilutions of PBMCs.
8. When making dilutions, it is important to be sure that the number of plated cells is sufficient for the development of a quantifiable number of spots, which can be easily counted manually or by using an automated ELISPOT plate reader.
9. When running ELISPOT experiments, it is extremely important to mix cells thoroughly before adding them into the wells. This may require shaking the tube with cells after filling every four wells to minimize well-to-well inconsistency.
10. The foil is used on the bottom of the plates to provide even heat distribution across the bottom of the ELISPOT plates while incubating the cells overnight. This also aids in well-to-well reproducibility across the plate.
11. Using a humidified incubator prevents the culture media from evaporating from the wells in the plate.
12. When incubating the plates with cells it is important to be sure that the shelves in the incubator are level, otherwise, the cells will move towards one side of the well, which will result in nonuniform distribution of spots across the bottom of the well.
13. It is also important to avoid disturbing (e.g., by slamming the door of the incubator) the plate with cells during the incubation.
14. After the wash steps, it is important to tap out the excess liquid in the well onto a paper towel to prevent diluting the sequential reagents added into the plate.

References

1. Butler, J. E., Ni, L., Nessler, R., Joshi, K. S., Suter, M., Rosenberg, B., et al. (1992) The physical and functional behavior of capture antibodies adsorbed on polystyrene. *J. Immunol. Methods* **150,** 77–90.
2. Butler, J. E. (2000) Solid supports in enzyme-linked immunosorbent assay and other solid-phase immunoassays (review). *Methods* **22,** 4–23.
3. Butler, J. E. (2004) Solid supports in enzyme-linked immunosorbent assay and other solid-phase immunoassays. *Methods Mol. Med.* **94,** 333–372.
4. Gazagne, A., Claret, E., Wijdenes, J., Yssel, H., Bousquet, F., Levy, E., et al. (2003) A Fluorospot assay to detect single T-lymphocytes simultaneously producing multiple cytokines. *J. Immunol. Methods* **283,** 91–98.
5. Czerkinsky, C., Moldoveanu, Z., Mestecky, J., Nilsson, L. A., and Ouchterlony, O. (1988) A novel two colour ELISPOT assay. I. Simultaneous detection of distinct types of antibody-secreting cells. *J. Immunol. Methods* **115,** 31–37.
6. Okamoto, Y., Abe, T., Niwa, T., Mizuhashi, S., and Nishida, M. (1998) Development of a dual color enzyme-linked immunospot assay for simultaneous detection of murine T helper type 1- and T helper type 2-cells. *Immunopharmacology* **39,** 107–116.

7. Okamoto, Y., Gotoh, Y., Shiraishi, H., and Nishida, M. (2004) A human dual-color enzyme-linked immunospot assay for simultaneous detection of interleukin 2- and interleukin 4-secreting cells. *Int. Immunopharmacol.* **4,** 149–156.
8. Kalyuzhny, A., and Stark, S. (2001) A simple method to reduce the background and improve well-to-well reproducibility of staining in ELISPOT assays. *J. Immunol. Methods* **257,** 93–97.

19

Fluorospot Assay
Methodological Analysis

Agnes Gazagne, Wolf Malkusch, Benoit Vingert, Wolf H. Fridman, and Eric Tartour

Summary

The conventional enzyme-linked immunospot (ELISPOT) technique detects only one secreted cytokine, which constitutes a major pitfall for the accurate characterization of the various T-cell subpopulations. We have therefore developed a fluorospot assay, which is a modification of the ELISPOT and is based on the use of multiple fluorescent-labeled anticytokines detection antibodies. A special automated ELISPOT reader consisting of a light microscope with incident fluorescence illumination and an integrating digital color camera has been adapted for this technique. This technique has been applied for the analysis of subpopulations of T-cells and polarized antigen-specific T-cells.

Key Words: ELISPOT; cytokines; T-lymphocytes; fluorospot.

1. Introduction

Upon activation, T-lymphocytes up-regulate expression of and secrete a number of cytokines *(1)*. The enzyme-linked immunospot (ELISPOT) assay is based on solid-phase immunoenzyme technology *(2)*. This test allows the detection of functionally specific T-cells secreting cytokines at a single cell level. When T-cells are incubated in the plates, the cytokines released are directly bound by capture antibodies; therefore, cytokines are not diluted in the culture supernatant or bound by cytokine receptors present in the supernatant or adjacent cells; this may explain the higher sensitivity of this technique compared with ELISA *(3,4)*. However, in most cases the ELISPOT procedure detects only one secreted cytokine; this constitutes a major drawback for the characterization of the various T-cell populations, which are identified by their profile or co-expression of cytokines *(5,6)*. Recent studies indicate that the

From: *Methods in Molecular Biology, vol. 302: Handbook of ELISPOT: Methods and Protocols*
Edited by: A. E. Kalyuzhny © Humana Press Inc., Totowa, NJ

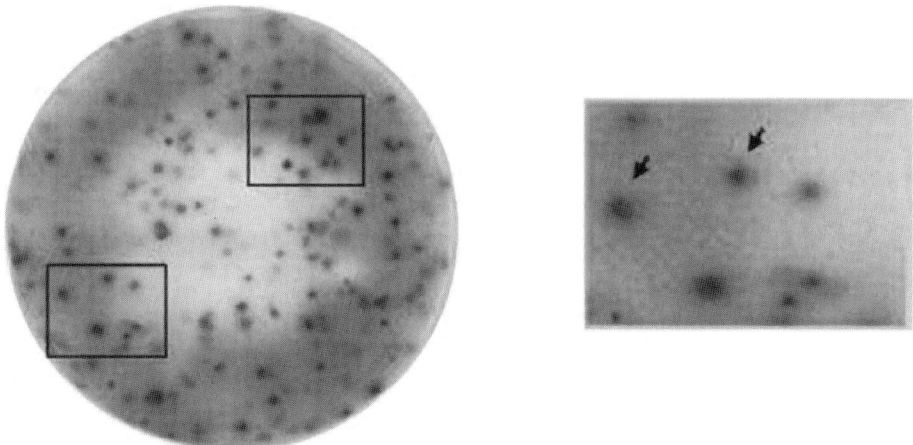

Fig. 1. Dual-color immunoenzymatic ELISPOT for the detection of IFN-γ and IL-4 producing cells. Left, B.EBV cells and a TH2 T-cell clone were mixed and activated with PMA–ionomycin. IFN-γ and IL-4 derived from secreting cells were detected using substrates specific for horseradish peroxidase (3-amino-9-ethylcarbazole, $C_{14}H_{14}N_2$) or alkaline phosphatase (5-bromo-4-chloro-3-indolylphosphate/Nitroblue tetrazolium chloride), respectively. Enzymes were linked to detection antibodies for IFN-γ and IL-4. Red spots corresponded to IFN-γ secreting cells, whereas blue spots belong to IL-4 producing cells. Right: Greater enlargement of a quadrant from the left. The arrows showed the difficulties in the interpretation of mixed color spots. *See* **Color Plate 5**, following page 50.

detection of double interferon (IFN)-γ–interleukin (IL)-2-producing T-cells provides additional clinical information regarding the prognosis of patients with human immunodeficiency virus than enumeration of IFN-γ- or IL-2-secreting T-cells alone *(7–10)*.

As others and we have experienced, attempts to develop an immunoenzymatic dual-color ELISPOT failed because of difficulties in the interpretation of mixed color spots (**Fig. 1**; **refs.** *11,12*). Therefore we have developed a fluorospot assay, which is based on a modification of the ELISPOT. The fluorospot assay is based on the use of multiple fluorescent-labeled anticytokine detection antibodies. This assay clearly provides better discrimination and characterization of double cytokine-producing cells than does an enzymatic reaction. This technique allows for the detection of regulatory T-cells and polarized type 1- and type-2 specific tetanus toxoid T-cells *(13)*. The availability of a large range of fluorophores should permit the extension of this technique to multiparameter analysis.

2. Materials

1. Ethanol (Merck-Eurolab-Polylabo, Denmark).
2. 96-Well polyvinylidene diflouride flat bottom plates (Millipore, Molsheim, France).
3. Cells were maintained in AIM V medium (Gibco-Life Technologies, Paisley, Scotland).
4. Tween 20 (Merck, Schardt, Germany).
5. Phosphate-buffered saline (PBS), pH 7.2–7.4 (Gibco-Life Technologies).
6. Anticytokine antibodies (*see* **Note 1**): Mouse IgG1 anti-IFN-γ (clone B-B1), mouse IgG1 anti-IL-2 (clone B-G5), fluorescein-labeled mouse anti-IFN-γ (clone BG-1), and biotinylated rabbit polyclonal anti-IL-2. All the antibodies were obtained from Diaclone (Besançon, France).
7. Secondary reagents for amplification step: anti-fluorescein rabbit IgG labeled with Alexa Fluor 488 (Molecular Probes, Eugene, OR), biotinylated goat anti-rabbit IgG (Southern Biotechnology), and phycoerythrin-conjugated streptavidin (Dako, Trappes, France). Fluorophores labeled conjugates are stored in aliquots undiluted at 4°C and protected from light.
8. Equipment required to read the fluorescence spots. For evaluation, we used an automated ELISPOT reader (KS ELISPOT, Carl Zeiss Jena) equipped with a light microscope (Axioplan 2 imaging mot, Carl Zeiss Jena) with incident fluorescence illumination (double band path fluorescence filter for FITC and Rhodamine), motor stage and automatic focusing unit.

3. Methods

To illustrate this technique, we describe a double-color fluorospot to detect IFN-γ and IL-2, but this technique also can be applied for monocolor fluorospot or for the detection of other cytokines.

1. Ninety-six well polyvinylidene diflouride (PVDF) flat-bottom plates (Millipore, Molsheim, France) were first treated with ethanol 70% for 10 min at room temperature (*see* **Notes 2** and **3**).
2. The plates then were washed three times with PBS; this step increases the binding efficiency of the plates.
3. The plates were then coated overnight at 4°C with 100 µL of mouse monoclonal anti-IL-2 and anti-IFN-γ antibodies in PBS at 10 µg/mL.
4. The plates were then blocked with 2% milk in PBS for 2 h at room temperature and washed twice with PBS containing 0.05% Tween-20. Less background was obtained when milk was used compared with medium that contained serum.
5. Cells diluted in AIMV medium in a volume of 100 µL were then added in serial dilutions in triplicate and were incubated for various times ranging from 18 h to 48 h at 37°C in a humidified atmosphere of 5% CO_2 in air (*see* **Notes 4** and **5**). We demonstrated that for human peripheral blood mononuclear cells, the plates were saturated at a concentration 2×10^5 cells and, therefore, it is not recommended to exceed this number of cells in the plates.

6. Then cells were removed and the plates were incubated with PBS containing 0.05% Tween-20 for 10 min to lyze all the remaining cells. This step was followed by three washes with PBS containing 0.05% Tween-20.

7. For the detection of IFN-γ and IL-2, 100 μL of a fluorescein-labeled mouse monoclonal anti-IFN-γ antibody (2 μg/mL) and 100 μL of biotinylated rabbit polyclonal anti-IL-2 (1.5 μg/mL) were added for 1.5 h at 37°C in a place protected from light (*see* **Note 6**). In general, the concentrations of the detection antibodies are always lower than that used for the capture antibodies.

8. For IFN-γ the signal was amplified with 15 μg/mL of an anti-fluorescein rabbit IgG conjugated with Alexa Fluor 488 and, for IL-2, the reaction was detected with 15 μg/mL of phycoerythrin-conjugated streptavidin. All the incubations were performed in a place protected from the light. It is necessary to avoid using anti-fluorescein rabbit IgG conjugated to fluorescein to prevent the formation of auto-aggregates of this antibody.

9. Images were taken with an integrating digital color camera (Axiocam MRc, Carl Zeiss Jena; *see* **Note 7**), which allows one to record high-resolution images. This camera, an alternative to an analog color camera, can be adjusted to an optimum fluorescence exposure time directly from the evaluation software. Various settings for different experiments may be saved in their own configuration files and easily recalled when necessary. **Figure 2** illustrates a dual-color fluorospot for IL-2 and IFN-γ recorded using this equipment.

10. After the digitization, the KS ELISPOT software proceeded with automatic data processing. In the user interface, the user elements are reduced to a minimum. Only four buttons are needed for routine evaluation. Finally, all data were transferred directly into a spreadsheet program for further evaluation and graphical display. The system settings were tested before the evaluation of a complete plate with the "check well" function. In this mode, the spot recognition can be adapted with the "teach" function, or a new configuration setting where multiple evaluation patterns can be defined.

11. Finally, the evaluation of the entire plate was started. The stage was moved to the center of the first well position and the auto-focus initiated. Then, the well position was scanned in a meander mode. From all fields, a complete well image was generated and displayed in a reduced mode in the image field for user control. The result was displayed in the overlay (spot indication and spot diameter). The stage moved to the next well position and the sequence repeated until the last well was evaluated. All rejected positions were skipped. After the evaluation of the last well position, the stage will automatically return to the start position. A detailed description of the evaluation software is given in Chapter 8 in this volume.

4. Notes

1. The use of antibodies without azide and with low endotoxin levels may help to reduce background levels when plate is developed.

2. The PVDF flat-bottom plates gave the best results when compared to plastic plates. This technique did not require the use of black plates.

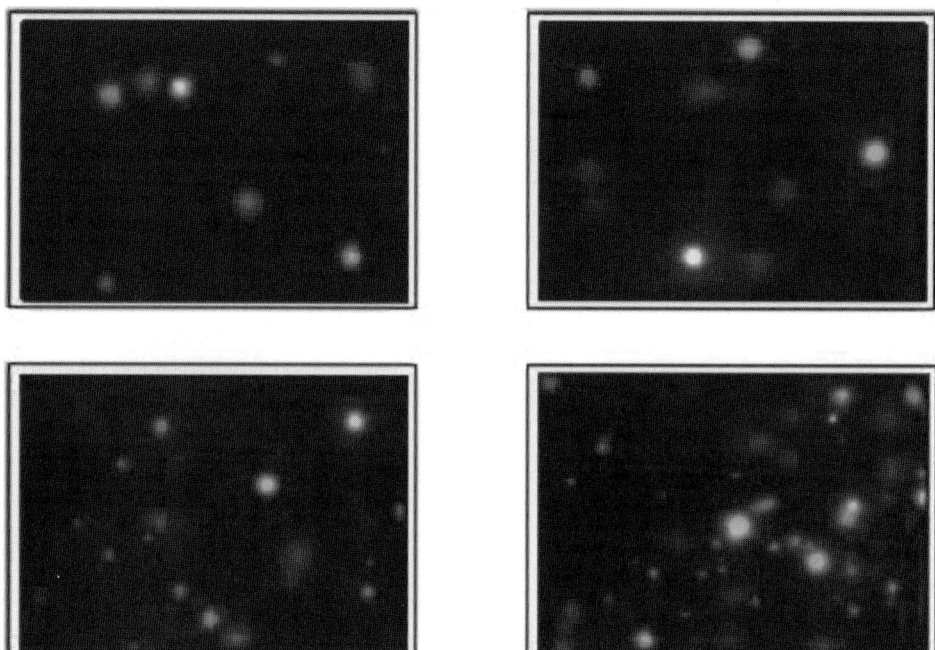

Fig. 2. Dual-color fluorospot for the detection of IL-2 and/or IFN-γ-producing cells. Peripheral blood mononuclear cells were stimulated with PMA and ionomycin in PVDF plates. IFN-γ- and/or IL-2-producing cells were characterized by a dual-color fluorospot assay. Green spots corresponded to IFN-γ secreting cells, whereas red spots belong to IL-2-producing cells. Yellow spots corresponded to cells coexpressing IFN-γ and IL-2. No spots were observed when non-stimulating cells were used for the dual-color fluorospot. *See* **Color Plate 6**, following page 50.

3. When dual-color fluorospot is performed, competition between capture antibodies may occur and this will introduce bias in the detection of spots. This difference may be related to the variable efficiency of antibodies binding to PVDF membrane. When antibodies are used for the first time, to ensure that the same concentrations of antibodies directed against IL-2 or IFN-γ were bound, we simultaneously tested the frequency of IL-2- and IFN-γ-producing cells with single color or double color fluorospot procedure. The number of cells producing IFN-γ and IL-2 in the single-color fluorospot assay has to match the sum measured with the double-color fluorospot.

4. Medium with fetal calf serum or human serum may non specifically activate the cells; therefore, we recommend medium without serum (either AIMV [Gibco] or X-Vivo [Cambrex]).

5. Cell incubation times vary depending on the cytokine to be detected. For IL-2 or IFN-γ, times range from 10 to 18 h. On the contrary, for the detection of IL-10 or IL-12, cells may need to be incubated up to 48 h.

6. Amplification system: when directly labeled anti-cytokine detection antibodies were used without amplification, fluorescence was not detectable. During dual-color fluorospot, one must be aware of crossreactivity between secondary and/or primary antibodies. Therefore, we recommend that one use species depleted antibodies and Fab'2 antibodies. For the detection of some cytokines a supplementary amplification step further increases the signal. For example, when detecting of IL-2, an amplification step with 100 μL of biotinylated goat anti-rabbit IgG for 1.5 h at 37°C could be added. During the double IFN-γ/IL-2 fluorospot assay, we first incubated the cells with the biotinylated goat anti-rabbit IgG and phycoerythrin-conjugated streptavidin, and the anti-fluorescein rabbit IgG conjugated with Alexa Fluor was added in a second step, after three washes, to avoid cross reactivity between the anti-rabbit IgG and the anti-fluorescein rabbit IgG. For the detection of IL-10, we use mouse IgG1 anti-IL-10 antibodies for the capture, biotinylated mouse IgG 2b anti-IL-10 and, in that case, the amplification step may include a biotinylated rabbit anti-mouse IgG 2b (Rockland, Gilbertsville)

7. Because the fluorescence illumination does not cross the PVDF membranes, the light from illuminator has to come from the top of the plates.

Acknowledgments

This work was supported by grants from Fondation de France, Agence Nationale de Recherche sur le SIDA (RIVAC program), Ligue Nationale Contre le Cancer, Association pour la Recherche sur le Cancer, INSERM (ATC Biotherapy), and the Canceropole.

Benoit Vingert is a fellow of the Fondation de France.

References

1. Fridman, W. H., and Tartour, E. (1997) Cytokines and cell regulation. *Mol. Aspects. Med.* **18**, 3–90.
2. Czerkinsky, C., Andersson, G., Ekre, H. P., Nilsson, L. A., Klareskog, L., and Ouchterlony, O. (1988) Reverse ELISPOT assay for clonal analysis of cytokine production. I. Enumeration of gamma-interferon-secreting cells. *J. Immunol. Methods.* **110**, 29–36.
3. Tanguay, S. and Killion, J. J. (1994) Direct comparison of ELISPOT and ELISA-based assays for detection of individual cytokine-secreting cells. *Lymphokine. Cytokine. Res.* **13**, 259–263.
4. Mo, X. Y., Sarawar, S. R., and Doherty, C. (1995) Induction of cytokines in mice with parainfluenza pneumonia. *J. Virol.* **69**, 1288–1291.
5. Mosmann, T. R. and Sad, S. (1996) The expanding universe of T-cell subsets: Th1, Th2 and more. *Immunol. Today.* **17**, 138–146.
6. Roncarolo, M. G., Bacchetta, R., Bordignon, C., Narula, S., and Levings, M. K. (2001) Type 1 T regulatory cells. *Immunol.Rev.* **182**, 68–79.
7. Sieg, S. F., Bazdar, D. A., Harding, C. V., and Lederman, M. M. (2001) Differential expression of interleukin-2 and gamma interferon in human immunodeficiency virus disease. *J. Virol.* **75**, 9983–9985.

8. Boaz, M. J., Waters, A., Murad, S., Easterbrook, P. J., and Vyakarnam, A. (2002) Presence of HIV-1 Gag-specific IFN-gamma+IL-2+ and CD28+IL-2+ CD4 T-cell responses is associated with nonprogression in HIV-1 infection. *J. Immunol.* **169**, 6376–6385.

9. Day, C. L. and Walker, B.D (2003) Progress in defining CD4 helper cell responses in chronic viral infections. *J. Exp. Med.* **198**, 1773–1777.

10. Harari, A., Petitpierre, S., Valellian, F., and Pantaleo, G (2004) Skewed representation of functionally distinct populations of virus-specific CD4 T-cells in HIV-1-infected subjects with progressive disease: changes after antiretroviral therapy *Blood.* **103**, 966–972.

11. Okamoto, Y., Abe, T., Niwa, T., Mizuhashi, S., and Nishida, M. (1998) Development of a dual color enzyme-linked immunospot assay for simultaneous detection of murine T helper type 1- and T helper type 2-cells. *Immunopharmacology* **39**, 107–116.

12. Karulin, A. Y., Hesse, M. D., Tary-Lehmann, M., and Lehmann, P. V. (2000) Single-cytokine-producing CD4 memory cells predominate in type 1 and type 2 immunity. *J.Immunol.* **164**, 1862–1872.

13. Gazagne, A., Claret, E., Wijdenes, J., Yssel, H., Bousquet, F., Levy, E., et al. (2003) A Fluorospot assay to detect single T-lymphocytes simultaneously producing multiple cytokines. *J. Immunol. Methods.* **283**, 91–98.

20

A Gel-Based Dual Antibody Capture and Detection Method for Assaying of Extracellular Cytokine Secretion

EliCell

Lisa A. Spencer, Rossana C. N. Melo, Sandra A. C. Perez, and Peter F. Weller

Summary

A distinguishing feature of eosinophils is their ability to rapidly release preformed cytokines from intracellular pools. Cytokines are delivered to the cell surface from granule stores by transport vesicles and are released in small packets at discrete locations along the cell surface through a process termed "piecemeal" degranulation. The study of this process has been hindered by lack of an assay sensitive enough to register minute protein concentrations and the inability to visualize morphology of cytokine secreting cells. These hindrances have necessitated our development of the EliCell assay, an agarose-based dual cytokine capture and detection system through which cytokine secretion and cellular morphology may be analyzed in concert. Cells are embedded within capture antibody-containing agarose and stimulated under conditions of interest. Extracellularly released cytokine is captured within the matrix at the point of release from the cell and can be labeled with a fluorochrome-conjugated antibody. Cytokine release and cellular morphology are visualized in parallel by phase contrast and fluorescence microscopy, respectively.

Key Words: EliCell; agarose matrix; eosinophil; cytokine; piecemeal degranulation; vesicular transport; secretion.

1. Introduction

Eosinophils have long been noted for their content of cationic granule proteins, the deposition of which leads to tissue damage and cellular dysfunction *(1,2)*. A more recently noted characteristic of eosinophils is their internal stores of preformed cytokines and chemokines with a wide range of biological functions *(3)*. In most other cells (i.e., T-cells), the release of these cytokines depends

From: *Methods in Molecular Biology, vol. 302: Handbook of ELISPOT: Methods and Protocols*
Edited by: A. E. Kalyuzhny © Humana Press Inc., Totowa, NJ

upon *de novo* synthesis in response to a particular stimulus. The ability of eosinophils to release these potent immunomodulators in the absence of *de novo* synthesis highlights the potential of these cells to play key roles in rapidly affecting the initiation or course of an allergic response.

Most, if not all, of the cytokines found preformed within eosinophils have been localized to specific granules *(4–9)*. Strong evidence provided mainly through electron microscopic analyses suggests that the mobilization of these factors from granule stores and subsequent extracellular release follows a process termed "piecemeal" degranulation (PMD), where small cytokine-containing vesicles traffic from granules to the plasma membrane. This process results in small packets of material being released at discrete locations along the cell surface. Although exocytosis of entire granule contents may occur under rare conditions, the more measured, specific process of PMD predominates *(4,10–12)*. Despite colocalization of numerous factors within specific granules and the common use of PMD as a method of exiting the cell, cytokines and chemokines are likely released individually in a tightly regulated, stimulus-dependent manner.

An understanding of molecular mechanisms responsible for this high level of specificity has been hindered by the lack of appropriate assays to detect low levels of stimulus-dependent cytokine release. Enzyme-linked immunosorbent assay (ELISA) methods relying on measurement of cytokine/chemokine levels in culture supernatants have been successful in the detection of some eosinophil-derived cytokines, particularly when nonphysiological stimuli and long incubation times are employed *(13–15)*. However, the progressive nature of release by PMD in response to more physiological stimuli cannot be fully appreciated because the released quantities are below the sensitivity of the assay. An additional obstacle is the potential of eosinophils to recapture or sequester released product with specific cytokine receptors at the cell surface, giving the false impression of lack of output *(16)*.

Enzyme-linked immunospot (ELISPOT) techniques avoid many of these disadvantages and provide the potential to detect minute quantities of product at the individual cell level. However, this approach provides little information concerning morphology, activation status, or even the viability of individual cytokine-producing cells. This can be especially important when considering eosinophil populations which, unlike most other cell types, do not require stimulus-induced de novo synthesis of the cytokine in question. Therefore, a damaged or permeabilized cell may appear as a positive spot because of the artificial release of its preformed contents or availability of intracellular stores to detecting antibodies.

In light of the unique difficulties introduced in the study of eosinophil PMD, we have developed a new approach to visualize the release of specific cytokine and chemokine products. Based upon the dual antibody capture and detection system of ELISPOT, the EliCell assay uses an avidin-conjugated agarose matrix to bind biotinylated cytokine-specific capture antibody. Cells are embedded within (rather than atop) this matrix and stimulated under conditions of interest. Released cytokine is captured within the matrix at the point of release from the cell and can be detected by a fluorochrome-labeled detection antibody. Viable cells remain embedded within the agarose substrate throughout the procedure and can be visualized under high magnification by phase contrast for morphological analysis in parallel with detection of released product by fluorescence microscopy *(17,18)*.

This approach provides substantial advantages, namely the ability to microscopically observe the cytokine-secreting cell in parallel with detection of the secreted product, providing information on viability and potential polarization of factor release. In addition, the procedure may be easily modified to allow for simultaneous detection of multiple cytokines, observance of surface markers in conjunction with cytokine release, detection of intracellular products, and other cytochemical analyses.

2. Materials

2.1. Agarose Activation

1. Low-melting point agarose (mp 65.5°C, gelling point 24°C; Promega, Madison, WI; cat. no. V2111).
2. 10 mM NaIO$_4$ dissolved in 100 mM sodium acetate buffer, pH 5.5. Store at 4°C and protect from light.
3. 70°C Water bath.

2.2. Coupling of Activated Agarose to Avidin

2.2.1. Coupling of Streptavidin to Agarose

1. Streptavidin-hydrazide (Pierce Chemical, Rockford, IL; cat. no. 21120).

2.2.2. Coupling of NeutrAvidin to Agarose

1. NeutrAvidin biotin-binding protein (Pierce; cat. no. 31000).
2. Quenching buffer (pH 7.4; may be purchased, along with sodium cyanoborohyde in **step 3**, in the AminoLink Plus Immobilization Kit [Pierce, cat. no. 44894]): 1 M Tris-HCl; 0.05% NaN$_3$.
3. Sodium cyanoborohydride* (32 mg; Pierce; cat. no. 44892) dissolved in 5 mL of 10 mM NaOH.
   Because of the high toxicity of CNBH$_3$, this step should be performed in a fume hood.
4. ImmunoPure HABA (Pierce; cat. no. 28010).

---

*Reagents may be purchased together in AminoLink Plus Immobilization Kit (Pierce, cat. no. 44894.

2.3. EliCell Assay

1. 37°C Water bath.
2. 10X RPMI-1640 (Sigma-Aldrich).
3. 1X RPMI 1640 (Sigma-Aldrich) supplemented 0.1% ovalbumin.
4. Biotinylated capture antibody (0.1 mg/mL diluted in RPMI + 0.1% ovalbumin).
5. Ultra-microtips with elongated tip (USA Scientific, Ocala, FL; cat. no. 1111-4000).
6. Microscope slides and cover slips (22 × 50 mm).
7. CoverWell perfusion chambers (Grace Bio-Labs, Bend, OR, cat. no. PC1L-0.5).
8. Hanks Buffered Salt Solution without calcium chloride and magnesium chloride (hereafter HBSS –/–).
9. Paraformaldehyde (diluted to 2% in HBSS –/–) (Electron microscopy Sciences, Ft. Washington, PA; cat. no. 15710). Dilutions should be made in fume hood and fresh dilutions of PFO should be used in each experiment. Protect from light.
10. Detection antibody pre-labeled with fluorochrome (*see* **Note 1**).
11. Aqua Poly/Mount (Polysciences, Inc., Warrington, PA; cat. no. 18606).
12. Acridine orange: ethidium bromide solution (recipe for solution adapted from Becton Dickinson *Immunocytometry Systems Cytometry Source Book*). Prepare 100X stock: 50 mg of ethidium bromide; 15 mg of acridine orange; dissolve in 1 mL of 95% ethanol. Add 49 mL of dI water. Mix well, divide into 1-mL aliquots and freeze. 1X working solution (prepare fresh): Dilute 100X stock with PBS and mix well. Store at 4°C for up to 1 mo protected from light.
13. Fast Green (Sigma).
14. Hematoxylin (Sigma).
15. Hema 3 Staining kit (Pierce).
16. AX-70 Provis Olympus with FITC and TRITC filter sets (HBO burner). Objectives used: 40X 1.00 Ph3 UplanApo and 100x 1.35 Ph3 UplanApo

3. Methods

3.1. Activation of Agarose

Incorporation of capture antibody into the support matrix requires prior activation of the agarose to generate functional aldehydes. Oxidation of *cis*-vicinal hydroxyl groups of the agarose by sodium metaperiodate is detailed below and in **Fig. 1A**.

1. Weigh 0.125 g of low-melting point agarose into a 250-mL Erlenmeyer flask (to increase surface area) and dilute to 2.5% by adding 5 mL sterile dI water
2. Mix well, but avoid swirling to prevent agarose binding to flask wall.
3. Solubilize agarose in 70°C water bath for 15 min with gentle agitation.
4. Solidify agarose at 4°C for 20 min.
5. Cover surface of solidified agarose with 5 mL of periodate solution (10 mM NaIO$_4$ in 100 mM sodium acetate buffer, pH 5.5) to activate the agarose, generating reactive aldehyde groups.
6. Protect from light and store overnight at 4°C.
7. Discard periodate solution.

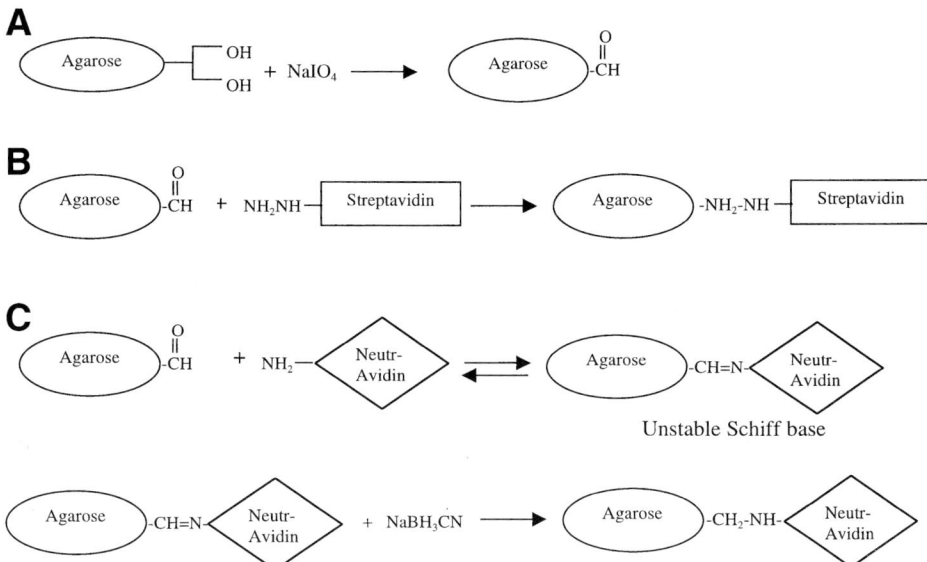

Fig. 1. Chemistry of agarose activation and conjugation to avidin derivatives. (**A**) Sodium meta-periodate oxidation of agarose to generate functional aldehydes. (**B**) Reaction of hydrazide groups of streptavidin–hydrazide with functional aldehydes of activated agarose. (**C**) Reaction of primary amine of NeutrAvidin with functional aldehydes of activated agarose to form unstable Schiff base, which is stabilized by reduction with NaBH₃CN in a reductive amination reaction.

8. Wash oxidized agarose for 10 min with 20 mL sterile dI water at 4°C
9. Repeat wash step 49 times, using a total of 1 L of water. Do not allow agarose to dry out.

3.2 Conjugation of Activated Agarose Substrate to Avidin

To uniformly incorporate biotinylated capture antibody into the oxidized matrix, agarose must be conjugated to avidin. We developed the assay using streptavidin, a biotin-binding analogue of egg-white avidin, gaining its prefix from its bacterium source, *Streptomyces avidinii*. The process of coupling agarose to streptavidin is detailed in **Subheading 3.2.1.** and in **Fig. 1B**. Streptavidin has been demonstrated to compete with extracellular matrix (ECM) proteins for binding to cell surface receptors of the integrin family through an RYD-containing sequence. This RYD-containing sequence mimics the RGD sequence common to several ECM proteins, such as fibronectin *(19–22)*. As potential uses for the EliCell system have evolved to include studies addressing eosinophil interactions with ECM proteins, we have begun to utilize alternatives to streptavidin. One such substitute is NeutrAvidin® (Pierce), a carbohydrate-free modified avidin derivative. The process of coupling NeutrAvidin to agarose is described in **Subheading 3.2.2.** and in **Fig. 1C**.

3.2.1. Chemical Coupling of Streptavidin to Agarose

The hydrazide group of streptavidin–hydrazide is reacted with the oxidized agarose, to stably couple streptavidin to the activated agarose support matrix.

1. Cover agarose layer with 5 mL of 0.2 mg/mL streptavidin–hydrazide (diluted in water).
2. Incubate overnight at room temperature.
3. Repeat washing (**steps 8** and **9** in Subheading **3.1.**) to remove any unbound streptavidin.
4. OPTIONAL: the degree of streptavidin conjugation may be determined by analyzing an aliquot of the gel in benzoic acid at 500 nm using HABA reagent (refer to **Subheading 2.**).
5. Streptavidin–agarose may be re-solubilized at 70°C and aliquoted before storage. Aliquots may be stored up to 6 mo at 4°C.

3.2.2. Chemical Coupling of NeutrAvidin to Agarose

Reactive aldehydes of the oxidized agarose are covalently coupled to available amine groups of neutravidin through a reductive amination reaction.

1. Cover agarose layer with sodium cyanoborohydride solution (32 mg $NaCNBH_3$ in 5 mL of 10 mM NaOH) plus 2 mg of NeutrAvidin (dissolved in 0.5 mL H_2O) and incubate overnight at 4°C. Because of the high toxicity of $NaCNBH_3$, this step should be performed in a fume hood.
2. To block unreacted groups, cover agarose with 5 mL of 1 M Tris and incubate for 2 h at room temperature.
3. Repeat washing **steps 8** and **9** from **Subheading 3.1.**
4. NeutrAvidin–agarose may be resolubilized at 70°C and aliquoted prior to storage.

3.3. EliCell Assay

Biotinylated cytokine-specific antibody is bound to avidin–agarose substrate in liquid state. Cells are exposed to stimulus, mixed with the liquid matrix and spread onto slides. As agarose solidifies, cells are immobilized in the capture antibody-containing substrate. The solid phase agarose–cell mixture is kept hydrated with stimulus containing medium and incubated for appropriate stimulation times. After stimulation, slides are incubated with fluorochrome-labeled cytokine-specific antibody to detect captured protein, and cell morphology and released product are analyzed in parallel by phase and fluorescence microscopy, respectively. The assay is illustrated in **Fig. 2**. *See* **Note 2** for a description of essential controls to be included in each experiment, and **Table 1** for sugges-

1. Agarose-avidin substrate

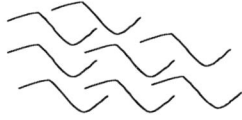

2. Biotinylated capture antibody is bound to agarose matrix

ᐯ Capture antibody

3. Eosinophils are combined with matrix

4. Cell-agarose mixture is spread onto slide

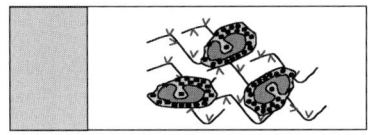

5. Chamber is affixed and stimulus added

6. Slides are incubated @ 37°C in humidified chamber. Secreted product is bound by capture antibody.

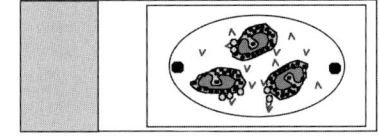

○ Released cytokine

7. Chamber is removed, and slides are fixed and incubated with flourochrome labeled anti-cytokine antibody.

ᖯ Fluorochrome-labeled detection antibody

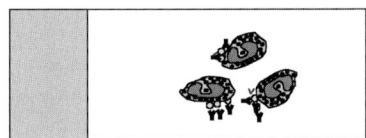

8. Slides are analyzed
Phase-contrast

Fluorescence

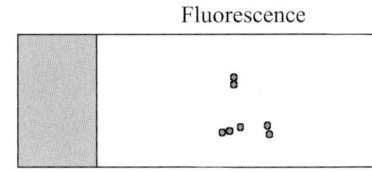

Fig. 2. EliCell assay. Schematic of EliCell assay to detect release of a single cytokine.

tions on specific capture-detection antibody pairs and stimulus concentrations. For adaptation of protocol to detect two cytokines simultaneously, to detect released product in parallel with surface marker staining or for intracellular detection of non-released products refer to **Notes 3–5**, respectively.

Table 1
Examples of Capture–Detection Antibody Pairs and Stimulus
Working Concentrations

Capture AB	Detection AB	Stimulus	Reference
Biotinylated goat polyclonal αhIL-4 (R&D Systems, cat. no. BAF204)	Mouse αhIL-4 (R&D Systems, Clone no. 3010.211)	Eotaxin (6 n*M*)	*15*
Biotinylated goat polyclonal αh RANTES (R&D Systems, cat. no. BAF278)	Mouse αhRANTES (R&D Systems, Clone no. 21418.211)	IFN-γ (500 U/mL)	*17*
Biotinylated goat polyclonal αh IL-12 (R&D Systems, cat. no. BAF219)	Mouse αhIL-12 (R&D Systems, Clone no. 24910.1)	αCD9 (2.5 µg/mL)	*18*

3.3.1. Binding of Capture Antibody to Substrate

1. Melt aliquot of streptavidin or NeutrAvidin-conjugated agarose at 70°C, then transfer to 37°C to maintain liquid state
2. Keeping at 37°C, combine the following in an Eppendorf tube (to determine final volume required multiply number of desired samples by 20 µL): 1 volume agarose, (+1/10 volume 10X concentrated RPMI-1640 medium), 1 volume biotinylated capture antibody (100 µg/mL in RPMI + 0.1% ovalbumin). Refer to **Note 6** for selection of appropriate capture-detection antibody pairs.
3. To above Eppendorf tube add 3 volumes of purified eosinophils (15×10^6 cells/mL in RPMI + 0.1% ovalbumin). Refer to **Note 7** for tips on eosinophil purification strategies.

3.3.2 Stimulation of Cells, Gelification of Substrate, and Incubation

1. For each stimulus, prepare in advance 1X and 10X working solutions. (Total volume required for 1X solution will be 400 µL × total number of slides, and for 10X solution will be 2 µL × total number of slides.)
2. Mix agarose/cell mixture (from **step 3** of **Subheading 3.3.1.**) thoroughly by aspirating gently with a plastic pipet tip.
3. In small Eppendorf carefully combine 20 µL agarose/cell mixture with 2 µL of 10X stimulus.
4. Using Ultra-microtips, gently spread (using surface tension) 22 µL stimulated agarose/cell mixture onto microscope slide. Avoid contacting slide surface with tip, as this may lead to cell damage.

5. Cover sample with perfusion chamber and place slide on tray atop hydrated pad. (Once chamber has been affixed, slides may remain at room temperature for several minutes while additional slides are prepared.)
6. Carefully pipet 400 μL of appropriate 1X stimulus over sample through a chamber access port, ensuring that chamber area is uniformly saturated.
7. Place tray in humidified incubator (37°C, 5% CO_2) for desired incubation time.
8. OPTIONAL: while slides are incubating, centrifuge fluorochrome-labeled detection antibody (15,000g for 30 min) to pellet any precipitate that may have formed during storage. Use supernatant to prepare working dilutions (**step 3**, **Subheading 3.3.3.**).

3.3.3. Fixation and Detection

1. After incubation, remove slides from incubator, carefully remove perfusion chambers and fix slides by immersion in 2% paraformaldehyde for 5 min at room temperature.
2. After fixation, wash slides for 10 min in HBSS –/– at room temperature with gentle agitation.
3. Prepare working dilutions of fluorochrome-labeled detection antibody in HBSS –/– (*see* **step 8, Subheading 3.3.2.**). Optimal concentrations must be determined for each detection antibody but generally range from 1 to 10 μg/mL. To determine total volume of detection antibody required, multiply number of slides by 400 μL. For suggestions on fluorochrome labeling of detection antibody, refer to **Note 1**.
4. Dry area immediately around agarose film on slides to prevent seepage of antibody, and place slides on hydrated pad. (Alternatively, pap pen may be used to outline staining area. However, care must be taken to avoid contacting agarose with pap pen.) DO NOT ALLOW SLIDES TO DRY.
5. Cover agarose film with 400 μL of diluted detection antibody and incubate 45 min at room temperature, protected from light.
6. After incubation with detection antibody, wash slides 3 × 10 min. in fresh HBSS –/– with gentle agitation.
7. After final wash, quickly dip slides in water to remove salt residue and allow slides to dry at room temperature, protected from light. (Slides may be left overnight to ensure drying.)
8. Once dry, slides should be mounted using aqueous mounting medium, coverslip affixed, and stored at room temperature in the dark until analysis. (Allowable storage time will depend upon fluorochrome used with detection antibody).

3.4. Analysis

The major advantage provided by the EliCell system is the ability to observe morphology and viability of individual cytokine-secreting cells. The next three subsections discuss analysis of results using fluorescence, phase-contrast and light microscopy.

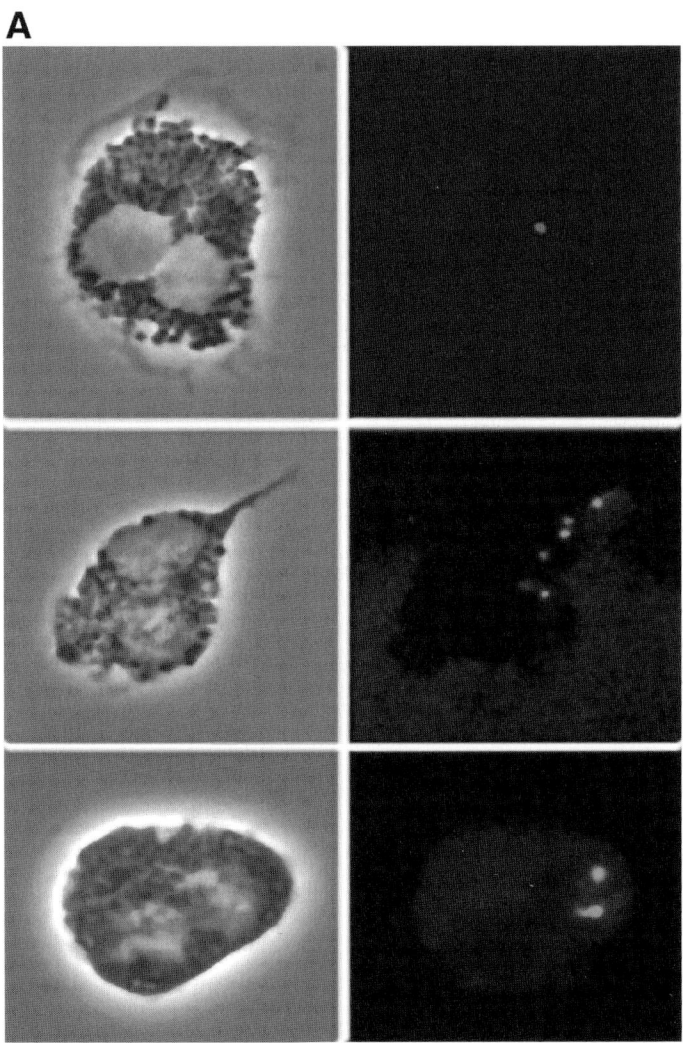

Fig. 3. Positive staining results using EliCell system. Detection of cytokine release from eosinophils stimulated with physiologic (**A**) or nonphysiologic (**B**) stimuli. In (**B**), eosinophils were stimulated with 0.5 µ*M* A23187. The bottom panel illustrates simultaneous detection of two cytokines labeled with Alexa 488 or Alexa 546. Digital pictures were taken using 100X magnification objective. *See* **Color Plate 7**, following page 50. *(Figure continues)*

3.4.1. Positive Staining Results

The detection of vesicle-released products using physiologic stimuli will appear as punctate spots at the cell surface (*see* **Fig. 3A**). A more diffuse staining pattern may indicate an insufficient concentration of capture antibody within the

B

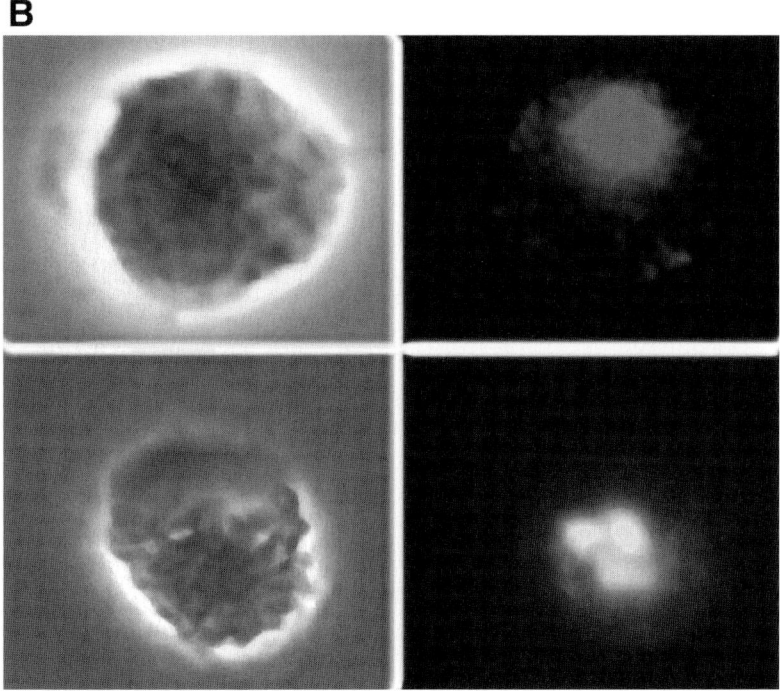

Fig. 3. *(continued)*

agarose matrix *(18)*. In contrast, detection of products released through cytolysis (i.e., following stimulation with a calcium ionophore) will appear as a much more robust staining, without discrete limits (**Fig. 3B**) and may be visualized in the absence of capture antibody, due to loss of membrane selectivity.

3.4.2. Staining Artifacts

It is essential that fluorescent product be analyzed in parallel with phase contrast analysis, to avoid counting staining artifacts as positive signal. In addition, appropriate controls must be included in each experiment (*see* **Note 2**). Common sources of staining artifacts arise from cellular autofluorescence exhibited by damaged or dying cells, non-specific antibody binding and permeabilization of cells. Generally, about 5% of the cells undergo degeneration and die by apoptosis during the EliCell assay. Examples of staining artifacts are shown in **Fig. 4**.

3.4.3. Cell Viability and Morphological Analysis

A major benefit of the EliCell system is the maintenance of cell viability (greater than 90%) even after fixation and detection steps. Viability of cells may

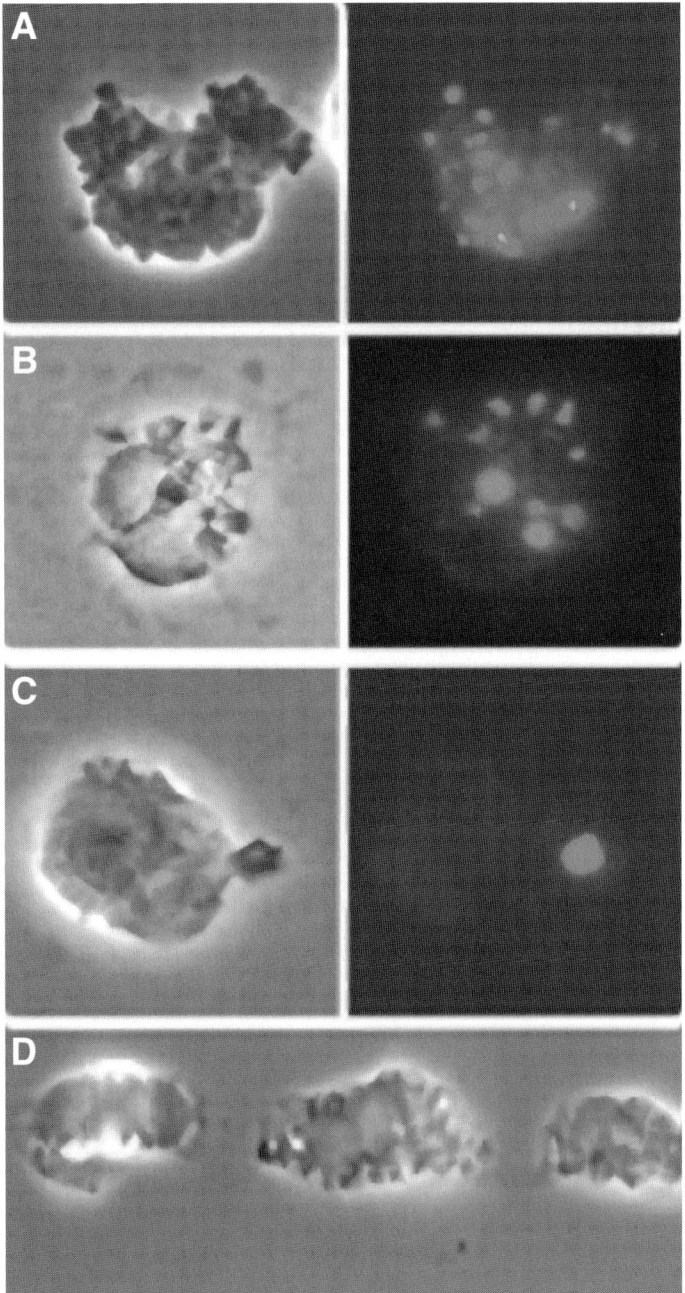

Fig. 4. Staining artifacts using EliCell system. Phase-contrast and fluorescence microscopy of identical fields of eosinophils incubated in EliCell preparations. Damaged (**A, B,** and **C**) or permeabilized **(D)** cells show nonspecific staining. In **D,** the image was overlaid. Digital pictures were taken using 100X magnification objective. *See* **Color Plate 8,** following page 50.

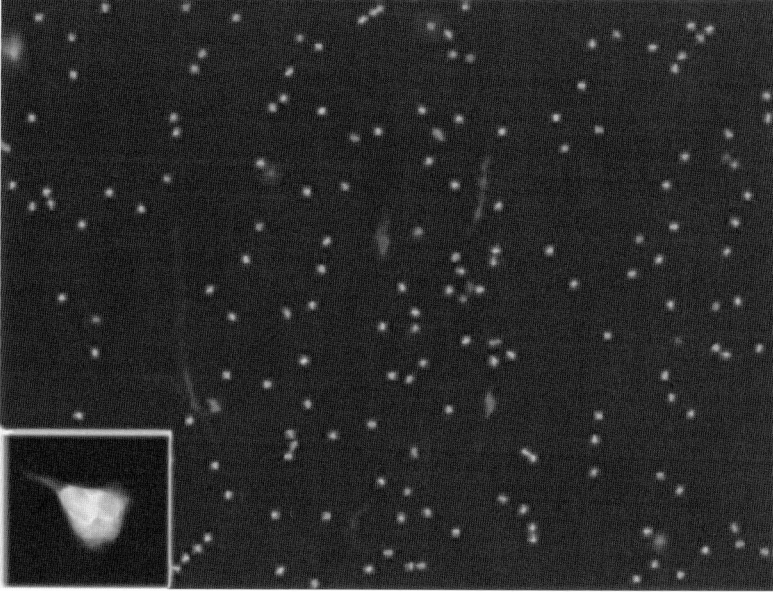

Fig. 5. Viability of cells after EliCell assay. EliCell preparation of eosinophils stained with acridine orange/ethidium bromide mixture after fixation. Most cells show green fluorescent nucleus indicative of cell viability. Digital pictures were taken using 100X magnification objective. *See* **Color Plate 9**, following page 50.

be monitored in the EliCell system using ethidium bromide staining as outlined in **Subheading 3.4.3.1.** Another key feature of the EliCell system is the ability to analyze cellular morphology throughout experimental manipulations. In addition to analysis by phase-contrast (described in **Figs. 3** and **4**), cell morphology may also be studied by bright-field microscopy. Preparation of cells for this purpose is described in **Subheading 3.4.3.2**.

3.4.3.1. Monitoring Cell Viability

1. After a 5-min fixation of slides in 2% PFO (**step 1, Subheading 3.3.3.**), wash once in HBSS –/–.
2. Add 300 μL of 1:1000 dilution of acridine orange:ethidium bromide mixture (*see* **Subheading 2.**).
3. Coverslip and analyze fluorescence using FITC and rhodamine filters without allowing cells to dry. Live cells will appear slightly red with green nuclear staining, while nuclei of dead cells will intercalate ethidium bromide and appear red (**Fig. 5**).
4. Alternatively, cell viability may also be analyzed adding the acridine orange:ethidium bromide solution directly into the chamber access port after incubation with stimulus. In this case, an inverted microscope will be necessary for slide analysis.

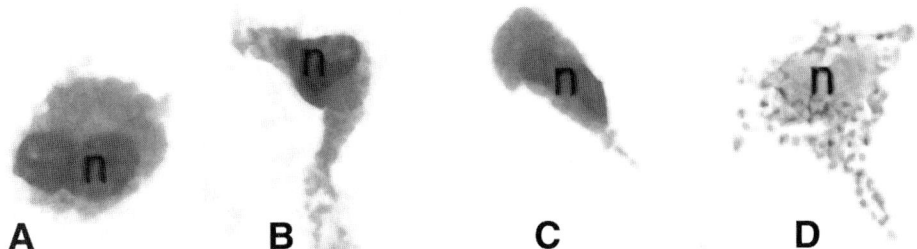

Fig. 6. Morphology of eosinophils during EliCell assay. Light micrographs of eosinophils observed in the EliCell system before (**A**) and after stimulation with eotaxin (**B–D**). Morphological changes characterized by cell elongation are clearly seen in stimulated cells. Cells were stained with Hema 3 (**A–C**) or fast green/hematoxylin (**D**). n, nucleus. Digital pictures were taken using 100X magnification objective. *See* **Color Plate 10**, following page 50.

3.4.3.2. MORPHOLOGICAL ANALYSIS

After the desired EliCell step to be analyzed, cells may be prepared as follows:

1. Fix the agarose film containing cells in methanol for 1 min. Fixation is accomplished by either slide immersion in the fixative (without chamber) or by adding the fixative directly into the chamber access port.
2. Stain the cells with acid/basic dyes such as fast green/hematoxylin (0.2% fast green in 70% ethanol for 20 min followed by hematoxylin for 5 s) or Hema 3® kit (Fisher Scientific).
3. Mount and analyze the slides using a bright-field microscope with 100X objective.
4. Alternatively, cells may be fixed with 2% paraformaldehyde for 5 min and stained with 2% chromotrope 2R for 5 min.

Eosinophils in the EliCell system may show different morphologies, ranging from round to highly polarized cells. Generally, after stimulation, a great proportion of eosinophils are seen as elongated cells (**Fig. 6**). This morphological change indicates that the cells are activated. Refer to **Note 8** for adaptation of protocol for additional cytochemical analyses.

4. Notes

1. When choosing a fluorochrome for detection antibody labeling, it is important to consider stability of the signal, as well as its staining intensity. We routinely use Alexa Fluor 546 (red) or Alexa Fluor 488 (green). Protein labeling kits are available from Molecular Probes (cat. no. A-10237 and A-10235 for Alexa 546 and Alexa 488, respectively). Allow approx 2 h for the labeling procedure.

2. Because of the staining artifacts discussed in **Subheading 3.4.2.** and to control for quality of purified eosinophils, the following controls must be included in each experiment, for each condition: (1) medium alone (no stimulus); (2) isotype control fluorochrome labeled detection antibody; (3) capture antibody omitted from agarose or replacement with irrelevant capture antibody.

3. The EliCell assay may be adapted to detect two cytokines simultaneously. In this case two capture antibodies will be combined with the avidin-agarose in **step 2** of **Subheading 3.3.1**. To maintain appropriate matrix consistency, the total volume added must be equivalent to 1 volume. Therefore, add 0.5 volume each of 2X stock biotinylated capture antibody. Fluorescently labeled detection antibodies should be chosen which do not cross-react with each other or either capture antibody, and whose emission spectra do not overlap. Diluted detection antibodies may be added simultaneously (**step 5, Subheading 3.3.3.**) in a total volume of 400 µL. (*See* **Fig. 3B** for illustration of positive signal.)

4. In addition, the EliCell assay may be adapted for the simultaneous detection of surface marker expression and released product. It is important to first determine whether fixation alters the antigenicity of the surface marker in question. If fixation does not alter ability of detection antibody to bind surface marker, then EliCell procedure may be followed as described, with the addition of a fluorochrome labeled antibody against surface marker added simultaneously with anti-cytokine detection antibody. Alternatively, slides may be incubated with fluorochrome-labeled antibody against the surface marker prior to fixation in 2% PFO (**step 1, Subheading 3.3.3.**). Slides should then be washed in HBSS –/– before continuing with fixation and subsequent staining with anti-cytokine detection antibody.

5. The EliCell system may be modified to immunolocalize intracellular products formed by eosinophils (i.e., the principal cysteinyl leukotriene LTC_4). In this case, the agarose is prepared without avidin and the cells must be permeabilized before detection with a specific fluorochrome-labeled antibody. For LTC_4 detection, cells are prepared as for conventional EliCell assay, with the following modifications *(23)*: (1) Eosinophils are embedded in the agarose matrix without binding of capture antibody to substrate. 1 volume of RPMI + 0.1% OVA should be added to tube to compensate for depleted volume (**step 2, Subheading 3.3.1.**). (2) After stimulation, cells are permeabilized and fixed with carbodiimide chemistry that induces cross-links between carboxyl group of newly synthesized LTC_4 to amines of adjacent proteins. (3) Fluorescently labeled detection antibody (i.e., Alexa488-labeled Ab specific for LTC_4) is used to detect and localize the immobilized leukotriene.

6. When choosing capture and detection antibody pairs, it is important to choose antibodies which react with unique epitopes of the target cytokine. We generally choose a polyclonal capture antibody and a monoclonal detection antibody to decrease the probability of antibody hindrance.

7. We routinely isolate eosinophils from human peripheral blood by Hypaque-Ficoll separation followed by negative selection using an immunomagnetic depletion protocol (Stem Cell Technologies, Vancouver, BC; cat. no. 14156). Much of the success of the EliCell assay will depend upon the initial quality of cells. Therefore, it

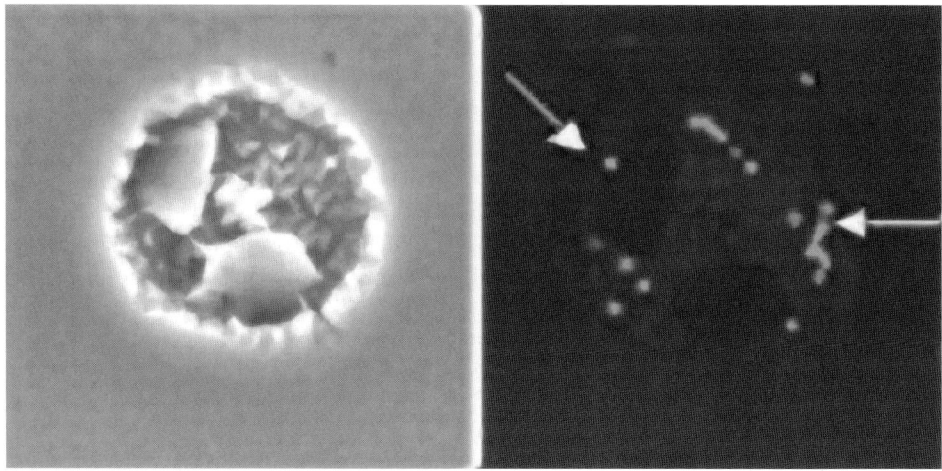

Fig. 7. Intracellular lipid body staining using EliCell system. Phase-contrast and fluorescence microscopy of identical field of chemokine-stimulated eosinophil in an EliCell preparation. Cytoplasmic lipid bodies are indicated (arrows). Cells were stained with BODIPY. Digital pictures were taken using 100X magnification. *See* **Color Plate 11**, following page 50.

is essential that great care be taken during the eosinophil purification process. Specifically, hypotonic saline solution should be used for red blood cell lysis (avoid using NH₄Cl solution). HBSS without calcium or magnesium should be used throughout purification process to avoid activation of eosinophils. In addition, it has been our experience that hypodense eosinophils are more fragile than normodense cells. Therefore if possible, care should be taken to recover only the normodense layer after Ficoll separation.

8. In addition to immunolocalization analyses, the EliCell assay may be useful for multiple cytochemical studies. For instance, viable eosinophils embedded in the agar matrix (prepared without avidin incubation) may be stimulated to induce lipid body formation, which may be detected by Nile Red or BODIPY® (Molecular Probes) staining (**Fig. 7**).

Acknowlegments

Supported by NIH grants AI20241, AI22571, AI51645, and HL70270. Dr. Sandra A.C. Perez and Dr. Rossana C. N. Melo were supported in part by a Fellowship from CNPq (Brazil).

References

1. Gleich GJ, Adolphson, C. R., and Leiferman, K. M. (1992) Eosinophils, in *Inflammation: Basic Principles and Clinical Correlates* (Gallin, J. I., Goldstein, I. M., and Synderman, R., eds.), Raven Press, New York. pp. 663–700.

2. Kita H, A. (1998) Biology of eosinophils, in *Allergy: Principles and Practice* (Adkinson, N. F., Busse, W. W., Ellis, E. F., Middleton, E., Jr., Reed C. E., and Yunginger, J. W., eds.) Mosby, St. Louis. pp. 242–260.

3. Lacy, P. and Moqbel, R. (1997) Eokines: synthesis, storage and release from human eosinophils. *Mem Inst Oswaldo Cruz.* **92 Suppl 2**, 125–33.

4. Weller, P. F., and Dvorak, A. M. (1994) Human eosinophils—development, maturation and functional morphology, in *Asthma and Rhinitis* (Busse,W. M., and Holgate, S., eds.), Blackwell Scientific Publications, Boston. pp. 225–274.

5. Gleich, G. J., Adolphson, C. R., and Leiferman, K. M. (1993) The biology of the eosinophilic leukocyte. *Annu Rev Med.* **44**, 85–101.

6. Beil, W. J., Weller, P. F., Tzizik, D. M., Galli, S. J., and Dvorak, A. M. . (1993) Ultrastructural immunogold localization of tumor necrosis factor-alpha to the matrix compartment of eosinophil secondary granules in patients with idiopathic hypereosinophilic syndrome. *J. Histochem. Cytochem.* **41**, 1611–1615.

7. Moller, G. M., de Jong, T. A., van der Kwast, T. H.,, et al. (1996) Immunolocalization of interleukin-4 in eosinophils in the bronchial mucosa of atopic asthmatics. *Am. J. Respir. Cell Mol. Biol.* **14**, 439–443.

8. Moqbel, R. (1996) Synthesis and storage of regulatory cytokines in human eosinophils. *Adv Exp Med Biol.* **409**, 287–294.

9. Ying, S., Meng Q., Taborda-Barata, L., et al. (1996) Human eosinophils express messenger RNA encoding RANTES and store and release biologically active RANTES protein. *Eur J Immunol.* **26, 70**–76.

10. Dvorak, A. M., Estrella, P., and Ishizaka, T. (1994) Vesicular transport of peroxidase in human eosinophilic myelocytes. *Clin. Exp. Allergy.* **24**, 10–18.

11. Dvorak, A. M., and Ishizaka, T. (1994) Human eosinophils in vitro. An ultrastructural morphology primer. *Histol Histopathol.* **9**, 339–374.

12. Dvorak, A. M., Ackerman, S. J., and Weller, P. F. (1990) Subcellular morphology and biochemistry of eosinophils, in *Blood Cell Biochemistry: Megakaryocytes, Platelets, Macrophages and Eosinophils* (H. J. R., ed), Plenum Publishing, London. pp. 237–344.

13. Dunzendorfer, S., Feistritzer, C., Enrich, B., and Wiedermann, C. J. (2002) Neuropeptide-induced inhibition of IL-16 release from eosinophils. *Neuroimmunomodulation.* **10**, 217–223.

14. Schmid-Grendelmeier, P., Altznauer, F., Fischer, B., et al. (2002) Eosinophils express functional IL-13 in eosinophilic inflammatory diseases. *J Immunol.* **169**, 1021–1027.

15. Bandeira-Melo, C., Sugiyama, K., Woods, L. J., and Weller, P. F. (2001) Cutting edge: eotaxin elicits rapid vesicular transport-mediated release of preformed IL-4 from human eosinophils. *J Immunol.* **166**, 4813–4817.

16. Sabin, E. A., Kopf, M. A., and Pearce, E. J. (1996) Schistosoma mansoni egg-induced early IL-4 production is dependent upon IL-5 and eosinophils. *J Exp Med.* **184**, 1871–1878.

17. Bandeira-Melo, C., Gillard, G., Ghiran, I., and Weller, P. F. (2000) EliCell: a gel-phase dual antibody capture and detection assay to measure cytokine release from eosinophils. *J. Immunol. Methods.* **244**, 105–115.

18. Bandeira-Melo, C., Perez, S. A., Melo, R. C., Ghiran, I., and Weller, P. F.(2003) EliCell assay for the detection of released cytokines from eosinophils. *J. Immunol. Methods.* **276**, 227–237.

19. Alon, R., Bayer, E. A., and Wilchek, M. (1992) Cell-adhesive properties of strep- tavidin are mediated by the exposure of an RGD-like RYD site. *Eur. J. Cell Biol.* **58,** 271–279.

20. Alon, R., Bayer, E. A., and Wilchek, M. (1993) Cell adhesion to streptavidin via RGD-dependent integrins. *Eur. J. Cell Biol.* **60,** 1–11.

21. Alon, R., Hershkoviz, R., Bayer, E. A., Wilchek, M., and Lider, O.. (1993) Streptavidin blocks immune reactions mediated by fibronectin-VLA-5 recognition through an Arg-Gly-Asp mimicking site. *Eur. J. Immunol.* **23,** 893–898.

22. Ferguson, T. A., Mizutani, H., and Kupper, T. S. (1991) Two integrin-binding pep- tides abrogate T-cell-mediated immune responses in vivo. *Proc. Natl. Acad. Sci. USA.* **88**, 8072–8076.

23. Bandeira-Melo, C., Phoofolo, M., and Weller, P. F. (2001) Extranuclear lipid bod- ies, elicited by CCR3-mediated signaling pathways, are the sites of chemokine- enhanced leukotriene C4 production in eosinophils and basophils. *J. Biol. Chem.* **276**, 22779–22787.

Index